页岩气开发理论与实践

(第三辑)

贾爱林 位云生 编著

石油工业出版社

内 容 提 要

页岩气开发是非常规气开发的重要组成部分。本书分为综合类、开发地质类、气藏工程类、生产应用类，汇总了国内一批页岩气开发领域专家的最新研究成果与心得，以及部分国外页岩气开发的最新进展，可为国内页岩气开发提供理论参考和方法借鉴。

本书可供从事页岩气等非常规天然气开发的科研人员使用，也可以作为高等院校相关专业师生的参考用书。

图书在版编目（CIP）数据

页岩气开发理论与实践.第三辑/贾爱林等编著.—北京：石油工业出版社，2021.6
ISBN 978-7-5183-4396-6

Ⅰ.①页… Ⅱ.①贾… Ⅲ.①油页岩资源-油气田开发-文集 Ⅳ.①P618.130.8-53

中国版本图书馆 CIP 数据核字（2021）第 068581 号

出版发行：石油工业出版社
　　　　　（北京安定门外安华里 2 区 1 号　100011）
　　　　　网　址：www.petropub.com
　　　　　编辑部：（010）64523708　图书营销中心：（010）64523633
经　　销：全国新华书店
印　　刷：北京中石油彩色印刷有限责任公司

2021 年 6 月第 1 版　2021 年 6 月第 1 次印刷
787×1092 毫米　开本：1/16　印张：16.5
字数：400 千字

定价：150.00 元
（如出现印装质量问题，我社图书营销中心负责调换）
版权所有，翻印必究

前 言

中国页岩气资源丰富，技术可采资源量达 31.6 万亿立方米（据 EIA 预测），仅次于美国，同时中国也是世界上除北美地区之外，少数获得页岩气规模开发的国家之一，特别是近几年，中国页岩气发展迅猛，年产量从 2014 年的 13 亿立方米快速增长至 2020 年的 200 亿立方米，成为继美国、加拿大之后，世界第三大页岩气生产国。在国内非常规气藏开发中，页岩气产量已成为继致密气之后，第二大非常规气藏类型。

中国页岩气的跨越式发展，得益于四川盆地及其周缘长宁—威远、昭通、涪陵三个国家级页岩气示范区海相页岩气开发技术的长足进步。"十二五"期间，在政府大力投资和补贴政策的激励下，通过系列技术攻关和开发试验，形成了埋深 3500 米以浅海相页岩气藏储层评价、优快钻井、体积压裂、产能评价和开发技术优化五大主体技术系列：高演化强改造复杂山地地球物理资料解释及储层综合评价技术、水平井地质工程一体化的优快钻井技术、以大排量滑溜水 + 低密度支撑剂 + 可溶桥塞 + 拉链式压裂模式为主的水平井体积压裂工艺技术、页岩气体积压裂水平井概率性产能评价技术、水平井及裂缝参数综合优化技术。这些技术的试验与应用，提高了单井产量，降低了开发成本，奠定了中国页岩气规模开发的技术基础。"十三五"期间，中国埋深 3500 米以深的南方海相页岩气获得了多个井点突破，有望在"十四五"及未来的页岩气规模开发中发挥主力军作用，同时为北方大面积的陆相及海陆过渡相深层页岩气资源开发技术的突破提供了借鉴和示范作用。

《页岩气开发理论与实践》系列论文集，立足中国页岩气勘探开发所取得的成果，收集整理了近年来具有代表性的学术论文，按照综合类、开发地质类、气藏工程类、生产应用类进行分类编辑，接集出版，希望在总结页岩气勘探开发理论和技术成果的同时，对这些理论和技术成果的推广应用发挥积极作用，并对新理论、新技术的探索有所启发，从而对我国页岩气勘探开发技术的进一步发展起到推动作用。

目　录

一、综合类

中国页岩气勘探开发进展及发展展望 ················ 赵文智　贾爱林　位云生等（3）

未来十五年中国天然气发展趋势预测 ················ 贾爱林　何东博　位云生等（21）

二、开发地质类

地质工程一体化实施过程中的页岩气藏地质建模 ···· 舒红林　王利芝　尹开贵等（37）

页岩气储层关键参数评价及进展 ···················· 乔　辉　贾爱林　贾成业等（52）

基于叠后地震数据的裂缝预测与建模——以太阳—大寨地区浅层页岩气储层为例
··· 王建君　李井亮　李　林等（64）

蜀南地区五峰—龙马溪组页岩微观孔隙结构及分形特征
··· 朱汉卿　贾爱林　位云生等（80）

基于氩气吸附的页岩纳米级孔隙结构特征 ············ 朱汉卿　贾爱林　位云生等（92）

三、气藏工程类

页岩气压裂水平井控压生产动态预测模型及其应用
··· 贾爱林　位云生　刘　成等（107）

Transient pressure analysis for multi-fractured horizontal well with the use of multi-linear flow model in shale gas reservoir
··· Wang Guangdong　Jia Ailin　Wei Yunsheng et al（124）

页岩气井 EUR 快速评价新方法 ······················ 赵玉龙　梁洪彬　井　翠等（150）

Comprehensive optimization of managed drawdown for a well with pressure-sensitive conductivity fractures: workflow and case study
··· Wei Yunsheng　Jia Ailin　Wang Junlei et al（159）

页岩气水平井组产量递减特征及动态监测 ············ 谢维扬　刘旭宁　吴建发等（171）

四、生产应用类

地质—工程—经济一体化页岩气水平井井距优化——以国家级页岩气开发示范区宁209井区为例 ················· 雍　锐　常　程　张德良等（187）

基于多井模型的压裂参数—开发井距系统优化 ······ 王军磊　贾爱林　位云生等（198）

Optimization workflow for stimulation-well spacing design in a multiwell pad
·················· Wang Junlei　Jia Ailin　Wei Yunsheng et al（218）

四川盆地威远区块典型平台页岩气水平井动态特征及开发建议
························ 位云生　齐亚东　贾成业等（241）

页岩气水平井生产规律 ······················· 郭建林　贾爱林　贾成业等（248）

一、综合类

中国页岩气勘探开发进展及发展展望

赵文智，贾爱林，位云生，王军磊，朱汉卿

（中国石油勘探开发研究院）

摘 要：中国页岩气进入了快速发展的黄金时期，随着天然气在能源消费结构的比重不断攀升，以及在川南地区页岩气商业化开发的成功，页岩气成为中国未来最可靠的能源接替类型。文章系统梳理了近10年来中国页岩气在勘探开发实践中的理论认识和开发技术进展，总结了中国页岩气商业开发的成功经验，明确了页岩气在中国未来天然气发展中的前景与地位。中国页岩气资源潜力巨大，是未来天然气产量增长的现实领域；获得工业性页岩气资源的条件包括"两高"（含气量高、孔隙度高）、"两大"（高TOC集中段厚度大、分布面积大）、"两适中"（热演化程度适中、埋藏深度适中）和"两好"（保存条件好、可压裂性好）；中国海相页岩气最具勘探开发潜力，是目前页岩气上产的主体，已形成了适用于四川盆地及周缘奥陶系五峰组—志留系龙马溪组页岩气开发的六大主体技术系列（地质综合评价技术、开发优化技术、水平井优快钻井技术、水平井体积压裂技术、工厂化作业技术、高效清洁开采技术）。中国页岩气商业开发的成功经验可总结为四点：(1) 选准最佳水平井靶体层位；(2) 配套优快钻进和高效体积改造技术；(3) 促进地质工程一体化数据融合；(4) 探索先进组织管理模式。对中国页岩气未来发展的三点建议：(1) 加强非海相及深层海相页岩气低成本开发关键技术与装备攻关；(2) 重视提高区块页岩气采收率问题，实现整体规模效益开发；(3) 重视非资源因素对页岩气上产节奏的影响。

关键词：页岩气；勘探理论；开发技术；成功经验；发展前景；中国

页岩是指由黏土和极细粒矿物堆积并固化形成的岩石，多沉积于无波浪扰动的海洋、湖泊等稳定环境，一般含丰富有机物，是生油气母岩[1]。页岩气以游离和吸附方式存在于页岩微纳米级孔隙，需要人工改造才能释放出工业性天然气，故页岩气藏又称"人工气藏"，具有初期产量高、衰减快、后期低产、时间较长的特点[2-3]。随着页岩气工业技术水平的不断进步，美国页岩气产量大幅提升，产量从2005年的$204\times10^8m^3$增长到2018年的$6072\times10^8m^3$，占天然气总产量的71%，年增长率20%。美国依靠"页岩气革命"，预计将在2022年实现能源独立。以页岩气为代表的非常规油气资源的成功开发，标志着油气工业理论和技术的重大突破和创新，极大地拓展了油气勘探开发的资源领域。

近年来，中国页岩气勘探开发取得重大突破，成为北美之外第一个实现规模化商业开发的国家[4]。加快页岩气勘探开发，提高天然气在一次能源消费中的比重，是加快建设清洁低碳、安全高效的现代能源体系的必由之路，也是化解环境约束、改善大气质量、实现绿色低碳发展的有效途径。中国目前已在上扬子区五峰组—龙马溪组4个"甜点区"建成涪陵、长宁—威远等千亿立方米级的海相页岩大气田，2018年产量达到$108\times10^8m^3$，但仅占天然气总产量6.8%，远低于美国页岩气在天然气总产量中的比重。目前除了3500m

以浅海相页岩气资源得到了有效动用外，中国在海相深层、陆相、海陆过渡相等页岩层系中存在巨量资源，随着页岩气勘探开发理论技术配套成熟，未来中国页岩气产量将会大幅攀升。本文通过科学梳理中国页岩气近10年来在勘探、开发、工程等理论技术领域取得的新进展和新思路，系统总结勘探理论认识和开发关键技术，剖析中国页岩气实现商业开发的成功经验，并进行发展前景的预测，为中国页岩气未来大规模发展提供有益借鉴。

1 中国页岩气勘探开发现状

1.1 资源潜力

中国页岩气资源总量大，但基于评价方法和认识的不同，各家研究机构的资源评价预测结果有较大的出入。根据中国石油天然气股份有限公司（以下简称中国石油）第四次最新资源评价结果，中国陆上页岩气可采资源量为 $12.85 \times 10^{12} m^3$，其中，海相页岩气可采资源量为 $8.82 \times 10^{12} m^3$，占比69%；海陆过渡相页岩气可采资源量 $2.37 \times 10^{12} m^3$，占比18%；陆相页岩气可采资源量 $1.66 \times 10^{12} m^3$，占比13%。

与北美海相页岩气相比，中国页岩气形成的资源基础具有多样性。中国陆上沉积盆地内广泛发育海相、海陆过渡相以及陆相3种类型的富有机质页岩（表1）[5-7]。其中，海相富有机质页岩主要沉积于早古生代，主要分布在四川盆地周缘等广大南方地区以及塔里木盆地、羌塘盆地等西部地区，总面积为 $(60 \times 90) \times 10^4 km^2$；海陆过渡相—煤系页岩沉积于石炭纪—二叠纪，主要分布于华北以及西北部盆地，面积约为 $(15\sim20) \times 10^4 km^2$；陆相富有机质页岩沉积于中—新生代，主要分布于东部松辽盆地、渤海湾盆地以及中部的鄂尔多斯盆地等，面积约为 $(20\sim50) \times 10^4 km^2$。

中国海相页岩形成时代老，平面上分布稳定，优质页岩连续厚度大，有机碳含量高，以 I 型干酪根为主，II$_1$ 型有机质为辅，热演化程度较高，有机质成熟度通常高于2.0%，以原油裂解生气为主，有机质孔的发育为气体的赋存提供了大量的储集空间，脆性矿物含量高，可压性好，黏土矿物以伊利石为主，成岩演化程度高，页岩气成藏条件优越[1]。目前四川盆地及其周缘五峰—龙马溪组海相页岩气勘探开发已取得重大突破，形成了工业产能。

与海相页岩相比，中国海陆过渡相页岩及陆相页岩成藏条件和潜力相对较差。海陆过渡相页岩干酪根类型以 II$_2$—III 型为主，有机质成熟度为1.0%~2.5%，以干酪根热解气为主，普遍处于生气高峰。纵向上多层分散分布，含砂岩夹层，平面上横向变化大，有机质孔发育程度低，这些因素都制约了海陆过渡相页岩气藏的成藏和有效开发。在海陆过渡相页岩的前期勘探评价中，要优选埋深适中、连续厚度较大、构造稳定、气体保存条件较好的区块作为有利区。目前中国海陆过渡相页岩总体处于勘探评价阶段，近期在鄂东大宁—吉县区块取得了重要发现，个别探井见工业气流，显示具备良好的开发前景。陆相富有机质页岩形成时间较晚，页岩总厚度较大，集中段相对发育，是中国陆上大型产油区的主力烃源岩。干酪根为 I—II$_2$ 型，有机质热演化程度较低，普遍处于生油阶段，有机质孔不发育，页岩储集空间有限，黏土含量较高，储层可压性差，且陆相沉积受高频旋回的控制，岩性变化较快，页岩层系连续性较差[8]，这些特点都加大了陆相页岩气勘探开发的难度。目前仅在四川盆地下侏罗统[9-10]和鄂尔多斯盆地南部三叠系延长组[11]获得了工业气流，资源前景存在较大不确定性。

表1 中国3类富有机质页岩气藏特征简表[7]

类型	有利区范围	集中段特征	生气潜力	含气性	可压裂性
海相页岩	面积大 [(10~20) ×10⁴km²]	厚度大连续 (30~80m)	生气量大（I—II₁，R_o=2.0%~5.0%，油裂解气为主）	含气量高（有机质孔发育，比表面积大，含气量1.0~6.0m³/t）	好（脆性矿物>40%，黏土以伊利石为主）
海陆过渡相页岩	面积较大 [(5~10) ×10⁴km²]	厚度小不连续 (<15m)	生气量偏小（II₂—III，R_o=1.0%~2.5%，热解气为主）	含气量低（有机质孔不发育，比表面积小，含气量多数<1m³/t）	一般（脆性矿物30%~60%，黏土以伊/蒙混层为主）
陆相页岩	分布局限 (<5×10⁴km²)	厚度较大变化快 (20~70m)	生气量小（I—II₂，R_o=0.5%~1.3%，生油为主）	含气量偏低（有机质孔不发育，比表面积小，含气量0.5~2.2m³/t）	差（脆性矿物20%~50%，黏土以蒙皂石为主）

1.2 勘探开发历程

中国页岩气勘探起步相对较晚，自2005年开始，国土资源部油气资源战略咨询中心联合国内石油公司和高等院校开展了规模性的页岩气前期资源潜力研究和选区评价工作，根据中国海相页岩气的勘探开发历程，可以将其划分为4个阶段。

1.2.1 评层选区阶段（2007—2009年）

2007年，中国石油与美国新田石油公司合作，开展了威远地区寒武系筇竹寺组页岩气资源潜力评价与开发可行性研究。2008年，中国石油勘探开发研究院在川南长宁构造志留系龙马溪组露头区钻探了中国第一口页岩气地质评价浅井——长芯1井。2009年，原国土资源部启动了"全国页岩气资源潜力调查评价与有利区优选"项目，对中国陆上页岩气资源潜力进行系统评价[12]。与此同时，中国石油与壳牌石油公司在富顺—永川地区开展了中国第一个页岩气国际合作勘探开发项目。

1.2.2 先导试验阶段（2010—2013年）

2010年开始，中国页岩气勘探开发陆续获得单井突破。2010年4月，中国石油在威远地区完钻中国第1口页岩气评价井——威201井，压裂获得了工业性页岩气流[7, 13]。2011年原国土资源部正式将页岩气列为中国第172种矿产，按独立矿种进行管理。2012年4月，中国石油在长宁地区钻获第一口具有商业价值页岩气井——宁201-H1井，该井测试获得日产气$15×10^4m^3$[4]，实现了中国页岩气商业性开发的突破。2012年11月28日，中国石油化工股份有限公司（以下简称中国石化）在川东南焦石坝地区完钻的焦页1HF井在五峰—龙马溪组获得页岩气测试产量$20.3×10^4m^3$，正式宣告了涪陵页岩气田的发现[14-15]。

1.2.3 示范区建设阶段（2014—2016年）

2014年开始，中国页岩气产量呈现阶梯式快速增长的态势，2014年中国页岩气产量跃升至$13.1×10^8m^3$，2015年产量为$45.4×10^8m^3$，2016年产量为$78.9×10^8m^3$。2014年中国石化焦石坝区块提交中国首个页岩气探明地质储量 $1067.5 \times 10^8 m^3$，实现了中国页岩气探明储量零的突破[10]。2015 年在威远 W202 井区、长宁 N201—黄金坝 YS108 井区及涪陵页岩气田累计提交探明页岩气地质储量 $5441.3 \times 10^8 m^3$ [16]。中国石油西南油气田建成第一个日产气量超百万立方米的页岩气平台——CNH6 平台，在四川盆地及其周缘逐渐形成了涪陵、长宁、威远和昭通四个国家级海相页岩气开发示范区，页岩气探明地质储量及产量逐年增长迅速[17]。

1.2.4 工业化开采阶段（2017 年至今）

2017 年，涪陵页岩气田如期建成百亿立方米产能，相当于建成千万吨级的大油田，2017 年全年产量超过 $60 \times 10^8 m^3$。同年，中国石油西南油气田公司 CNH10-3 井单井产量突破 $1 \times 10^8 m^3$，中国石油 2017 年全年产量超过 $30 \times 10^8 m^3$。2017 年全年中国页岩气产量超过加拿大（$52.1 \times 10^8 m^3$），成为世界第二大页岩气生产国。截至 2018 年底，累计完钻井数 898 口，提交探明地质储量超过 $1 \times 10^{12} m^3$，2018 年全年页岩气产量超过 $108 \times 10^8 m^3$。

2 中国页岩气勘探理论及地质认识

2.1 有机质形成与天然气解析机理

随着实验测试技术的不断发展，国内外学者在富有机质页岩中发现大量的有机纳米孔隙[18-19]。这是页岩气的重要储集空间，并借助纳米 CT、氩离子抛光扫描电镜、气体吸附仪等先进测试分析仪器，实现了页岩储层纳米级孔隙结构的定性描述和定量表征[20-21]。有机质孔隙通常都是次生孔隙，原生有机质中的孔隙很难在沉积成岩过程中保存下来。次生有机质孔隙又可以分为两种成因，一是干酪根生烃过程中转化为流体而形成的次生孔隙[22-24]，二是在过成熟阶段残留的沥青裂解形成的有机质孔隙。有机质孔隙的发育与有机质的显微组分以及成熟度存在一定相关性[25]。有机质孔隙在Ⅰ型和Ⅱ型干酪根中较发育，在Ⅲ型干酪根中很少发育[18, 26]，腐泥组作为母质的有机质生烃潜力最大，是有机质孔隙发育的有利条件，中国海相页岩和陆相页岩以Ⅰ型和Ⅱ型干酪根为主，生烃潜力较大，在成熟度达到一定程度时有机质孔隙广泛发育；海陆过渡相页岩以Ⅱ$_2$—Ⅲ型干酪根为主，生烃潜力有限，有机质孔发育有限，以粒内溶蚀孔、粒间微孔为主[27]。

成熟度是影响页岩有机质孔隙发育的关键因素（图 1），国内外大量研究表明[28-31]，有机质孔隙在有机质成熟度达到一定程度时开始形成，然后随着有机质成熟度的升高，有机质孔隙数量增多，且在过成熟阶段存在二次裂解，有机质孔隙继续增加，当成熟度达到一定上限后，成熟度过高导致生烃衰竭，有机质发生炭化，有机质孔隙被充填，孔隙体积大幅度减小。

页岩气主体由游离气和吸附气组成，吸附气主要吸附在有机质和黏土矿物的表面。大量实验结果表明[32]，在高成熟度海相富有机质页岩中，随着有机碳含量的增大，页岩吸附能力增强，有机碳含量是影响页岩吸附能力的最主要因素。中国南方海相页岩气开发实践表明，川南页岩气中游离气比例约为 70%，吸附气比例为 30%[33]。在页岩气井开发早期，人工裂缝沟通的孔隙和天然裂缝中的游离气最先被开采出来，游离气占主导；随着地层压力的降低，当达到临界解吸压力时，吸附气开始大量解吸，产气比重逐渐增加，但由于解吸过程相对缓慢，可以一定程度上弥补气井产量递减。

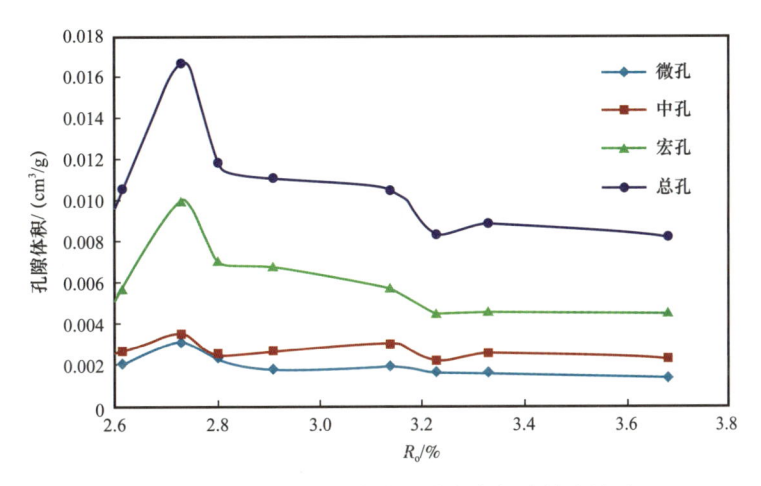

图 1 海相富有机质页岩有机孔发育与成熟度关系

2.2 有机质"接力成气"与页岩气最佳勘探窗口

赵文智等[34]于 2005 年提出了有机质"接力成气"模式，其内涵是指成气过程中生气母质的转换和生气时机与贡献的交替。生气母质的转换是指干酪根降解生气和液态烃裂解生气两者的转换，干酪根降解形成的液态烃一部分排出烃源岩，形成常规油藏，另一部分液态烃呈分散状仍滞留在烃源岩内，在高—过成熟阶段发生热裂解，烃源岩仍具有较好的生气潜力（图 2）。常规油气勘探阶段，R_o 值介于 0.6%～1.6% 被视为最佳勘探窗口，当

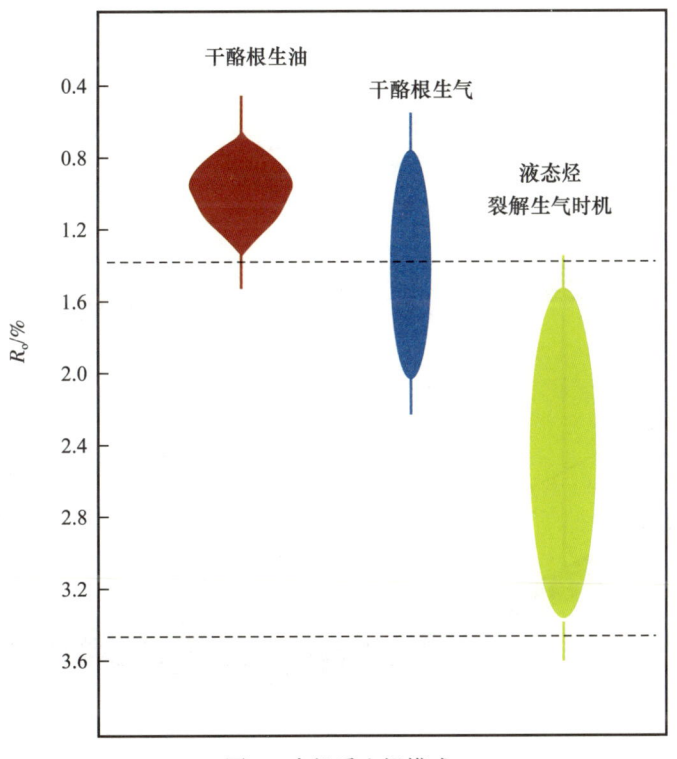

图 2 有机质生烃模式

R_o>1.6% 之后，视为此时有机质已经耗尽生烃能力，长期缺少勘探，正是由于忽略了滞留烃成藏的潜力。排烃效率模拟实验表明[23]，一般烃源岩的排烃效率在40%～60%，说明仍然有40%～60%的液态烃滞留在烃源岩中，成为天然气规模成藏的有效气源灶，当有机质达到高—过成熟阶段之后，滞留烃数量又急剧减小，发生二次裂解和排烃。有机质接力成气模式虽然关注的是滞留烃裂解气源外成藏，与页岩气自生自储的源内成藏有一定的差异性，但是接力成气的理论能够很好地解释中国南方海相页岩高—过成熟阶段高含气性的原因，强调了烃源岩内部滞留烃裂解生气的重要性，指明了页岩气勘探的最佳成熟度窗口R_o为1.35%～3.5%。

2.3 连续性油气聚集与"甜点"勘探战略

与常规油气圈闭聚集理论不同，非常规油气强调连续性油气聚集[33, 35]。页岩气属于典型的连续性油气聚集成藏，其大范围连续分布于盆地中心和斜坡区的页岩地层中（图3），源储共生，没有明显的圈闭界限，突破了常规油气"藏"的概念，勘探战略转向"甜点"。对于页岩气勘探来说，"甜点"平面上是指具有工业开采价值的非常规油气高产富集"甜点区"，剖面上是指富有机质黑色页岩层段内，经过人工改造可形成工业价值的页岩气高产"甜点段"。川南五峰—龙马溪组海相页岩气勘探开发实践表明，工业性页岩气资源形成需满足以下条件：（1）含气量高，通常含气量要求大于$3m^3/t$，以保证页岩气藏有较好的物质基础；（2）孔隙度高，高产区孔隙度大于4%，且孔隙类型以有机质孔隙为主；（3）高TOC集中段连续厚度大，优质页岩段有机碳含量普遍大于4%，集中段厚度通常大于15m；（4）分布面积大，大于100～130km^2，足够的面积和连续厚度是工业性页岩气的资源基础；（5）热演化程度适中，以R_o值2.0%～3.0%最佳，此时处于液态烃裂解生气最佳窗口内；（6）埋藏深度适中，一般介于2000～4500m；（7）保存条件好，自身具有良好的顶底板保存条件，且不发育大的断裂构造，地层超压，压力系数通常大于1.2；（8）可压裂性好，包括储层脆性高，水平地应力差值小等，使得页岩地层能够得到充分改造，从而获得高产。

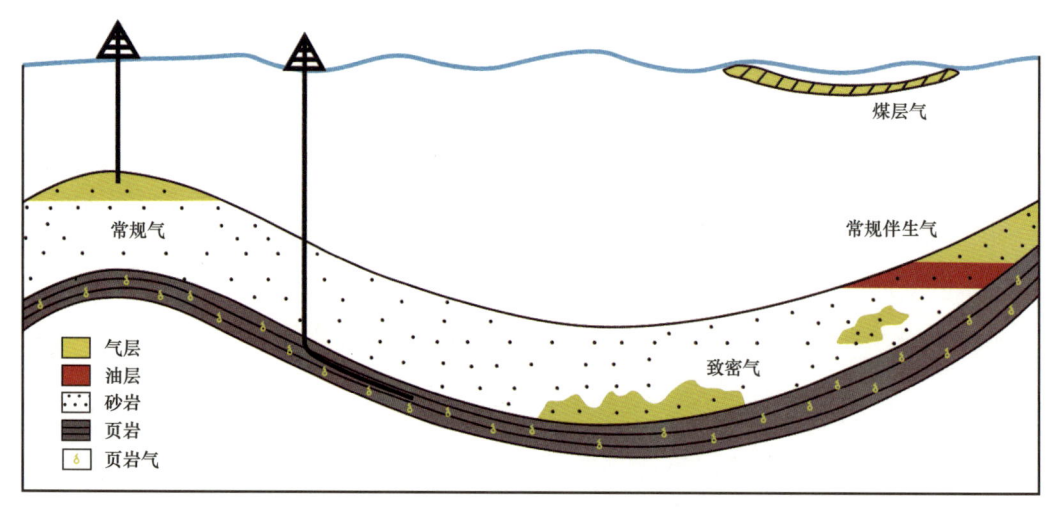

图3 非常规油气与常规油气剖面分布示意图

3 开发关键配套技术

在近10年来的页岩气勘探开发过程中，中国形成了埋深3500m以浅的页岩气六大开发主体技术系列（图4），掌握了储层特点及生产规律，提高了单井产量、有效控制了开发成本。

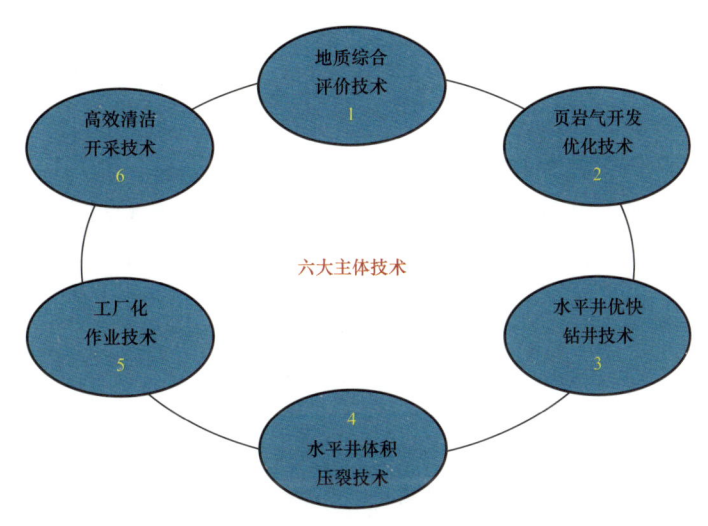

图4 中国页岩气开发主体技术

3.1 地质综合评价技术

在勘探评价阶段，提前开展气藏精细描述工作，创新形成了多期构造演化、高—过成熟页岩气地质综合评价技术。通过开发小层划分、岩心纹层刻画、高精度地球物理预测和沉积相成因分析等手段，形成了页岩储层描述的理论基础。主要内容包括：（1）测井与岩心分析相结合，明确了龙马溪组的地层结构；（2）海相深水陆棚沉积环境，决定了优质页岩储层的"大甜点"分布特征；（3）地质与工程因素综合分析，构建了优质页岩储层表征关键参数。

目前形成了以复杂山地页岩气地震采集处理及系列参数解释技术、水平井存储式测井系列设备及评价技术、主力开发小层划分技术为主体的地质综合评价技术。将开发层段研究尺度从几十米精细到几米，优质靶体位置明确为龙马溪组龙一$_1^1$小层，同时建立以综合系数（联合地质和工程指标以整体反映页岩气井生产能力[36]）为评价标准的页岩气井综合分类标准（表2）和以动态储量为标定的地质储量计算方法。地质综合评价技术为页岩气地质—工程"甜点区"优选和有效开发提供了地质依据。

表2 中国南方海相页岩气井综合分类标准[36]

气井类型	综合分类系数	试气产量/($10^4 m^3/d$)	单井累计产量/$10^4 m^3$	内部收益率/%
Ⅰ	>1.603	>9.62	>8380	>12
Ⅱ	1.456～1.603	8.64～9.62	7651～8380	8～12
Ⅲ	<1.456	<8.64	<7651	<8

3.2 开发优化技术

在页岩气开发的不同阶段面临着不同的开发优化问题：开发部署阶段优化水平井井距（即巷道间距）、钻井阶段优化水平井箱体层位（即靶体位置优化）、完井阶段优化完井压裂方案（即水平段及压裂参数优化）、投产阶段优化单井生产制度。

（1）水平井靶体位置是水平井获得高产的重要地质保证，也有利于优质储层充分改造。开发实践表明，水平井开发效果（初始产量及累计产量）与优质储层钻遇长度相关性高，五峰组上部—龙一$_1^3$是优质导向窗口，其中长宁地区水平井测试产量与龙一$_1^1$+龙一$_1^2$小层钻遇长度相关度高，威远地区水平井测试产量与龙一$_1^1$小层钻遇长度相关度高[37]。

（2）水平井及压裂参数是水平井获得高产的重要工程保证。水平井产液剖面结果显示无产能贡献的射孔簇平均占比45%，无产能贡献的压裂级占比超过20%[38]。综合考虑储层品质和完井品质，结合地质条件和压裂施工条件限制，将储层物性和完井参数相近的层段划分为同一压裂段并优选射孔位置以降低压裂段内应力差异，打破"几何完井"设计方法的盲目性，最大程度增加储层压裂设计的均匀性和有效性。

（3）水平井间距是气藏提高储量动用程度的关键。EIA发表至2018年6月所有钻完井的平均情况，按水平段平均长度1370m取值，平均井距约315m。川南页岩气形成了以生产干扰和压裂干扰为判别技术的开发井距优化技术，长宁、威远、昭通等井区开发井距从最初的400～500m逐步缩小至目前300m，井间储量得到有效动用，区块采出程度可从25%提高至35%左右[39]。

（4）单井生产制度是保证气井EUR最大化的保障。2010年以前，美国Haynesville页岩气田基本采用大油嘴生产，2010年后考虑到该地区地层压力较高、应力敏感性显著，逐渐转变为控压限产方式。长宁—威远、昭通等页岩气示范区的地质条件与Haynesville相似，经过几年的现场试验后也开始推广使用控压限产的生产方式，大量开发实践证明了控压生产较放压生产可普遍提高28%的单井EUR[40-41]。

3.3 水平井优快钻井技术

经过多年页岩气钻采工程现场试验，通过优化钻井液性能、钻井参数和钻具组合，缩短钻井周期，提高机械钻速，中国基本形成了3500m以浅五峰组—龙马溪组海相页岩气水平井优快钻井技术[42-43]。通过不断技术优化，形成了成熟的配套技术（表3），包括：井身结构优化、直井段高效马达+个性化PDC钻头、造斜段和水平段旋转导向精确控制井眼轨迹。

表3 钻井优化前后参数对比

关键参数	优化前	优化后
井身结构	四开四完	三开三完
井眼轨迹	三维井	双二维井
钻井参数	常规参数	大排量，高钻压
钻井装备	35MPa泵及管汇	52MPa泵及管汇
定向工具	PDC+弯螺杆	PDC+旋转导向
钻井液体系	水基	油基

中国石化涪陵页岩气田焦页 22-S1HF 井创造了 2536m 的国内超长水平段"一趟钻"最长的钻井记录，为 2000m 以上超水平段页岩气开发积累了经验。同时"高性能水基钻井液"的成功试验，进一步降低了钻井周期和环保风险，长宁区块单井钻井周期由 139 天缩短至 69 天，最短 27.6 天，水平段最长达到 2810m。海相页岩气钻井经验表明，为了获得高产井要保证水平段龙一$_1^1$小层有利储层的钻遇率、控制优质井眼轨迹、确保优质的井身质量，同时要实现快速钻井，有效控制钻井作业成本。近两年来，中国页岩气在 3500m 以深主体钻井工艺也取得技术突破，钻井周期由 210 天下降至 120 天，Ⅰ类储层钻遇率由 50% 提高到 90% 以上。

3.4 水平井体积压裂技术

对页岩储层而言，气井无自然产能或自然产能极低，需要经过大规模压裂改造后才能有效投产，形成"人造气藏"。随着水平段长度的增加，相应的压裂强度不断增加，通过增加压裂级数、减小段间距、增加压裂簇数、提高支撑剂浓度、暂堵转向、加砂压裂和提高压裂液用量等一系列技术措施，可以有效增加储层改造强度，实现"超级缝网"，进而提高单井产能。

经过持续攻关试验，中国自主研发的体积压裂技术、工具已成熟配套，并实现了规模化应用，成为低成本开发的关键技术之一[44]。目前形成以"大排量低黏滑溜水 + 低密度高强度支撑剂 + 可溶桥塞 + 拉链式压裂模式"和"密切割分段分簇 + 高强度加砂技术 + 暂堵转向（多级）压裂"为主的 3500m 以浅的水平井体积压裂装备及工艺技术，"千方砂、万方液"的大规模体积压裂已经成为中国页岩气体积改造的标志。针对深层页岩高应力差、高闭合应力特征，初步形成的深层主体压裂工艺采用了"低黏滑溜水、大液量高排量泵注 + 大粒径支撑剂、高强度加砂"主体压裂工艺，实现了复杂缝网体积压裂。当前，分段更短、簇数更多、加砂强度更大的新一代改造技术正在川南页岩气开发区内积极推广应用，未来有信心将单井 EUR 提高到（1.5～2.0）×$10^8 m^3$。

3.5 工厂化作业技术

水平井 + 多级压裂的开发模式广泛应用于页岩气开采。通过改变压裂模式和优化压裂参数在地层内产生一定的诱导应力，促使地应力大小和方向发生变化，从而大幅度提高网络裂缝复杂程度、增加气藏的增产改造体积[45]。中国页岩气通过引进、消化吸收、再创新的方式，逐渐形成了适用于复杂山地海相页岩气开发的"井工厂"作业模式：双钻井作业、批量化钻井、拉链式压裂、统一供水供电。完全实现了钻井压裂"工厂化布置、批量化实施、流水线作业"和"资源共享、重复利用、提高效率、降低成本"的目标。通过加强对各技术环节的把控，提高了施工效率，降低了单井投资，采用页岩气水平井组"工厂化"作业技术，形成了川南页岩气工厂化技术指标体系，设备安装时间减少 70%，压裂作业效率提高 50%，钻井作业效率提高 50% 以上，钻井液压裂液回用率达 85%。

3.6 高效清洁开采技术

形成了以"标准化设计技术 + 组合式橇装技术 + 数据采集与数据集成技术 + 实时监测与远程控制技术 + 协同分析与辅助决策技术"为主体的地面采输与数字化气田建设技

术,达到了工厂化预制、模块化安装、快建快投、重复利用的目的,实现了平台无人值守、井区集中管控,远程支持协作。井均地面投资由1200万元降至700万元以内,信息化覆盖率超过90%,操作成本控制在220元/10^3m^3。形成了以"土地保护技术+地下水保护技术+地表水保护技术+重复利用技术"为主体的清洁开采技术,实现了井下无窜漏,钻井压裂废弃物无害化处理,压裂返排液回收重复利用,减少了土地占用率,有效保护了环境。实施后减少土地占用70%以上,压裂返排液重复利用率85%以上,废弃物无害化处理率100%,地表、地下水质监测未发现异常。

4 页岩气商业开发成功经验

中国已经落实了页岩气资源基础,证明了具有商业开发价值,经过多年的勘探开发实践,页岩气实现了从无效资源向单井有效产量的技术跨越,中国海相页岩气已逐步进入精细化开发阶段,既要产量也要效益[46-49]。通过系统总结上文所述的页岩气井高产地质和工程主控因素,将页岩气实现商业开发的经验总结为以下4条。

4.1 水平井靶体层位优选

受制于川南海相页岩特殊的地质和工程条件,靶体位置不仅决定着资源基础,也影响钻井压裂质量。由于四川盆地五峰组—龙马溪组海相页岩水平层理发育,限制了压裂裂缝缝高延伸,即使加大加砂强度及施工排量也无法实现对整个优质页岩段的充分改造[50-51],只有选准靶体才能保证井筒附近层位得到充分改造。

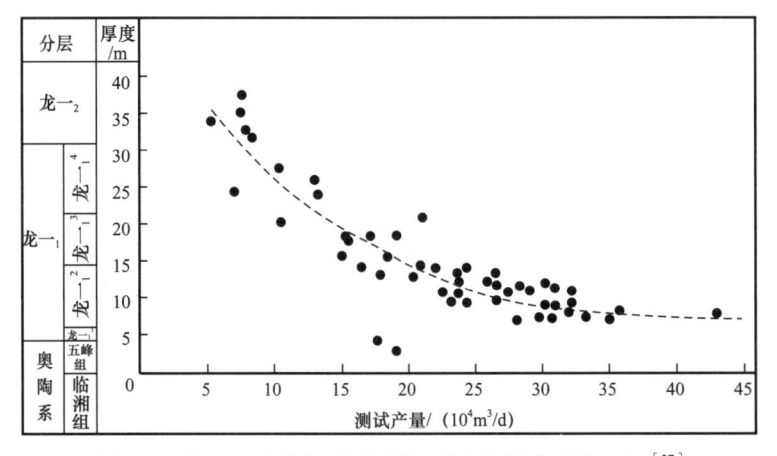

以长宁—威远页岩气示范区为例,中国石油不断探索水平井箱体位置,最终获得了优化的靶体层位。在第一轮开发过程中水平井靶体位于龙一$_1$亚段下部,距页岩底部较高,产能建设实施效果一般,长宁、威远井均测试日产气量分别为$10.9×10^4m^3$和$11.6×10^4m^3$,井均EUR分别为$0.53×10^8m^3$和$0.41×10^8m^3$;第二+三轮实施过程中,通过优选纵向上地质和工程"甜点",将水平井靶体下沉至五峰组—龙一$_1^1$层,产能建设效果明显好于第一轮,测试日产量、井均EUR均有所提高(图5)。在长宁区块,建产区最佳靶体位置优选到龙一$_1$亚段1~2小层,箱体高度5m,保证单井效果最好,从2014年Ⅰ+Ⅱ类井比例25%,提高到目前Ⅰ类井比例100%。

图5 长宁201井区水平井靶体层位与测试产量关系图[37]

4.2 优快钻进和高效体积改造

在确定最优靶体层位基础上，通过优化开发技术获得合理的完井压裂参数（压裂段数、射孔簇数、段间距、簇间距等）、压裂工艺参数（压裂液类型、支撑剂类型及参数、施工排量、加砂浓度等），进而使用"优快钻井＋高效体积压裂"技术实施最优设计方案，切实发挥技术组合优势，起到"1+1＞2"的协同效应。

中国页岩气开发经验表明，"长水平井＋超级压裂"工程技术直接推动页岩气增产增效[52-53]。长宁和焦石坝钻井实施效果证实，优快钻井技术能保证水平井按设计轨迹方位、方向钻进，在提高Ⅰ类储层钻遇长度同时提高井筒完整性，这是提高单井产能和EUR的基础（图6）。高效体积改造技术既能增加缝网产能也能降低套变、压窜风险。

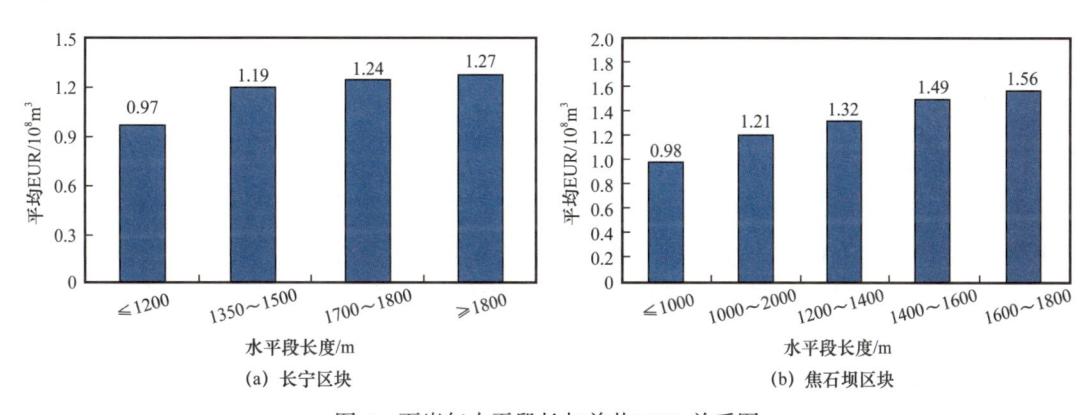

图6 页岩气水平段长与单井EUR关系图

4.3 地质工程一体化及数据优化

采用地质工程一体化研究、一体化设计、一体化实施，全面在作业公司推广精细化管理模式，消除组织上的工作障碍和技术上的人为切割，确保"定好井、钻好井、压好井、管好井"，实现储量整体动用，优化综合开发效益[53-55]。在地质工程一体化平台下对地下资源进行全生命周期的仿真模拟（图7），通过使用三维地质建模、地应力模拟和复杂缝网模拟打造"透明"页岩气藏，通过井位部署设计、钻井工艺设计、压裂工艺设计实现三维空间优化部署和设计，通过轨迹精确控制、压裂实时调整、生产动态管理获得最大的单井EUR。同时在一体化研究模式中，不同技术环节中的数据信息融合是关键，进一步利用大数据、云平台等先进信息技术，建立页岩气勘探开发关键数据库，打破"数据孤岛"，强化地质工程一体化研究[56-57]。

长宁区块在第三轮方案实施过程中全面推进地质工程一体化技术，页岩气井Ⅰ类储层钻遇率、井筒完整性、体积改造效果显著提升，单井产量、EUR大幅度提高，奠定了未来大规模建产的基础。同时通过构建页岩气全生命周期的数据湖和知识库系统平台，基于平台规则来甄选地质工程数据，实现多专业数据高效连通和参数交互优化，促进了一体化的项目组织和管理框架[37]。

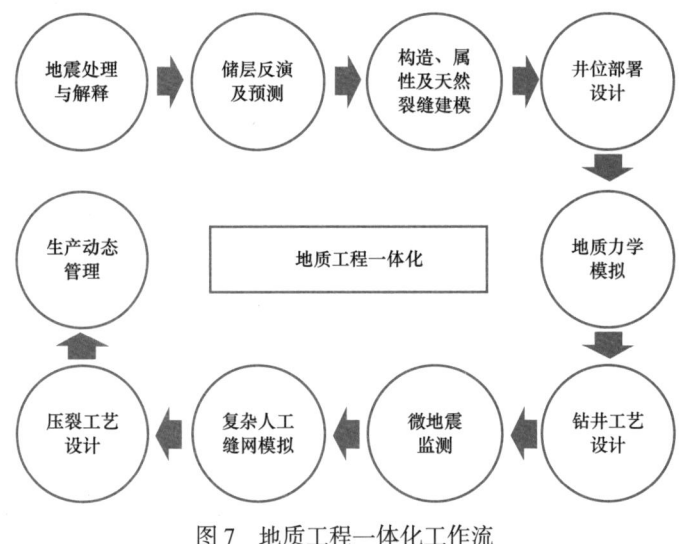

图 7　地质工程一体化工作流

4.4　先进生产组织管理模式

中国页岩气上产的资源基础是落实的，但每年页岩气上产工作量大，参与单位多，保工程施工质量、控降成本与组织难度大，需精心运筹，以保进度。2019—2035年页岩气快速上产年均需钻井500口以上，是2018年工作量的近2倍，保障质量和进度难度大。目前页岩气开发有企业自营、国内合资、风险作业、对外合作等多种经营模式、十几个主体参与，涉及管理部门和参与单位多，高效协调和有序组织难度较大[17]。同时企地关系、地质灾害等非资源因素存在较大不确定性，影响上产进度。良好的企地关系和利益共享机制，国家对页岩气的扶持政策为页岩气开发创造了良好的宏观环境。

中国石油建立了"总部领导小组、前线指挥部、地方油公司＋工程服务公司"的三级管理体制，采取了"国际合作、国内合作、风险作业、自营开发"四种生产作业机制，推广应用"井位部署平台化、钻井压裂工厂化、采输作业橇装化、工程服务市场化、生产管理数字化、组织管理一体化"的"六化"管理模式，支撑川南页岩气大规模快速建产。中国石化形成了一套包括"组织机构、运行计划、施工组织、企地关系"为主体的体制机制。通过整合各方资源和优势，提速、提质、提效页岩气开发，形成了成熟的管理组织模式，包括：（1）稳定队伍，实现最佳学习曲线；（2）甲乙方通力合作，实施统一计价方法、统一激励措施、统一资源配置的"三统一"降本增效配套措施，共同控降成本；（3）良好的企地关系与利益共享的机制，积极争取政府财政政策支持，合理劈分国家、公司、资源地三者利益，为页岩气开发创造良好的社会环境。

5　发展前景及展望

5.1　页岩气资源的现实性与资源基础

四川盆地及周缘奥陶系五峰组—志留系龙马溪组超压海相页岩气，埋深4500m以浅面积$3.09 \times 10^4 km^2$，可采资源量约为$3.78 \times 10^{12} m^3$（表4）。仅中国石油在川南地区5个区块埋深4500m以浅可工作面积$1.83×10^4 km^2$，可采资源量达到$1.9×10^{12} m^3$，因此四川盆地及周缘奥陶系五峰组—志留系龙马溪组海相页岩气是目前持续上产的最现实领域。

表4　四川盆地及周缘五峰—龙马溪组页岩气资源评价结果

目标区 名称	面积 /km²	有效厚度 /m	埋深 /m	可采资源 /10⁸m³
长宁	3980	40～80	2000～4500	3483
叙永	450	20～40	2000～3000	126
威远	2748	20～60	2600～4500	2198
富顺—永川	6446	80～100	3000～4500	14504
内江—大足	3799	40～80	2000～3500	3799
璧山—江津	3680	40～80	2600～4500	5520
江安—水富—屏山	3340	60～100	2000～3500	5010
犍为	2010	10～50	3500～4500	531
江津东线状区块	361	20～40	4000～4500	190
綦江北线状区块	280	20～30	4000～4500	123
江北—邻水	810	20～30	3500～4350	227
涪陵	416	60～90	2200～4500	510
巫山—巫溪	2560	20～60	4000～4500	1536
合计	30880			37757

四川盆地及周缘寒武系筇竹寺组海相页岩同属于深水陆棚沉积环境，埋深4500m以浅面积$1.1×10^4 km^2$，可采资源量$0.8×10^{12} m^3$，与龙马溪组相比，筇竹寺组物性和含气性略差，地质品质与工程品质较差（表5），是未来重要的接替领域。

表5　四川盆地及周缘龙马溪组与筇竹寺组页岩基础参数对比

层系	沉积环境	有机碳 /%	孔隙度 /%	含气量 /(m³/t)	脆性矿物 /%	高TOC集中段连续厚度 /m	R_o /%
龙马溪组	深水陆棚	2.2～4.8	3.1～7.2	1.7～8.4	40～80	20～80	2.0～3.0
筇竹寺组	深水陆棚	4.3～8.1	0.9～1.9	0.8～2.8	52～95	60～135	2.5～3.4

四川盆地侏罗系、上二叠统龙潭组、鄂东本溪—太原组等非海相页岩气（包括海陆过渡相+陆相），可采资源量$4.2×10^{12} m^3$（表6）。与海相页岩相比，孔隙度低（0.5%～3%）、脆性矿物含量低（石英含量10%～30%）、储层横向不连续，单井测试日产量仅为数万立方米，目前尚不能有效开发，但随着未来勘探开发技术进步，是潜在开发领域。

表 6 中国非海相页岩气可采资源量统计表

地区	可采资源量 /$10^{12}m^3$ 海陆过渡相	陆相
四川盆地	0.9	1.14
鄂尔多斯盆地	0.8	0.2
渤海湾盆地	0.1	0.1
中扬子盆地	0.5	—
下扬子盆地	0.2	0.2
松辽盆地	—	0.09
塔里木盆地	—	0.08
准噶尔盆地	—	0.05
吐哈盆地	—	0.03

5.2 页岩气产量发展趋势与地位

受消费需求驱动，中国天然气产量将持续快速增长，大力发展天然气是消费需求增长和能源结构转型的客观需求。页岩气资源潜力较大，是天然气倍增发展的生力军，未来将保持高峰增长态势，其对整个天然气产业链意义非常重大，是未来高质量发展的重中之重。

海相页岩气是未来中国页岩气上产的主体，目前已经形成了商业化开发。其中，3500m 以浅主体开发技术配套成熟，单井综合投资不断降低，具备快速上产的基础和条件；3500～4000m 开发技术逐渐成熟，4000～4500m 关键技术不断攻关，深层海相页岩气未来产量增长可期，其形成产能将主要用以弥补产量递减。按照"向烃源岩要产量"的思路，近期在非海相页岩气区块部署的几口评价井有望获得突破，未来也将具备一定的开发潜力。

需要强调的是，中国页岩气发展也面临着诸多现实挑战。四川盆地及周缘五峰组—龙马溪组页岩气井初期采用"几何均匀压裂、面积均匀布井、全程放压生产"的开发策略，忽视了井间地质品质和工程品质的差异性，导致 1/3 以上压裂段无产能贡献、大量储量难以动用，区块整体采收率仅有 20% 左右；相对于美国，中国页岩气开发井距较小，在不同批次水平井压裂投产过程中，井间压窜现象普遍存在，新井、老井开发效果均受严重影响，EUR 损失在 25% 以上。而且未来随着深层海相、非海相页岩气资源的进一步动用，储量品位更低、地下条件更复杂，现场工作量和开发效益将是页岩气发展的主要受限点。只有聚焦关键技术创新、探索"一井一藏"精细化开发模式、加强未动用储量和尾矿利用的政策支持、走低成本之路，中国页岩气未来方可持续。

6 结论及建议

经过近 10 年的科技攻关和勘探开发实践，中国页岩气在地质评价、开发评价、水平井钻完井、体积压裂及清洁开采等不同技术领域取得了长足进步，积累了宝贵经验，形成一套适用于 3500m 以浅的海相页岩气勘探开发理论体系和技术装备。以五峰组—龙马溪组为代表的海相页岩气已经实现了商业性开发，深层海相及非海相页岩气领域也不断取得突破，探明储量、年产量逐年增加，为未来形成更大产量规模提供了坚实的资源基础和技术保障。

页岩气勘探开发过程中多学科、多专业相互支撑、相互配合，实行一体化研究、一体化实施、一体化管理是页岩气未来实现快速上产和较长时间稳产的技术关键，既要产量更要效益。

虽然中国页岩气潜力较大，但面临的挑战多。为实现页岩气产业快速发展，提出以下建议：（1）加强深层页岩气低成本关键技术与装备攻关，加快深层体积压裂工艺、非海相薄层精准钻井技术、全可溶桥塞、旋转导向工具等关键技术与装备攻关，释放低品位页岩气潜在产能；（2）开发需关注提高区块采收率问题，应从立体井网、合理井距与生产制度多方面综合施策，从纵向、平面及改造区内部提高储量动用；（3）实行差异化税费政策，给予页岩气稳定的扶持政策，可考虑对应用长水平段、高性能水基钻井液等新技术，探索筇竹寺组、非海相层系等新层系，进一步提高页岩气开发补贴水平。

参 考 文 献

[1] 赵文智，李建忠，杨涛，等.中国南方海相页岩气成藏差异性比较与意义［J］.石油勘探与开发，2016，43（4）：499–510.

[2] Jenkins C D, Bayer C M. Coalbed-and shale-gas reservoirs［J］. Journal of Petroleum Technology, 2008, 60（2）: 92–99.

[3] Ozkan E. The way ahead for US unconventional reservoirs［J］. The Way Ahead, 2014, 10（3）: 37–39.

[4] 邹才能，董大忠，王玉满，等.中国页岩气特征、挑战及前景（二）［J］.石油勘探与开发，2016，43（2）：166–178.

[5] 张金川，姜生玲，唐玄，等.我国页岩气富集类型及资源特点［J］.天然气工业，2009，29（12）：109–114.

[6] 李玉喜，聂海宽，龙鹏宇.我国富含有机质泥页岩发育特点与页岩气战略选区［J］.天然气工业，2009，29（12）：115–120.

[7] 董大忠，王玉满，黄旭楠，等.中国页岩气地质特征、资源评价方法及关键参数［J］.天然气地球科学，2016，27（9）：1583–1601.

[8] 林腊梅，张金川，唐玄，等.中国陆相页岩气的形成条件［J］.天然气工业，2013，33（1）：35–40.

[9] 何发岐，朱彤.陆相页岩气突破和建产的有力目标——以四川盆地下侏罗统为例［J］.石油实验地质，2012，34（3）：246–251.

[10] 马永生，蔡勋育，赵培荣.中国页岩气勘探开发理论认识与实践［J］.石油勘探与开发，2018，45（4）：561–574.

[11] 王香增，高胜利，高潮.鄂尔多斯盆地南部中生界陆相页岩气地质特征[J].石油勘探与开发，2014，41（3）：294-304.

[12] 张大伟.《页岩气发展规划（2011—2015年）》解读[J].天然气工业，2012，32（4）：6-10.

[13] 邹才能，张国生，杨智，等.非常规油气概念、特征、潜力及技术——兼论非常规油气地质学[J].石油勘探与开发，2013，40（4）：385-399.

[14] 王志刚.涪陵页岩气勘探开发重大突破与启示[J].石油与天然气地质，2015，36（1）：1-6.

[15] 郭旭升.涪陵页岩气田的发现与勘探认识[J].中国石油勘探，2016，21（3）：24-37.

[16] 董大忠，王玉满，李新景，等.中国页岩气勘探开发新突破及发展前景思考[J].天然气工业，2016，36（1）：19-32.

[17] 马新华，谢军.川南地区页岩气勘探开发进展及发展前景[J].石油勘探与开发，2018，45（1）：161-169.

[18] Loucks R G, Reed R M, Ruppel S C, et al. Morphology, genesis, and distribution of nanometer-scale pores in siliceous mudstones of the Mississippian Barnett shale [J]. Journal of Sedimentary Research, 2009, 79 (12): 848-861.

[19] Loucks R G, Reed R M, Ruppel S C, et al. Spectrum of pore types and networks in mudrocks and a descriptive classification for matrix-related mudrock pores [J]. AAPG Bulletin, 2012, 96 (6): 1071-1098.

[20] Mastalerz M, Schimmelmann A, Drobniak A, et al. Porosity of Devonian and Mississippian new albany shale across a maturation gradient: Insights from organic petrology, gas adsorption, and mercury intrusion [J]. AAPG Bulletin, 2013, 97 (10): 1621-1643.

[21] 高凤琳，宋岩，梁志凯，等.陆相页岩有机质孔隙发育特征及成因——以松辽盆地长岭断陷沙河子组页岩为例[J].石油学报，2019，40（9）：1030-1044.

[22] Ross D J K, Bustin R M. Shale gas potential of the Lower Jurassic Gordondale Member, northeastern British Colunbia, Canada [J]. AAPG Bulletin, 2007, 55 (1): 51-75.

[23] 赵文智，王兆云，王红军，等.再论有机质"接力成气"的内涵与意义[J].石油勘探与开发，2011，38（2）：129-135.

[24] 罗小平，吴飘，赵建红，等.富有机质泥页岩有机质孔隙研究进展[J].成都理工大学学报（自然科学版），2015，42（1）：50-59.

[25] Curis M E, Cardott B J, Sondergeld C H, Rai C S. Development of organic porosity in the Woodford shale with increasing thermal maturity [J]. International Journal of Coal Geology, 2012, 103: 26-31.

[26] Jarvie D M. Unconventional shale-gas systems: The Mississippian Barnett Shale of north-central Texas as one model for thermogenic shale gas assessment [J]. AAPG Bulletin, 2007, 91 (4): 475-499.

[27] 朱汉卿，贾爱林，位云生，等.低压气体吸附实验在页岩孔隙结构表征中的应用[J].东北石油大学学报，2017，41（6）：36-47.

[28] 孙健博，孙兵华，赵谦平，等.鄂尔多斯盆地富县地区延长组7湖湘页岩油地质特征及勘探潜力评价[J].中国石油勘探，23（6）：29-37.

[29] 郭旭升，胡东风，魏志红，等.涪陵页岩气田的发现与勘探认识[J].中国石油勘探，2016，21（3）：24-37.

[30] 武恒志, 熊亮, 葛忠伟, 等. 四川盆地威远地区页岩气优质储层精细刻画与靶窗优选[J]. 天然气工业, 2019, 39（3）: 11-20.

[31] Reed R M, Loucks R, Milliken K L. Heterogeneity of shape and microscale spatial distribution in organic-matter-hosted pores of gas shales[C]//AAPG Annual Convention and Exhibition, Abstract. Long Beach: CA, 2012: 22-24.

[32] 朱汉卿, 贾爱林, 位云生, 等. 蜀南地区富有机质页岩孔隙结构及超临界甲烷吸附能力[J]. 石油学报, 2018, 39（4）: 391-401.

[33] 端祥刚, 胡志明, 高树生, 等. 页岩高压等温吸附曲线及气井生产动态特征实验[J]. 石油勘探与开发, 2018, 45（1）: 119-127.

[34] 赵文智, 王兆云, 张水昌, 等. 有机质"接力成气"模式的提出及其在勘探中的意义[J]. 石油勘探与开发, 2005, 32（2）: 1-7.

[35] 贾爱林, 位云生, 金亦秋. 中国海相页岩气开发评价关键技术进展[J]. 石油勘探与开发, 2016, 43（6）: 949-955.

[36] 谢军, 赵圣贤, 石学文, 等. 四川盆地页岩气水平井高产的地质主控因素[J]. 天然气工业, 2017, 37（7）: 1-12.

[37] Miller C, Waters G, Rylander E. Evaluation of production log data from horizontal wells drilled in organic shales[C]//SPE-144326-MS presented at the SPE North American Unconventional Gas Conference and Exhibition, Texas, USA, 14-16 June 2011.

[38] 位云生, 王军磊, 齐亚东, 等. 页岩气井网井距优化[J]. 天然气工业, 2018, 38（4）: 129-137.

[39] 贾爱林, 位云生, 刘成, 等. 页岩气压裂水平井控压生产动态预测模型及其应用[J]. 天然气工业, 2019, 39（6）: 71-80.

[40] 焦方正. 非常规油气之"非常规"再认识[J]. 石油勘探与开发, 2019, 46（5）: 803-810.

[41] 熊小林. 威远页岩气井EUR主控因素量化评价研究[J]. 中国石油勘探, 2019, 24（4）: 532-538.

[42] 钟城, 秦启荣, 胡东冈, 等. 川东南丁山地区五峰组—龙马溪组页岩气藏"六性"特征[J]. 油气地质与采收率, 2019, 26（2）: 14-23.

[43] 梁兴, 徐进宾, 刘成, 等. 昭通国家级页岩气示范区水平井地质工程一体化导向技术应用[J]. 中国石油勘探, 2019, 24（2）: 226-232.

[44] 习传学, 高东伟, 陈新安, 等. 涪陵页岩气田西南区块压裂改造工艺现场试验[J]. 特种油气藏, 2018, 25（1）: 155-159.

[45] 位云生, 齐亚东, 贾成业, 等. 四川盆地威远区块典型平台页岩气水平井动态特征及开发建议[J]. 天然气工业, 2019, 39（1）: 81-86.

[46] 潘继平. 非常规天然气资源开发政策困境及对策建议[J]. 石油科技论坛, 2019, 38（1）: 1-7.

[47] 康玉柱. 中国非常规油气勘探重大进展和资源潜力[J]. 石油科技论坛, 2018, 37（4）: 1-7.

[48] 侯启军, 何海清, 李建忠, 等. 中国石油天然气股份有限公司近期油气勘探进展及前景展望[J]. 中国石油勘探, 2018, 23（1）: 1-13.

[49] 金之钧, 蔡勋育, 刘金连, 等. 中国石油化工股份有限公司近期勘探进展与资源发展战略[J]. 中国石油勘探, 2018, 23（1）: 14-25.

[50] 位云生, 贾爱林, 何东博, 等. 中国页岩气与致密气开发特征与开发技术异同[J]. 天然气工业,

2017，37（11）：43-52.

[51] 刘乃震，高远文，王延瑞，等.提高四川盆地页岩气开发效益的工程方案［J］.天然气工业，2019，39（5）：93-98.

[52] 谢军.长宁——威远国家级页岩气示范区建设实践与成效［J］.天然气工业，2018，38（2）：1-7.

[53] 谢军，鲜成钢，吴建发，等.长宁国家级页岩气示范区地质工程一体化最优化关键要素实践与认识［J］.中国石油勘探，2019，24（2）：174-185.

[54] 李国欣，王峰，皮学军，等.非常规油气藏地质工程一体化数据优化应用的思考与建议［J］.中国石油勘探，2019，24（2）：147-152.

[55] 刘乃震，王国勇，熊小林.地质工程一体化在威远页岩气高效开发中的实践与展望［J］.中国石油勘探，2018，23（2）：59-68.

[56] 刘合，孟思炜，苏健，等.对中国页岩气压裂工程技术发展和工程管理的思考与建议［J］.天然气工业，2019，39（4）：1-7.

[57] 贾爱林.中国天然气开发技术进展及展望［J］.天然气工业，2018，38（4）：77-86.

未来十五年中国天然气发展趋势预测

贾爱林,何东博,位云生,李易隆

(中国石油勘探开发研究院)

摘　要: 国内天然气产量是中国天然气产业布局的基础,预测中国天然气中长期产量趋势对天然气产业发展具有重要意义。对比中美两国天然气发展特点,常规与非常规天然气并重发展的趋势是一致的,而中国非常规天然气对常规天然气主动接替的格局更加明显,更加主动。本文详细梳理了中国天然气发展在资源特点、开发技术、开发效益、组织与管理及安全环保等5个方面面临的挑战,探讨勘探开发理论技术进展与发展方向,评价常规与非常规天然气勘探增储和开发上产潜力,预测2035年常规天然气产量维持在$1350×10^8m^3$,非常规天然气产量上升至$1160×10^8m^3$,加上溶解气$90×10^8m^3$,国内天然气总产量达到$2600×10^8m^3$,为中国能源结构调整及进口管道和储气库战略布局提供决策基础。

关键词: 天然气;发展趋势;主动接替;技术进展;产业布局;决策基础

　　天然气作为优质燃料和清洁化石能源,在中国能源结构调整中发挥着越来越重要的作用[1]。鉴于中国天然气的资源禀赋和经济发展阶段,国内天然气产量与消费量的增速差距不断加大,保障国内天然气稳定供应已经成为中国天然气产业的重要使命。李剑等[2]主要从消费和需求侧预测了2030年中国天然气产业发展前景,宏观上提出了加大天然气勘探开发力度、加快互联互通体系建设、理顺价格机制和市场管控4个方面的产业发展建议。陆家亮等[3]、王建良等[4]以公开报道的资源量和储量数据为基础,根据广义翁氏模型、灰色—哈伯特组合模型、产量构成法、储采比控制法等方法,主要从资源的角度,预测了中国天然气产量的发展趋势和峰值,不足之处在于:(1)未考虑理论技术的进步和局限;(2)未细化到具体气田和不同开发阶段气藏的分类预测。本文对比中美天然气的发展特点,分析中国天然气勘探开发面临的主要挑战,依据探明储量增长和技术进步,分已建成气田、在建上产气田和勘探新区3种类型预测常规天然气产量趋势,分致密气、页岩气、煤层气3种类型预测非常规天然气产量趋势,进一步降低影响产量预测的不确定性因素,提升中国天然气相关产业中长期发展布局的可靠性。

1　中国天然气消费及供给格局

1.1　中国天然气消费量

　　进入"十三五",中国GDP增速保持6%以上,一次能源消费总量由2015年的

43×10^8t 标煤增长至 2019 年的 48.6×10^8t 标煤，伴随着国家经济持续快速发展和能源结构转型的快速推进，天然气在中国一次能源消费中的比例由 2015 年的 6.0% 上升至 2019 年的 8.4%，年均增长 $283\times10^8m^3$，年均增速达到 12% 以上，2019 年天然气消费量已达到 $3067\times10^8m^3$[5]。

1.2 中国天然气消费结构

从天然气消费结构来看，随着城镇化、农村地区煤改气以及环保驱动工业用气不断增加，天然气消费增量将继续释放，2019 年城市燃气、工业和发电用气仍是天然气消费增长的主要动力，分别占天然气消费总量的 38%、34%、20%，化工用气占比 8%。由于"煤改气"涉及的北方地区冬季清洁取暖计划目前已规划至 2021 年，预计未来两年中国天然气需求增长动力将主要来自城市燃气。

1.3 中国天然气供给格局

2006 年中国成为天然气净进口国，从 2007 年开始，中国的天然气进口量以年均 60% 以上的速度高速增长，逐步由国产气和东南沿海进口 LNG，发展到国产气、沿海进口 LNG 和中亚 A/B/C 管线、中缅管线、中俄管线进口管道气的多元化供给格局[6]。通过形式的多元化（进口 LNG 和管道气）、路线的多样化（建立西北、西南、东北和沿海四大进口通道）、国家的多样化（土库曼斯坦、乌兹别克斯坦、哈萨克斯坦、俄罗斯、缅甸、卡塔尔等国家）进行境外天然气的进口，分散进口风险，提高了进口气的供给安全[7]。

截至 2019 年底，中国投产的中亚 A/B/C 管线、中缅管线、中俄东线多条进口管道最大输气能力 $1050\times10^8m^3$/a，广西、海南、广东、福建、浙江、上海、江苏、山东、天津、河北、辽宁等地的 22 座进口 LNG 接收站设计接收能力 7285×10^4t/a，宏观上为弥补中国国产气的消费缺口、提升中国天然气供给安全发挥了极其重要的作用。2019 年中国国产天然气 $1777\times10^8m^3$，进口天然气 $1322\times10^8m^3$，其中，进口管道气 $508\times10^8m^3$，进口 LNG 5856×10^4t（$814\times10^8m^3$），对外依存度 43.1%[8-9]，有力支撑了中国经济发展和人民生活改善。

2 中美天然气勘探开发历程对比

2.1 美国天然气勘探开发历程

美国是世界上最早发展现代天然气工业的国家，1821 年，美国纽约州弗里多尼亚钻了一口 9m 深的气井，开始利用天然气，标志着美国现代天然气工业的诞生[10]。二战以来，美国天然气生产快速发展，大致可分为 3 个阶段（图 1）。

第一阶段（1945—1970 年），常规气快速增长期。美国常规天然气田主要集中在墨西哥湾沿岸、西部内陆和二叠三大盆地，主要储层集中分布在古近—新近系的砂岩、白垩系、二叠系、石炭系砂岩与石灰岩交互沉积储层中，以中小型气田为主，原始储量大于 $1000\times10^8m^3$ 的 17 个大气田中，13 个气田的埋深小于 3500m。特别是 1945—1970 年，美国天然气从陆上发展到墨西哥湾近海，开发技术和储产量快速增长，一直处于世界领

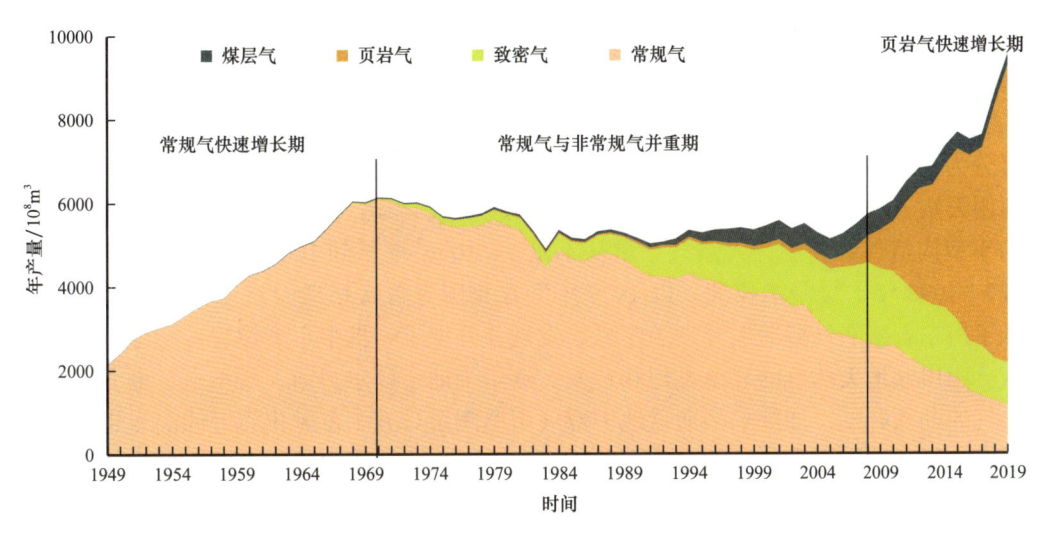

图 1 近 70 年美国天然气发展历程

先地位，1970 年常规气产量达到高峰，年产 $6086\times10^8m^3$，致密气和煤层气刚刚起步，产量仅有 $50\times10^8m^3$。

第二阶段（1970—2008 年），常规气与非常规气并重发展期。1970 年以后，美国开始对致密气、页岩气、煤层气开发出台税收抵免政策，大大提高了非常规气产量。西部和西南部新墨西哥州 San Juan 盆地、科罗拉多州 Piceance 盆地、怀俄明州 Greater Green River 盆地是美国致密气的主要产区，2008 年美国致密气产量达到高峰，年产 $1913\times10^8m^3$。同时，2007—2008 年，东部 Black Warrior 盆地浅煤层和 San Juan 盆地深煤层的规模开发，煤层气产量也达到高峰，年产 $540\times10^8m^3$。米歇尔能源公司经过多年的摸索和实践，将 3D 地震成像、水平井钻井、大型水力压裂和微地震压裂成像四项关键技术引入并完善到 Barnett 页岩气勘探开发实践中，推动 Barnett 区块页岩气规模上产至 $400\times10^8m^3$，助推美国页岩气产量达到 $600\times10^8m^3$ 以上，年增速超过 50%。2008 年非常规气产量首次超过常规气产量，美国天然气正式进入非常规时代。

第三阶段（2008 年至今），页岩气快速增长期。2008 年以后，随着页岩气开发关键技术的成熟与推广，美国页岩气呈现爆发式增长，TX-LA-MS Salt 盆地 Haynesville 区块、东部 Appalachian 盆地 Marcellus 区块、Utica 区块、Western Gulf 盆地 Eagle Ford 区块、Permian 盆地 Delaware 区块及 Arkoma 盆地 Fayetteville 区块等主力页岩气区块相继投入规模开发，在全球掀起了"页岩气革命"的热潮，2015 年页岩气产量突破 $4000\times10^8m^3$，在美国天然气总产量中的比例首次超过 50%，标志着美国进入页岩气时代。2019 年美国页岩气产量再创新高，达到 $7140\times10^8m^3$，占美国天然气总产量 75%，引领美国成为天然气净出口国[11]。

2.2 中国天然气勘探开发历程

中国是最早利用天然气的国家，西汉时期已有临邛火井，16 世纪开发的四川自流井气田，是大规模开发的第一个气田，天然气生产仅用作采卤制盐的燃料，地域局限于四川自贡市自流井地区，钻井深度均小于 200m。1835 年在四川自贡市大安区阮家坝山下由人工钻凿的燊海井，成为世界上最早挖掘的天然气深井，井深1001.42m，既产卤，又产气。早期发现和利用的天然气田均为浅层构造灰岩气藏。新中国成立后，中国现代天然气工业进入新的发展阶段，从四川地质勘探开发，延伸至陕甘宁、塔里木盆地及沿海地区，大致可分为3个阶段[12]（图2）。第一阶段（1949—1976年）发展起步期。依托四川盆地开发利用天然气的经验，新中国天然气工业在四川发展起步，主要以四川盆地二叠系、三叠系的小型构造气藏为主，天然气产量从不足 $1 \times 10^8 m^3$ 增长到 $100 \times 10^8 m^3$ 以上。第二阶段（1977—1998年）缓慢增长期。以1977年底在相国寺首次钻遇石炭系气藏为标志，由此揭开了川东石炭系气藏勘探开发的序幕，发现并开发了一批中小型裂缝—孔隙型背斜构造气藏。四川盆地天然气的规模发现和开发，逐步带动了鄂尔多斯、塔里木、渤海湾等盆地天然气的勘探开发，如鄂尔多斯盆地中部气田奥陶系马五1古风化壳溶蚀白云岩气藏、塔里木盆地柯克亚、渤海湾盆地黄骅坳陷板桥中渗砂岩边底水凝析气藏等，奠定了中国中浅层构造型边底水气藏的勘探开发技术基础，同时随着原油产量的增加，溶解气产量也快速增加，1998年天然气产量增长到 $220 \times 10^8 m^3$ 以上。第三阶段（1999年至今），快速增长期。1998年7月1日中国石油天然气总公司与中国石油化工总公司重组，形成中国石油与中国石化两大上中下游一体化的油气公司，两者的竞争关系推动了中国天然气勘探开发进程。2005年前，前陆冲断带构造天然气成藏理论、大面积源储共生砂岩天然气成藏理论的突破，相继指导了以克拉2、迪那2为代表的库车深层构造型大气区、以苏里格、大牛地、榆林、子洲、神木为代表的鄂尔多斯上古生界低渗透—致密砂岩型大气区、以徐深、克拉美丽为代表的火山岩气藏的系列重大发现，奠定了中国以中浅层碳酸盐岩为主的单一气藏类型向多种复杂类型气藏转变的发展格局。2005年以后，有机质接力生气理论、全过程生烃模式及连续型"甜点区"油气聚集理论的提出，指导了安岳、普光、元坝深层古老碳酸盐岩大气田、克拉苏超深层碎屑岩气藏及涪陵、长宁—威远等页岩气藏的发现和突破，引领中国天然气工业形成常规与非常规并重的勘探开发格局，按照"发现一类、攻关一类、成熟配套一类"的开发理念，攻关形成高精度三维地震、水平井、大规模压裂改

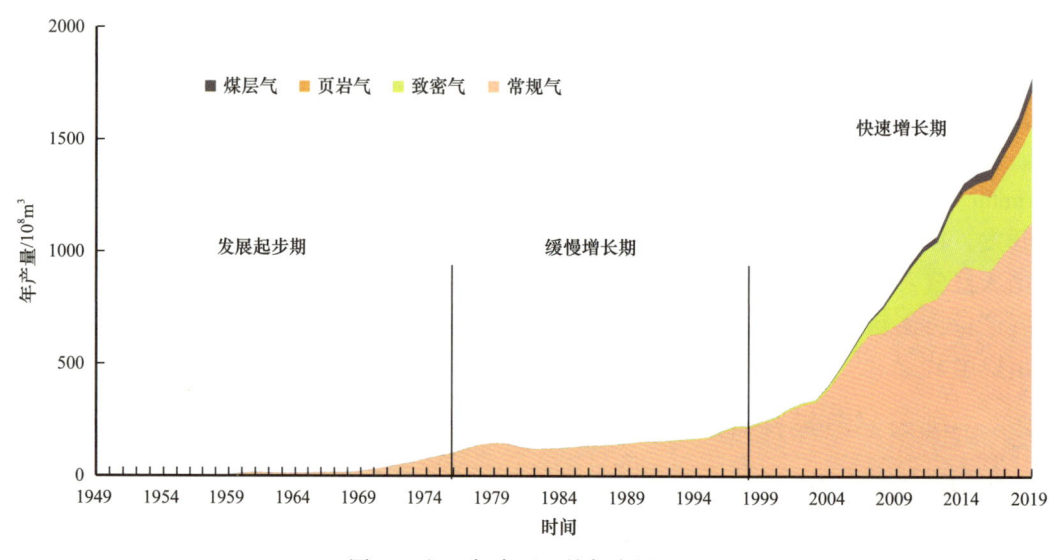

图2 近70年中国天然气发展历程

造3项核心关键技术利器，建成了中国陆上鄂尔多斯、四川、塔里木3个超大型天然气生产基地，其中，鄂尔多斯和四川2个生产基地探明储量超过$5×10^{12}m^3$、年产量超过$500×10^8m^3$，塔里木盆地探明储量超过$2×10^{12}m^3$、年产量超过$300×10^8m^3$，厚植了中国天然气未来持续上产的基础和潜力。

2.3 中美天然气发展对比与启示

中美天然气资源禀赋均差于中东和俄罗斯，因此天然气开采强烈依赖技术进步，美国天然气勘探开发技术目前处于世界领先地位，技术发展规律对中国天然气产业具有指导意义。美国常规气产量于1970年达到了峰值后，由于非常规气技术准备不足，天然气产量出现了较长时期的下滑，经过30多年的探索和实践，技术进步推动了非常规气对常规气的产量接替，美国天然气产量再次出现快速增长局面。美国非常规气产量快速增长为中国天然气发展带来的最大启示是，技术进步是推动天然气产业变革的最大动力，中国在天然气勘探开发方面作为"后发"国家，近年来通过引进和学习美国先进的非常规技术，实现了非常规气和常规气并驾齐驱快速发展的局面。随着一大批主力常规气田进入稳产中后期和非常规天然气的快速增长，中国非常规气"主动接替"常规气的趋势愈加明朗，非常规勘探开发技术进步将成为决定中国天然气生产的重要因素，技术自主与技术创新将成为未来天然气勘探开发的主旋律。

3 天然气勘探开发面临的挑战

3.1 资源劣质化成为常态

2005年以来，新发现气田以深层—超深层、复杂碳酸盐岩和非常规气为主[13]，近15年发现的常规优质大气田仅有库车山前克深构造带克深气田群、川中古隆起寒武系—震旦系安岳气田和川东上二叠统—下三叠统元坝气田，优质大气田发现难度和发现周期进一步加大。2006—2019年新增探明储量$11.5×10^{12}m^3$，其中，常规低品位（含低渗透强非均质碳酸盐岩、超深层致密含水碎屑岩、火山岩、基岩等）和非常规气储量占比超过70%，特别是近5年，占比逐年增加，2019年超过80%，其中页岩气占比53%、致密气占比19%、常规低品位占比8.6%。探明储量高峰增长与探明储量劣质化将成为常态。

3.2 深层—超深层等复杂气藏开发技术不成熟或效率低

常规气藏和非常规气深层化是必然趋势[14]。对于超深层气藏的勘探开发，目前技术发展空间较大，需求非常迫切。如超深层高精度三维地震成像技术、安全快速钻井技术、裂缝型致密储层改造技术、高温、高压、高含硫堵水、控水、排水技术、高抗硫大排量增压技术等。

非常规气开发技术和效益与国外差距较大，如中国页岩气开发在钻完井速度、施工质量、单位成本和实施效果等方面全面落后于北美（表1），除储层条件本身具有一定的差异外，主要原因是旋转导向等核心技术和组织方式等方面与北美存在较大差距，北美全部采用油公司主导的日费制，重要技术环节均采用全球招标，优选最先进的技术服务公司，保障实施进度、施工质量和实施效果，而国内以大包为主，成本、技术和质量受制于承包方，且难以改变。

表 1　中美页岩气关键开发参数对比表

参数指标	北美	中国	对比（北美/中国）
生产厚度 /m	15～20	10～15	基本相当
单台钻机每年钻井 /口	25～40	3～4	8～10 倍
平均水平段长度 /m	2000～2200	1500～1600	1.3 倍
一趟钻进尺 /m	2500	250	10 倍
水平段钻速 /（m/d）	1000～1500	30～50	30 倍
每压裂段簇数	7～8	3～5	基本相当
簇间距 /m	5～6	8～10	
单井 EUR/$10^8 m^3$	3～5	0.9～1.2	3～4 倍

3.3　常规低品位及非常规气开发效益差

截至 2019 年，中国天然气累计探明地质储量 $16.0 \times 10^{12} m^3$，其中，常规低品位及非常规气累计探明储量约 $8.0 \times 10^{12} m^3$，占比 50%，这一比例未来将进一步增大。按照不同气区不同气价和不同生产成本，评价 2019 年中国非效益产量约 $70 \times 10^8 m^3$，占比 3.9%，主要包括低品位碳酸盐岩、火山岩Ⅱ+Ⅲ类区、川中须家河组和鄂尔多斯盆地上古生界高含水区致密气、低品位煤层气、超深层低丰度气藏、海相页岩气Ⅲ类区和陆相及海陆过渡相页岩气低效区等。如川中须家河组、鄂尔多斯盆地上古生界高含水区致密气仅有 25% 的气井单井累计产量达到了效益边界；页岩气井不考虑补贴情况下，按照目前单井投资 7570 万元条件下，要实现 8% 的基准收益率，井均 EUR 需达到 $1.1 \times 10^8 m^3$，存在较大挑战[15]。

3.4　稳产与上产工作量大，工作量组织与管理难度大

以中国石油为例，保持中国石油 $1200 \times 10^8 m^3$ 以上的年产量规模持续上产至 2035 年达到产量规模（1600～1800）$\times 10^8 m^3$，每年需新建产能 $300 \times 10^8 m^3$ 以上，是"十三五"年均新建产能规模的 1.5 倍；年均钻井数量 4000 口，每年工作量是"十三五"钻完井总工作量的 2 倍以上。特别是塔里木超深层和川南页岩气钻井数量多，钻井周期长，对 70 型以上大功率钻井的需求量大，工作协调组织难度大。巨大工作量的组织过程中，还需要考虑和协调非资源非技术因素，如采矿权登记、采矿权内用地审批、泥石流、滑坡、地震等地质灾害等，这些因素多，且部分因素不可控，组织管理存在较大挑战。

3.5　安全环保风险大

建设美丽中国，构建绿色和谐家园，是中国现阶段发展的主基调。新的《安全法》和《环保法》对油气开发的要求更高，安评和环评更加细化，审批手续多。钻井、压裂施工对构造敏感区的影响、大量钻井废液和岩屑处理等难度不断增加。部分与保护区重叠的矿权，虽然国家多部门协调后，可以进行油气开发，但对安全环保的要求更为苛刻。

以四川盆地为例，中国石油矿权区与保护区重叠面积为 $2.2\times10^4 km^2$，占14%，资源量 $3.9\times10^{12}m^3$。中国石化也存在矿权与保护区重叠的类似情况。同时川渝地区人口稠密、环境敏感，安全环保和生产建设管控难度大。

4 天然气勘探开发理论技术及发展方向

伴随着风险与挑战，天然气勘探开发理论技术不断突破，引领储量和产量的持续增长。新理论、新技术、新工艺的更新换代将不断加快，逐步向精细化、一体化和智能化转变[16]。

4.1 天然气勘探理论技术及发展方向

新中国成立后，中国天然气工业从四川起步发展，1979年前主要油型气理论指导勘探发现，发现以中小型气藏为主。1979年煤成气理论的提出，奠定了中国陆相天然气地质理论基础。20世纪80—90年代，大中型气田勘探开发实践中形成的理论体系逐步成熟，2000年前后，前陆冲断带构造天然气成藏理论、大面积岩性气藏形成理论、深层碳酸盐岩成藏理论、叠合盆地油气富集理论、有机质接力生气理论指导了中国以常规气藏为主的持续发现，2000—2008年年均新增探明储量 $5000\times10^8m^3$ 以上。2010年前后，全过程生烃模式、古老碳酸盐岩成藏理论、连续型"甜点区"油气聚集理论指导了深层、超深层碳酸盐岩气藏、海相页岩气等非常规天然气的战略突破[17-18]，2009年至今，年均新增探明储量 $8000\times10^8m^3$ 以上，预计未来10~15年仍将保持探明储量高峰增长。

在系列理论技术推动下，鄂尔多斯、四川、塔里木三大盆地成为陆上最具成长性的天然气生产基地，特别是鄂尔多斯和四川盆地逐步呈现出含气层系多、埋藏深度跨度和累积厚度大、常规和非常规气叠置分布，天然气资源异常丰富的超级盆地的特点。鄂尔多斯盆地主力含气层系从古生界到中生界均有发育[19-20]，地层厚度可达1000m以上，资源类型包括常规气、致密气、煤层气、页岩气等多种类型，以低渗透和致密砂岩气为主，成为中国最大的天然气生产基地。四川盆地含气层系较为古老，从前寒武系到中生界数千米厚的地层中发育数十套含气层系[21]，常规气、页岩气、致密气、煤层气资源富集，特别是近年碳酸盐岩大型气藏和页岩气的连续突破，天然气累计探明储量近 $4\times10^{12}m^3$，其中页岩气累计探明储量近 $1.8\times10^{12}m^3$，成为中国天然气探明储量最大的盆地，常规气和非常规气均具有很好的增长潜力[22]。打造纵向多层系、常规与非常规多类型油气藏一次性立体勘探和整体勘探模式，避免或减少单层系多轮次勘探的重复投入，是未来超级盆地勘探的主要发展方向。

4.2 天然气开发理论技术及发展方向

中国天然气藏按照"发现一类、攻关一类、配套一类"的思路，逐步形成了常规无水气藏自然衰竭开发理论、常规边底水气藏控水均衡开发理论、致密气分级降压开发理论、页岩气人工气藏理论、煤层气排水解气理论等天然气开发理论体系[23]，打造了以高精度三维地震、水平井和大规模压裂改造为核心的复杂气藏开发技术利器[24]，推动了中国天然气近20年产量年均增速超过10%，引领中国成为世界第六大产气国。

近十年，中国天然气开发在鄂尔多斯、四川、塔里木陆上三大盆地取得重大进展，未来开发潜力可期[25]。鄂尔多斯盆地是中国最大的致密气生产基地，探明并开发了中国规模最大的气田——苏里格致密大气田（探明地质储量 $1.98\times10^{12}m^3$，2019年产量 $254\times10^8m^3$），2019年天然气产量 $520\times10^8m^3$，攻关多层系立体开发与大幅度提高采收率技术[26-27]，支撑鄂尔多斯盆地上产 $600\times10^8m^3$ 并保持长期稳产。四川盆地是中国最大的海相碳酸盐岩（高含硫）和海相页岩气的生产基地，发现并开发了中国单体规模最大的整装深层海相碳酸盐岩气田——安岳大气田（探明地质储量 $9441\times10^8m^3$，2019年产量 $135\times10^8m^3$）[28]，发现并开发了中国埋藏最深的生物礁气田——元坝大气田（探明地质储量 $2199\times10^8m^3$，2019年产量 $37\times10^8m^3$）[29]，高效建成了中国第一个 $100\times10^8m^3$ 高含硫大气田——普光大气田（探明地质储量 $3763\times10^8m^3$，2019年产量 $62\times10^8m^3$）[30]，高效建成了中国第一个 $100\times10^8m^3$ 大型页岩气田——涪陵大气田（探明地质储量 $6008\times10^8m^3$，2019年产量 $62\times10^8m^3$）[31]，2019年天然气产量 $486\times10^8m^3$，攻关完善深层碳酸盐岩气藏和深层海相页岩气有效开发技术，支撑四川盆地年产千亿立方米国家战略生产基地建设[32]。塔里木盆地是中国超深层气田开发的典范，发现并开发了中国最深（平均产层中深超过7000m）的陆相碎屑岩大气田——克深大气田（探明地质储量 $6690\times10^8m^3$，2019年产量 $92\times10^8m^3$），2019年天然气产量 $300\times10^8m^3$，持续攻关超深层气藏高效开发技术[33]，不断拓展中国超深层气藏新领域。

5 中国天然气产量趋势预测

中国天然气地质资源量 $173\times10^{12}m^3$（不含天然气水合物），累计探明率小于20%，总体处于勘探早中期[34]，预计2035年前年均新增探明储量 $7000\times10^8m^3$ 左右，累计新增探明 $11.5\times10^{12}m^3$，其中，常规气、非常规气分别新增探明 $4\times10^{12}m^3$ 和 $7.5\times10^{12}m^3$。随着致密气和页岩气的规模开发，中国天然气已进入常规与非常规并重的发展阶段，因此产量趋势预测按照常规气和非常规气两类进行论证。

5.1 不同类型气藏产量预测方法与原则

本文选用产量构成法进行产量预测，即以气田/区块为基本单元，依据不同类型气藏在不同开发阶段的生产规律、剩余可采储量的可靠性等指标，测算每个基本单元的产量潜力，进而按时间叠加得到不同年份的总产量规模。该方法结合了地质与气藏工程专业研究认识和开发规划方法，预测结果比较可靠。

目前中国已开发的常规气田540多个，开发阶段差异较大，故将常规气田分为已建成气田、在建上产气田和勘探新区三类。已建成气田根据已探明储量和气藏生产动态，预测产量趋势，进入递减期后，产量按照6%~12%的年综合递减率下降，并结合不同类型气藏经验采收率加以约束。在建上产气田，首先复算探明地质储量、动态储量和单井指标，与开发方案进行对比，若与方案基本一致，以方案产量剖面为准，否则根据实际储量和指标情况，分析上产趋势。勘探新区，根据气藏类型，参照天然气开发管理纲要标准，确定采气速度及不同产量规模的稳产年限，递减期产量类比同类气藏开发特征进行预测。

近年，中国非常规天然气探明储量和产量增长迅猛，特别是致密气和页岩气，未来整体仍处于储产量快速增长期（图4），因此3种非常规气单独进行预测。致密气以苏里格气田年产量综合递减率20%计算年弥补递减和净新增产量，页岩气以川南页岩气田年产量综合递减率35%计算年弥补递减和净新增产量，煤层气按年新增产量（2~3）× 10^8m^3 计算。

中国溶解气产量较为稳定，2020—2035年产量为（86~90）× 10^8m^3/a。总体，天然气年产量预测公式如下：

$$Q_{总} = Q_{常规} + Q_{非常规} + Q_{溶解}$$
$$= (Q_{常规已建} + Q_{常规在建} + Q_{常规新区}) + (Q_{致密气} + Q_{页岩气} + Q_{煤层气}) + Q_{溶解}$$

其中，$Q_{常规} = \sum_{i=1}^{n} Q_i$，$n$ 为常规气田个数，常规已建气田、常规在建气田和常规新区气田产量预测方法有所不同；

$Q_{致密气}$ = 上年致密气产量 ×（1–20%）+ 当年新投产井产量；

$Q_{页岩气}$ = 上年页岩气产量 ×（1–35%）+ 当年新投产井产量；

$Q_{煤层气}$ = 上年煤层气产量 +（2~3）× 10^8m^3。

5.2 常规气产量趋势预测

中国已开发气田中，大部分气田属于已建成方案设计规模的气田，2020年产量 $820 \times 10^8m^3$，占全国总产量的45%左右，总体处于开发中后期，其中榆林、靖边、元坝气田通过层系接替或扩边保持稳产，其他主力气田年综合递减率5%~8%，预测2035年产量维持 $330 \times 10^8m^3$。中国在建上产气田16个，主力气田主要包括四川盆地震旦系等深层、川东北高含硫、致密气、塔里木克拉苏构造带和海域5个领域，2020年计划产量 $215 \times 10^8m^3$，预计2035年上产 $435 \times 10^8m^3$，动用新增探明储量 $1 \times 10^{12}m^3$。以陆上三大盆地和海域为重点，突出四川下古生界—震旦系、塔里木库车、南海等重点领域区带实施集中勘探，加大新区新领域风险勘探与甩开预探，2020—2035年预计勘探新区可新增探明储量 $3 \times 10^{12}m^3$。预计2025年、2030年、2035年分别贡献产量 $240 \times 10^8m^3$、$395 \times 10^8m^3$ 和 $590 \times 10^8m^3$。勘探新区产量影响因素较多，具有较大的不确定性。3种常规天然气产量预测见图3。

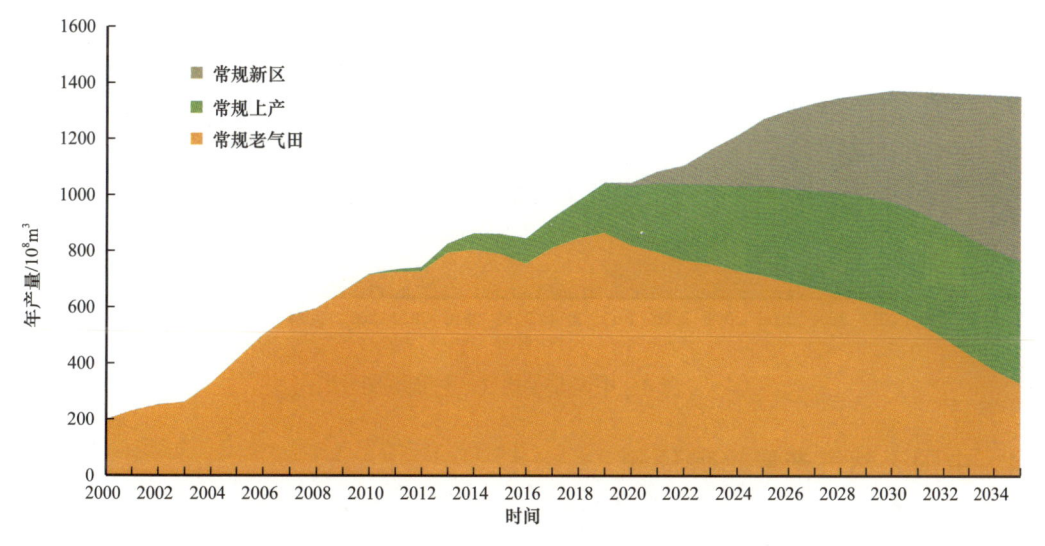

图3 中国常规气产量预测图

5.3 非常规气产量趋势预测

致密气产量规模大，保持稳定增长[35]。中国致密气开发时间较早，2006年之前产量较低，一直归到低渗透砂岩气藏类型。2006年后鄂尔多斯盆地苏里格、大牛地等致密气田的规模发现和成功开发，储量产量迅速增长，截至2019年底，探明地质储量$5×10^{12}m^3$，预计2020年产量$475×10^8m^3$，通过持续扩大建产区域和提高采收率技术的推广应用，2025年达到$525×10^8m^3$，稳产至2035年，动用新增探明储量$2.5×10^{12}m^3$，主要分布在鄂尔多斯盆地东部、南部上古生界和四川盆地沙溪庙组致密气。

页岩气是产量快速增长的主力[35-36]。中国页岩气建产起步于2012年，依靠水平井与体积压裂技术的突破，川南长宁201-H1井和川东涪陵焦页1HF井两口水平井获高产页岩气流，拉开中国海相页岩气开发的序幕。2014—2019年，累计提交探明储量$1.79×10^{12}m^3$，投产水平井超过1000口，以3500m以浅储量开发为主，2020年计划产量$205×10^8m^3$，依靠深层海相页岩气以及部分海陆过渡相页岩气的突破和建产，2035年持续上产至$550×10^8m^3$，动用新增探明储量$4.5×10^{12}m^3$，主要分布在四川盆地泸州、渝西等海相深层和鄂东海陆过渡相区块。

煤层气产量保持缓慢增长。中国煤层气开发利用大致可划分为井下抽放—试验勘探—技术引进—规模开发4个阶段。2005年以后，中国才开始大规模对煤层气进行商业开发。以沁水盆地、鄂尔多斯盆地东缘为主要建产区的中高阶煤层气率先获得突破，开发技术逐步成熟配套，2020年预计产量$42×10^8m^3$，依靠中高煤阶完善井网、措施增产及排采优化技术，2035年可缓慢上产至$85×10^8m^3$，动用新增探明储量$0.5×10^{12}m^3$。低煤阶开发初步在二连盆地吉尔格朗图获得局部井点突破，一旦技术成熟配套，煤层气产量有望进一步增加（图4）。

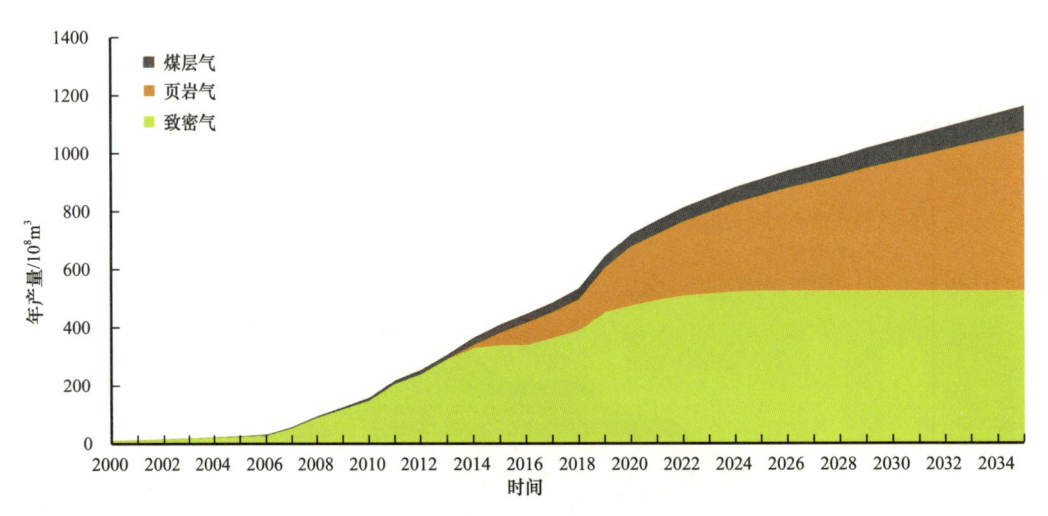

图4 中国非常规气产量预测图

5.4 中国天然气产量趋势预测

2035年前，中国天然气持续上产的趋势不会改变，常规气与非常规气并重发展的格局不会改变，但由于常规气优质接替资源的发现难度越来越大，新增储量中非常气资源比例不断增加，常规气产量由增速放缓转向稳定或小幅递减，非常规气保持快速增长，预计2035年中国天然气产量达到$2600×10^8m^3$，其中非常规气产量占比45%以上（表2），2035年后中国天然气将进入真正的非常规时代。国内天然气产量的预测，为中国合理布局进口管道和沿海LNG接收站建设和规模提供了关键依据。

表2 中国天然气产量预测表　　　　　　　　　　单位：10^8m^3

类型	2020年	2025年	2030年	2035年
常规气	1044	1270	1370	1350
非常规气	720	910	1040	1160
溶解气	86	90	90	90
合计	1850	2270	2500	2600

6　结论

随着能源结构调整和绿色环保理念的不断深入，中国天然气消费需求持续旺盛，天然气产业发展主要取决于供给侧。国内天然气产量是中国天然气产业发展的根基，也是产业布局的基础。

对比中美天然气发展特点，中国非常规气发展时机把握好，发展速度快，已形成对常规气"主动接替"的宏观态势。认清目前资源劣质化、开发技术效率低、新开发气田开发效益差、工作量组织与管理难度大、安全环保风险大等挑战，梳理勘探开发理论技术进展与发展方向，评价常规与非常规天然气勘探增储和开发上产潜力，预测2035年常规天然气产量维持在$1350×10^8m^3$，非常规天然气产量上升至$1160×10^8m^3$，加上溶解气$90×10^8m^3$，国内天然气总产量达到$2600×10^8m^3$，为中国能源结构调整及进口管道和储气库战略布局提供决策基础。

参 考 文 献

［1］邹才能，潘松圻，赵群.论中国"能源独立"战略的内涵、挑战及意义［J］.石油勘探与开发，2020，47（2）：416-426.

［2］李剑，佘源琦，高阳，等.中国天然气产业发展形势与前景［J］.天然气工业，2020，40（4）：133-142.

［3］陆家亮，赵素平，孙玉平，等.中国天然气产量峰值研究及建议［J］.天然气工业，2018，38（1）：1-9.

［4］王建良，刘睿.中国天然气产量中长期走势预测研究［J］.煤炭经济研究，2019，39（10）：41-47.

［5］国家发改委，经济运行局.2019年天然气运行简况［EB/OL］.（2020-1-21）.https：//www.ndrc.gov.cn/fggz/jjyxtj/mdyqy/202001/t20200121_1219615.html.

［6］何润民，李森圣，曹强，等.关于当前中国天然气供应安全问题的思考［J］.天然气工业，2019，39（6）：123-131.

［7］刘立.我国能源供应体系建设的思考［J］.国土资源情报，2019，（12）：58-63.

[8] 陆家亮，唐红君，孙玉平．抑制我国天然气对外依存度过快增长的对策与建议［J］．天然气工业，2019，39（8）：1-9．

[9] 刘剑文，杨建红，王超．管网独立后的中国天然气发展格局［J］．天然气工业，2020，40（1）：132-140．

[10] 宋岩．美国天然气分布特点及非常规天然气的勘探［J］．天然气地球科学，1990，1（1）：14-16．

[11] 梁涛，常毓文，许璐，等．北美非常规油气蓬勃发展十大动因及对区域油气供需的影响［J］．石油学报，2014，35（5）：890-900．

[12] 马新华，陈建军，唐俊伟．中国天然气的开发特点与对策［J］．天然气地球科学，2003，14（1）：15-20．

[13] 焦方正．非常规油气之"非常规"再认识［J］．石油勘探与开发，2019，46（5）：803-810．

[14] 李熙喆，郭振华，胡勇，等．中国超深层大气田高质量开发的挑战、对策与建议［J］．天然气工业，2020，40（2）：75-82．

[15] 胡文瑞．开发非常规天然气是利用低碳资源的现实最佳选择［J］．天然气工业，2010，30（9）：1-8．

[16] 马新华，胡勇，何润民．天然气产业一体化发展模式研究与实践［J］．技术经济，2019，38（9）：65-72．

[17] 邹才能，杨智，何东博，等．常规—非常规天然气理论、技术及前景［J］．石油勘探与开发，2018，45（4）：575-587．

[18] 马永生，蔡勋育，赵培荣．中国页岩气勘探开发理论认识与实践［J］．石油勘探与开发，2018，45（4）：561-574．

[19] 杨华，付金华，刘新社，等．鄂尔多斯盆地上古生界致密气成藏条件与勘探开发［J］．石油勘探与开发，2012，39（3）：295-303．

[20] 刘丹，张文正，孔庆芬，等．鄂尔多斯盆地下古生界烃源岩与天然气成因［J］．石油勘探与开发，2016，43（4）：540-549．

[21] 李鹭光．四川盆地天然气勘探开发技术进展与发展方向［J］．天然气工业，2011，31（1）：1-6．

[22] 马新华．四川盆地天然气发展进入黄金时代［J］．天然气工业，2017，37（2）：1-10．

[23] 邹才能，郭建林，贾爱林，等．中国大气田科学开发内涵［J］．天然气工业，2020，40（3）：1-10．

[24] 位云生，齐亚东，贾成业，等．四川盆地威远区块典型平台页岩气水平井动态特征及开发建议［J］．天然气工业，2019，39（1）：81-86．

[25] 李海平，贾爱林，何东博，等．中国石油的天然气开发技术进展及展望［J］．天然气工业，2010，30（1）：5-7．

[26] 武力超，朱玉双，刘艳侠，等．矿权叠置区多层系致密气藏开发技术探讨——以鄂尔多斯盆地神木气田为例［J］．石油勘探与开发，2015，42（6）：826-832．

[27] 冀光，贾爱林，孟德伟，等．大型致密砂岩气田有效开发与提高采收率技术对策——以鄂尔多斯盆地苏里格气田为例［J］．石油勘探与开发，2019，46（3）：602-612．

[28] 谢军．安岳特大型气田高效开发关键技术创新与实践［J］．天然气工业，2020，40（1）：1-10．

[29] 刘成川，柯光明，李毅．元坝气田超深高含硫生物礁气藏高效开发技术与实践［J］．天然气工业，2019，39（S1）：149-155．

[30] 孔凡群，王寿平，曾大乾．普光高含硫气田开发关键技术［J］．天然气工业，2011，31（3）：1-4．

[31] 王志刚．涪陵大型海相页岩气田成藏条件及高效勘探开发关键技术［J］．石油学报，2019，40（3）：370-382．

[32] 马新华，胡勇，王富平．四川盆地天然气产业一体化发展创新与成效［J］．天然气工业，2019，39（7）：1-8．

[33] 贾爱林，唐海发，韩永新，等．塔里木盆地库车坳陷深层大气田气水分布与开发对策［J］．天然气地球科学，2019，30（6）：908-918．

[34] 邹才能，赵群，陈建军，等．中国天然气发展态势及战略预判［J］．天然气工业，2018，38（4）：1-11．

[35] 邱中建，赵文智，邓松涛．我国致密砂岩气和页岩气的发展前景和战略意义［J］．中国工程科学，2012，14（6）：4-8．

[36] 赵文智，贾爱林，位云生，等．中国页岩气勘探开发进展及发展展望［J］．中国石油勘探，2020，25（1）：31-44．

二、开发地质类

地质工程一体化实施过程中的页岩气藏地质建模

舒红林[1]，王利芝[2]，尹开贵[1]，李庆飞[1]，张 卓[1]，罗瑀峰[1]

（1.中国石油浙江油田公司；2.斯伦贝谢中国公司）

摘 要：地质工程一体化强调地学研究与作业的互动，在中国地质条件复杂的非常规油气田，地质建模在数据基础和应用需求上体现出了独特的挑战。一方面，数据以水平井为主、数据量大且数据类型冗杂；另一方面，作业进程要求模型快速迭代甚至达到"适时"建模。因此如何充分利用多种数据快速建立高质量的地质模型至关重要，本文关注以页岩气为代表的非常规油气藏建模的独特性，并提出了具体流程和方法。其一，系统阐述了水平井地质建模流程，通过真厚度域旋回对比、二维导向剖面及井震数据融合从一维到三维解决水平井构造和属性建模难题。其二，以蚂蚁追踪为例介绍了天然裂缝预测与建模方法，非常规储层裂缝普遍发育，在理解裂缝发育背景的前提下通过成像测井、钻井、微地震等多学科资料的交叉验证有助于实现合理的裂缝建模。其三，不同应用需求下的地质建模流程与应用，如多学科集成的井位部署优化、适时建模支持地质导向以及压裂工程应用等。

关键词：地质工程一体化；地质建模；页岩气；水平井；天然裂缝

近年来，非常规致密油气藏逐步成为油气开发的热点领域，为接替常规油气能源、支撑油气革命做出了重要贡献[1]，页岩气更是成为非常规天然气发展的"热点"[2]。由于页岩气藏储层致密，一般无自然工业产能，而需依赖平台式水平井和体积压裂来提高产能，这就造就了其独特的开发特点和资料基础。相比于北美页岩油气分布稳定、地面条件好等特点，中国页岩气藏大多经历了多旋回的构造演化，这就决定了中国页岩气开发的独特性。为了更高效地开发页岩气资源，必须探索、实践和发展适合其独特性的、以地质工程一体化为核心的高效开发之路[3]。在地质工程一体化场景下，储层表征与建模技术的资料基础、应用场景和应用目的都发生了深刻的变化。相应地，地质建模技术也应当经历"非常规"的蜕变。本文以四川盆地南部五峰组—龙马溪组页岩气藏为例，探讨地质建模的技术特点，其中的研究思路也可借鉴到其他非常规油气藏。

1 页岩气地质建模的现状与挑战

常规油气藏地质建模技术已广泛应用于油气田储量评估及油气藏管理，基于地质、地球物理、测井等多种数据，通过地质统计学建立的三维地质模型已成为多学科团队沟通协作的重要桥梁[4-6]。以页岩气为代表的非常规油气藏地质建模相对于常规油气藏而言发生了深刻的变化。

第一，以水平井为主的开发方式对传统的构造和储层属性建模方法提出了挑战。通常页岩气水平井地质导向深度窗口较小，如四川盆地南部龙马溪组导向窗口已经从初期的15~20m降低到2~5m，水平轨迹在地层中穿行所反映的储层变化是垂向非均质性和平面非均质性的综合响应，采用在水平段设置虚拟直井的方式可以为水平段增加更多的构造控制点[7-8]，但通常仅限于特征明显的地层界面，如何有效地将水平段进行精细的构造地层归位已成为致密油气藏开发需要解决的首要问题。除此之外，在属性建模时，由于平台式布井资料存在采样偏误（在优质层段存在更多的采样点且采样点平面分布不均），而属性建模算法需要一个能反应三维体属性分布特征的分布模型（直方图）[5]，因此数据统计时需特别注意。

第二，以页岩为代表的致密油气藏对储层品质和完井品质参数建模提出了特殊的要求。致密油气储层模型需包括"六性"相关的属性[9]，即烃源岩特性（如总有机碳含量TOC）、岩性（如矿物含量）、物性（如孔隙度、渗透率、饱和度）、脆性（如杨氏模量、泊松比、脆性指数）、含油气性（如含气量）及地应力特性（如主应力大小、方位），另外还要考虑电性参数（如伽马、密度）。

第三，天然裂缝的关注度空前提高，且对天然裂缝的关注点发生了深刻的变化。传统上，对裂缝性油气藏主要关注天然裂缝对流体渗流的影响，从而分析裂缝与油气成藏、产能特征以及注水特征的相互关系[10]，即裂缝对渗透率各向异性和量值的贡献，裂缝是否开启、裂缝的开度参数成为裂缝性油气藏研究的重点[11]。对四川盆地的页岩气藏而言，由于复杂的多旋回构造演化导致断裂系统十分发育[2]，在水力压裂时，裂缝及其受力状态直接影响了水力裂缝的扩展以及缝网的复杂性[12-13]。即使是已发生矿物充填的裂缝，因其导致岩石强度降低，压裂时也会对水力裂缝扩展产生重要影响[12]。

第四，地质工程一体化应用场景下，对一体化建模的管理和应用方式提出了新的需求。地质工程一体化强调地学研究与作业的互动，地质模型对钻井、压裂等工程作业提供支持是建模工作的重要目的，因此建模思路上要适应地质力学模拟及压裂模拟等非常规研究需求。尤其在开发早期，地质模型的不确定性较高，这就要求地质模型根据工程作业进度不断获取的新数据和新资料对模型进行质控和更新，做到"适时建模"[14]，将多学科资料所反映的地质特征定量化表征到模型中去，从而支持井位部署、钻井、地质导向、压裂设计和后评估等工作，实现从一维到三维、从定性到定量以及从局部（单井）到整体（平台区、建产区）的跨越，最终实现油气田的高效开发。

2 页岩气藏地质建模技术对策

页岩气藏地质建模的目的是地质工程一体化应用，结合页岩气田以水平井开发为主、多学科资料丰富的特点，从构造、属性和天然裂缝三大方面表征影响页岩气开发的储层品质和工程品质参数。关键技术流程包括水平井地层归位与建模、井—震趋势约束的页岩气属性建模、地震蚂蚁追踪驱动的天然裂缝建模及力学稳定性分析，详见图1。

2.1 水平井地层归位及构造建模

2.1.1 真厚度域地层对比

由于水平井轨迹相对于地层产状而言，可以存在多个上切与下切段，通过垂厚（TVT）

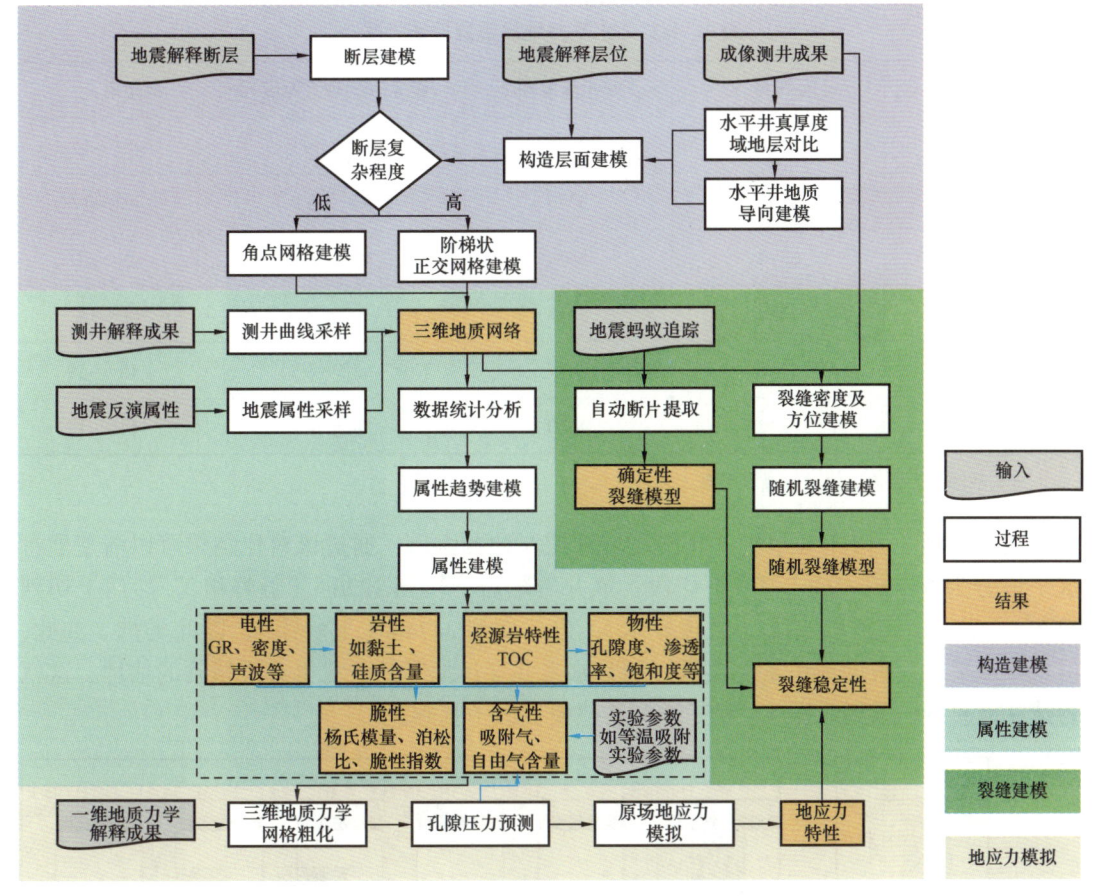

图 1 页岩气藏地质建模流程图

对比储层旋回特征的方法虽然在直井影响不大，但受轨迹重复和构造起伏的影响，在水平段使用会存在较大的问题。而在真厚度（TST）域对比是消除构造起伏后将上切或下切段"垂化"，其曲线形态可与直井保持较好的可对比性[15]，对比的认识是三维构造建模的重要输入。

页岩储层由于其非均质性以垂向为主，可以将同一平台储层横向厚度稳定展布作为假设条件，结合真厚度域曲线特征对比反推地层倾角[16-17]，具体流程见图2。由于采集随钻成像资料的井较少且随钻成像拾取的倾角往往存在一定误差，多数井需要根据地震解释提供的构造图计算井周的地层产状作为真厚度计算的初始输入。受地震资料精度的限制，对地层产状的初始认识往往需要不断地调整，调整的依据是水平段单一上切或下切段地层旋回厚度是否与直井及斜井段一致。通过不断地调整地层产状、更新真厚度计算结果，以达到水平段测井曲线与直井的合理对比。

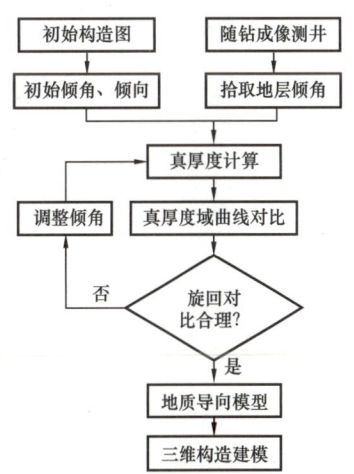

图 2 水平井真厚度域地层对比与构造建模流程图

表1 H1-2水平井地层重复段统计表

水平分段	起始深度/m	终止深度/m	距龙一$_1^3$顶真厚度/m	视倾角/(°)
下切段1	0	3226	<15.9	−5~−17
上切段1	3231	3370	8.8~15.9	−9~−17
下切段2	3370	3474	8.8~16.5	−9~−20
上切段2	3550	3718	−3.5~16.1	0~−13
下切段3	3719	3887	−3.5~8.2	−15~−19
上切段3	3889	4030	6.7~8.3	−14~−15

注：视倾角上倾为负值，下倾为正值。

表1和图3为H1-2水平井真厚度域地层对比的案例，通过该对比结果可以清楚地将水平段分为6个较大的上切/下切段：该井着陆后下切到3226m（五峰组），然后上切到3370m（龙一$_1^2$，该段曲线已倒转显示），再下切到3474m（五峰组），以此类推。如此，可将该水平井反映的地层重复再现，而且水平井穿过地层的产状也有了精确的估计，这就为建立地质导向模型（图4）和三维构造建模提供了基础。

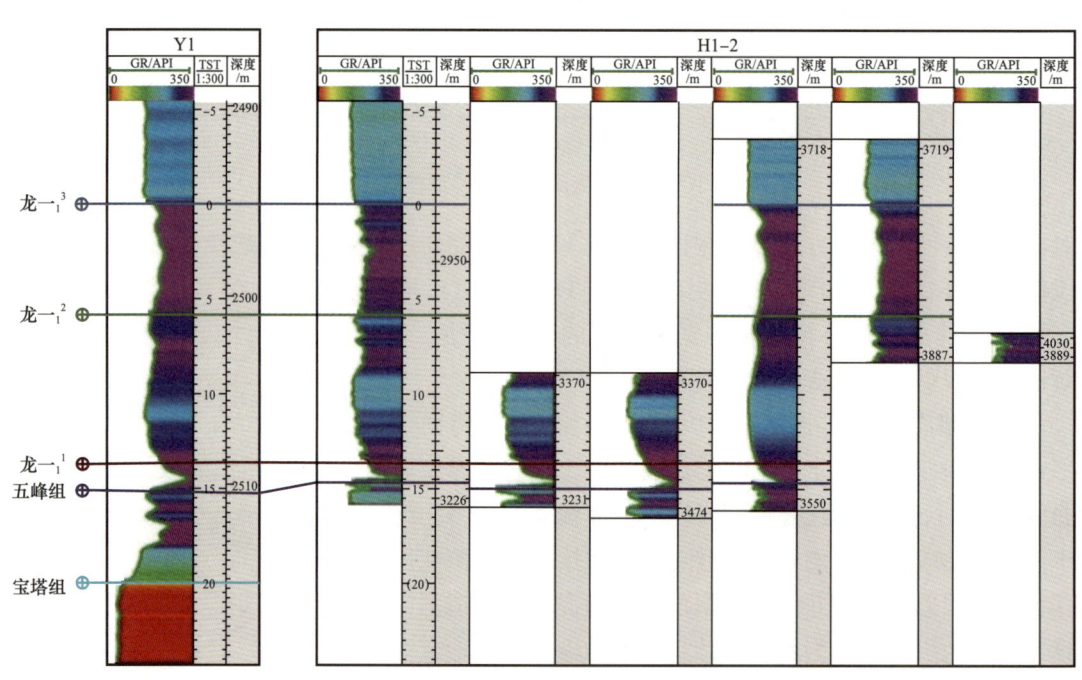

图3 页岩气真厚度域直井（Y1）—水平井（H1-2）地层对比图

2.1.2 地质导向模型的三维建模应用

根据资料基础的不同，地质导向技术存在3个互补的技术方法，第一种方法称之为"建模—对比—更新"方法，其思路是根据初始模型和井轨迹正演模拟曲线响应，并将其与实时测井响应（LWD）对比，进而开展曲线拟合并更新修正模型[18]；第二种方法是实时地层倾角探测方法，通过实时随钻井壁成像测井（如伽马成像、电阻率成像等）判断轨迹相对于地层的上、下切关系，其至拾取地层界面倾角，建立地层模型；第三种方法为边界探测技术，通过深电阻率探测等手段反演地层内的电阻率界面，如流体界面的随钻探测。页岩气常用的地质导向方法为前两种。

地质导向及其建模技术可以建立过水平井的二维导向模型，推断井筒上下的构造和地层发育特征，从而解决了水平井曲线的一维局限性，其结果可以作为地质建模的输入。与真厚度域地层对比相同，页岩二维构造导向模型假设同一平台内储层横向厚度和属性稳定展布，将导眼井的分层界面和电性曲线（如伽马）均质地推广到水平段，然后通过调整水平段构造倾角，来拟合水平井的模拟曲线和实测曲线以达到水平段二维构造建模的目的（图4）。由于二维地质导向模型可以精细地反映过井剖面的构造，在进行井区三维构造建模时可以将其作为输入。

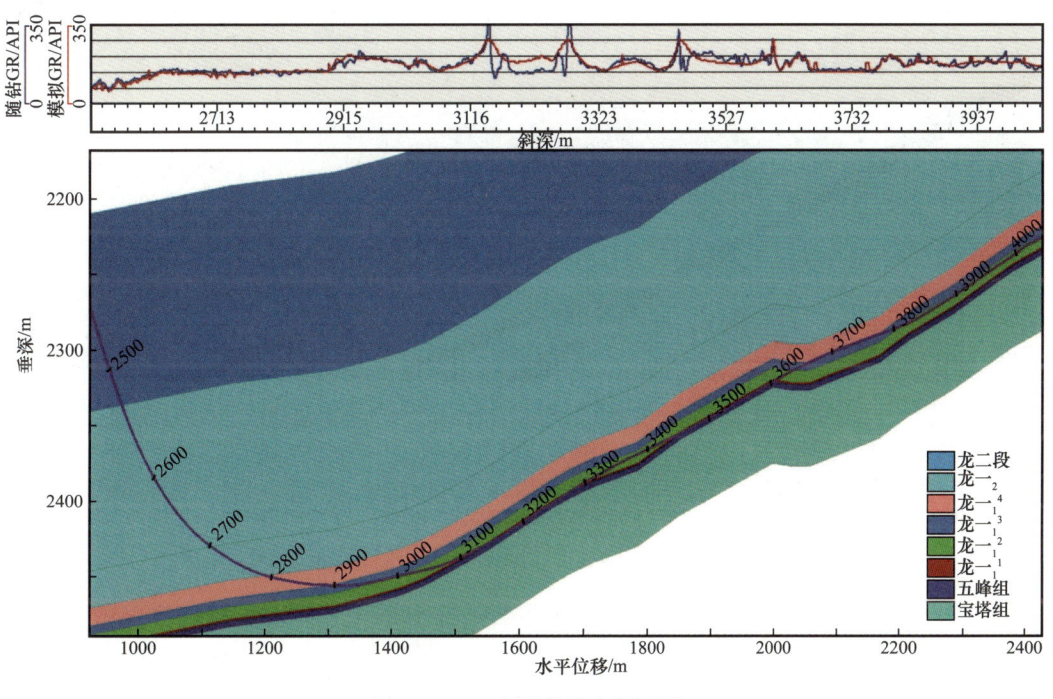

图4 H1-2水平井导向剖面图

2.1.3 三维构造建模

结合单井分层、二维导向剖面模型和地震构造解释可以建立可靠的三维构造层面模型，在建模时以上3种构造信息可以赋予不同的权重，如单井分层和二维导向剖面作为"硬数据"，地震解释的精度较低，故可作为"软数据"进行趋势控制。

在建立三维网格时，可考虑不同断层复杂程度来选择网格建模方法，在断层复杂程度较低时建议采用角点网格，在断层复杂程度较高时建议采用阶梯状正交网格。三维网格平面精度需兼顾地震面元的大小以及工程应用的精度需求，以YS井区为例，地震面元为20m×40m，水平段压裂设计时簇间距通常在20～50m，故该区平面网格精度定义为

40m×40m以兼顾精度与计算效率，通过角点网格方法建立的构造模型如图5所示。垂向网格劈分时，在优质储层段采用高精度的垂向网格精度（平均0.5m），远离优质储层段垂向网格精度逐渐变为1～10m，以合理控制网格数量。在进行地应力模拟时，可对网格的平面精度适度粗化（如80m×80m）或按照I、J索引提取平台模型，考虑页岩具有明显的垂向非均质性，垂向网格精度建议保持与精细地质模型一致。

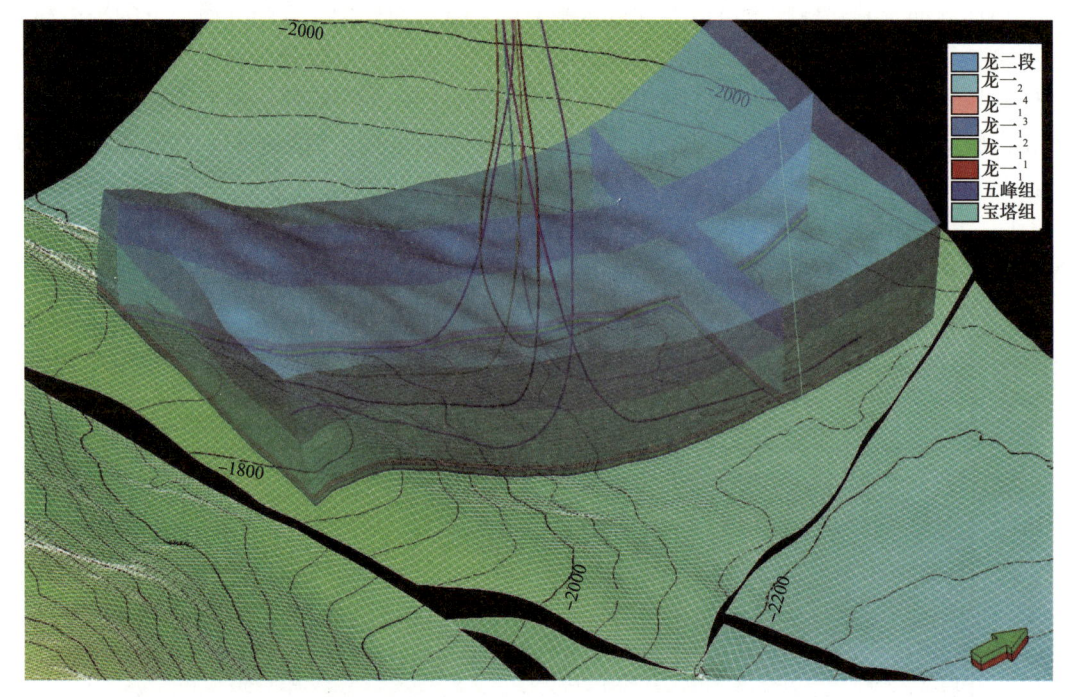

图5 YS井区三维构造层面模型（五峰组底面，40m×40m）及H1平台构造模型

2.2 水平井属性建模

2.2.1 采样偏误

采样偏误在油气田勘探开发领域十分常见，由于布井时倾向于在"甜点区"，从而使得"甜点区"采样数据更多。非常规油气藏以水平井开发为主，该现象变得更加凸显。水平段主要在优质储层中穿行，这就导致了测井数据在空间上不均匀的采样，如果直接使用水平井的数据分布进行属性预测，往往将带来乐观的预测结果。因此，当直井和水平井共存时，如果直井分布较为均匀，可以借鉴直井的测井数据分布特征[8]，如果直井及水平井分布不均，则需在数据空间插值基础上通过三维网格体获取更为合理的数据分布特征。

图6统计了同一区块龙一$_1$亚段测井有效孔隙度的分布特征，其中3口直井统计的平均值为3.6%，标准差为0.9%，3口直井与11口水平井一起统计的平均值为4.2%，标准差为1%。由于水平井在"甜点"层段钻井，导致统计的数据分布特征出现了偏误，呈现了乐观的统计特征，3口直井较为均匀地分布在不同的平台，其统计的数据分布特征则更为合理，建模时应采用直井的数据分布。

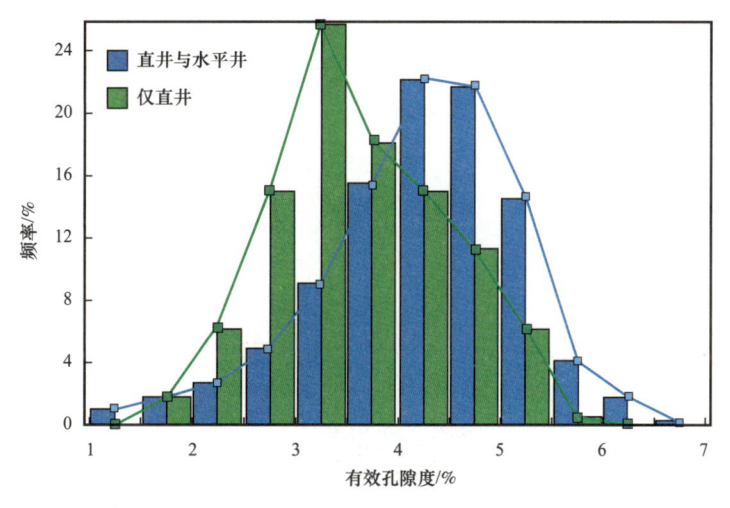

图6 龙一₁亚段直井与水平井测井有效孔隙度分布特征对比图

2.2.2 趋势建模

在使用地质统计学进行属性参数的空间插值时，需要保证其满足平稳性假设，如果数据体现出了系统性的趋势，则需要对其表征，并在变差函数分析和属性建模之前移除这种趋势[5]。以页岩为例，其垂向非均质性较强，即使是水平井资料，在完成地层归位之后仍然可以看到与直井相似的垂向趋势（图7a、b），对于平面趋势则可以由地

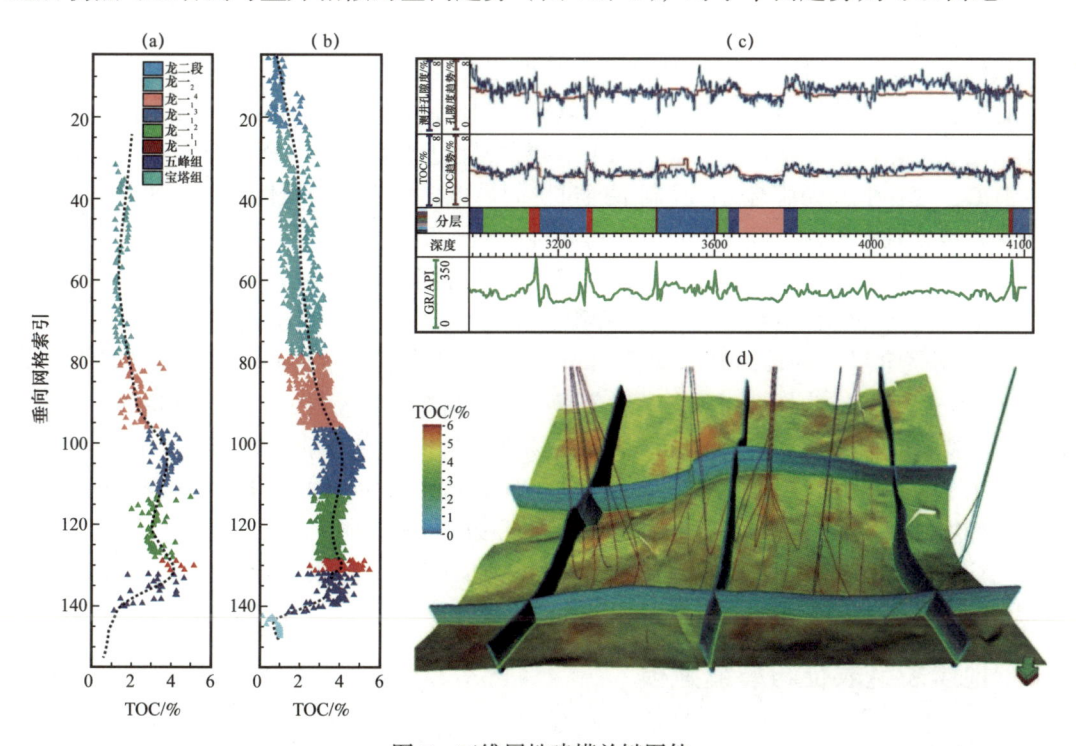

图7 三维属性建模关键图件

（a）3口直井反映的TOC垂向趋势；（b）11口水平井反映的TOC垂向趋势；
（c）单井提取的趋势模型与测井曲线对比图；（d）三维TOC模型

- 43 -

震资料（如反演）来提供，通过线性计算可以将垂向趋势与平面趋势结合为三维趋势体。图7b中值得注意的一点是，水平井测井对于薄层的响应相对于直井而言具有更好的分辨率，如五峰组顶部存在约0.5m厚的观音桥段生物介壳灰岩，具有低TOC特征，可见水平井对该段的响应更明显。图7c将水平井轨迹穿过的趋势模型网格值与测井曲线进行了对比，可见二者具有较好的相关性，在趋势模型控制下进行属性建模更能反映页岩储层的变化特征（图7d）。

2.2.3 属性建模

常规油气藏属性建模离不开沉积相建模，沉积相建模是反映地质概念模式和控制属性模型的重要手段。应用沉积相建模时需要重点考虑两个方面：首先，不同沉积相之间具有明显的物性差异，如孔隙度、渗透率、饱和度等；其次，沉积相具有直观的空间变化模式，如特定的形态（可指导变差函数设置）、沉积相之间的接触关系等。对于页岩而言，在岩心微观尺度[20]或通过矿物含量（硅质、钙质、黏土矿物）三端元法[21]可以将其进一步细分为不同的岩相，但由于目前缺乏不同岩相的空间变化模式，难以通过沉积相建模方法将其合理的三维表征以至于控制属性分布。

因此本文推荐的属性建模方法为趋势模型约束下的随机连续属性建模，如序贯高斯模拟，在不同属性参数之间存在明显的相关性时通过协同克里金模拟以保证三维属性参数之间的相关性。

通过分析地震反演属性和直井测井曲线可以获取横向和垂向变程，在YS井区，横向变程约1000m，垂向变程在2~3m。

属性建模时，由于不同参数之间存在物理相关性，需要考虑不同属性的先后模拟次序，比如，烃源岩特性对物性和含油气性具有控制作用、岩性对脆性具有控制作用等，如图1所示。因此，首先在地震反演的约束下建立岩性及烃源岩特性等基础属性，如黏土含量、TOC。然后在这些基础属性控制下通过协同模拟建立其他的属性参数，如黏土含量控制脆性参数，TOC控制孔隙度和饱和度等物性参数。

页岩含气量需在三维孔隙压力预测基础上，结合网格属性参数（孔隙度、饱和度、密度、TOC、天然气体积系数等），通过容积法在三维网格分别计算吸附气量和游离气量获得[5]。吸附气含量计算时，郎氏体积、郎氏压力需结合实验室岩心测试获取的与TOC的回归关系进行计算。由图8可见，总含气量一方面与烃源岩特性（TOC）正相关，另一方面也与孔隙压力正相关。这与页岩气赋存的物理特征相吻合，即TOC越高吸附能力越强，同时，由于有机孔为自由气赋存提供了空间，TOC越高储存自由气的孔隙体积也越大；孔隙压力的增加对吸附气和自由气含量增加都有直接的贡献作用。

地应力特性则在岩石力学属性（杨氏模量、泊松比等）建模基础上通过三维地质力学模拟器进行原场地应力模拟[22-23]。

2.3 地震裂缝表征与裂缝建模

2.3.1 地震裂缝表征

地震裂缝预测的方法很多[24]，本文选取蚂蚁追踪方法[25]作为重点探讨。蚂蚁追踪常被用来自动识别和追踪地震属性体中的异常和不连续性。川南页岩气独特的地层特征为

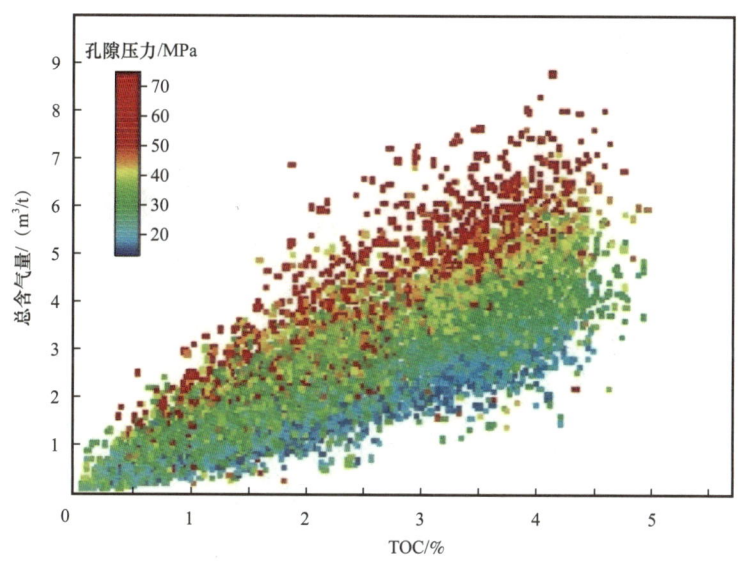

图8 三维模型中TOC、总含气量及孔隙压力属性交会图

该方法的应用提供了良好的前提条件，页岩本身横向展布稳定，五峰组—龙马溪组页岩与下伏的宝塔组石灰岩之间为明显的岩性界面，从而产生了强波阻抗特征，方差属性的变化主要来自构造信息，在方差属性基础上开展蚂蚁追踪可以很好地反映断层及裂缝的特征。该方法在川南页岩气区块取得了较好的裂缝预测效果[26-27]。

蚂蚁追踪算法中涉及的参数较多，比如初始蚂蚁边界、方向、步长等，在实际应用中不同参数的使用往往会得出不同的结果，保证蚂蚁追踪结果的合理性是应用该方法的重要前提。这就要求对蚂蚁追踪结果进行验证，页岩气田采集的成像测井、钻井、录井、微地震监测等资料为此奠定了很好的基础。

合理的蚂蚁追踪裂缝预测应该从以下几个方面开展合理性验证：首先，蚂蚁追踪结果不应包含地层信息和地震采集脚印等非裂缝信息；其次，蚂蚁追踪结果应与其他地震属性相互验证，方差、曲率、相干等属性都不同程度地反映了断层和裂缝的信息，蚂蚁追踪应与这些属性中的异常带吻合；第三，蚂蚁追踪结果应与单井资料吻合，如微地震监测中的高震级事件，钻井漏失、气测异常以及成像测井反映的主要裂缝方位等；最后，蚂蚁追踪结果应与区域断裂系统和构造背景吻合。

2.3.2 裂缝建模

在蚂蚁追踪结果的基础上开展裂缝建模较为直接，通过确定性建模和随机建模两种方式可分层次的建立断裂—裂缝模型。所谓确定性建模即通过断层片提取技术[25]将蚂蚁体中的断层或者裂缝带提取成面，而随机建模方式是通过离散裂缝建模建立裂缝片网络（DFN）。DFN建模的两个关键参数是裂缝产状和裂缝密度，裂缝的走向即为蚂蚁追踪时的主方位，裂缝密度可在井点统计的裂缝密度标定的基础上对蚂蚁追踪的量值进行回归转换。从图9案例可见，H1平台发育两组裂缝。H1平台北支天然裂缝走向以平行于东北侧的走滑断层为主，多呈北西走向，高震级的微地震事件也反映了这一特点。H1平台南支裂缝则以北东走向为主，呈带状分布，平行于南侧呈北东走向的逆冲断层，经实钻证实，这些裂缝带所在的位置都伴有构造倾角的突变（图4中H1-2井3400m左右、3600m左右），因此认为其为褶皱相关的裂缝。

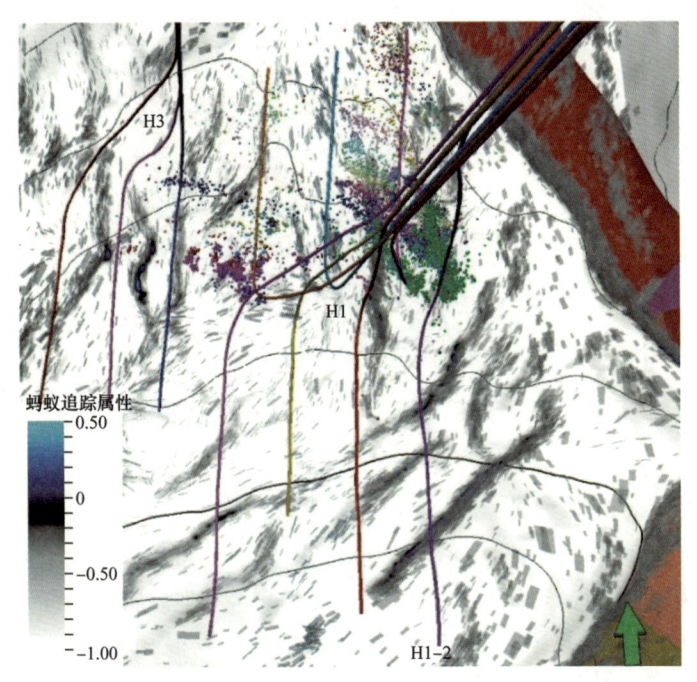

图9 龙一₁亚段蚂蚁追踪、微地震（震级>-1）以及天然裂缝模型综合图

天然裂缝的受力状态影响了天然裂缝的稳定性及其在压裂作业时与水力裂缝的相互作用。钻井实践表明，受力不稳定的裂缝容易在钻井时造成井漏以及井壁失稳，在压裂时容易造成压裂液的漏失进而引起砂堵。裂缝受力状态可分解为正应力和剪应力，根据摩尔—库伦准则可判定处于极限应力状态而倾向于滑动的裂缝，从而在工程作业之前提前预警。

以H1平台为例，其现今地应力状态为走滑应力状态，最大水平主应力73MPa左右，最小水平主应力57MPa左右，垂直主应力69MPa左右，孔隙压力37MPa左右，最大水平主应力方位为110°。图10为DFN裂缝模型在现今地应力条件下的滑动可能性。从图10中可见，北东（75°左右）和北西（330°左右）走向的裂缝在现今地应力条件下较不稳定，容易滑动，从而可能对钻完井造成不利影响。

图11为H20平台南支裂缝稳定性分析的平面图，该平台南支实钻4口井。其中，H20-4井周裂缝发育程度低，该井钻井顺利，钻进过程中未发生钻井液漏失等钻井复杂事件；H20-2井井周预测存在裂缝发育带，但裂缝走向为近南北向，裂缝稳定性较好，该井钻进过程也较为顺利；H20-6井井周裂缝由着陆时的近南北向逐渐转变为南部的北西走向，裂缝稳定性也逐渐变差，该井在3658m和3811m发生了两次钻井液漏失，其中3811m处漏失严重，迫使该井提前完钻；H20-8井井周裂缝与H20-6井类似，该井南部北西走向裂缝的稳定性差，该井在3480m、3622m和3751m发生了3次钻井液漏失。通过该平台的钻井分析可见，裂缝对井壁稳定性的影响除了与裂缝密度有关之外，裂缝的力学稳定性也十分关键，这一认识也被用于指导相邻平台的布井优化。

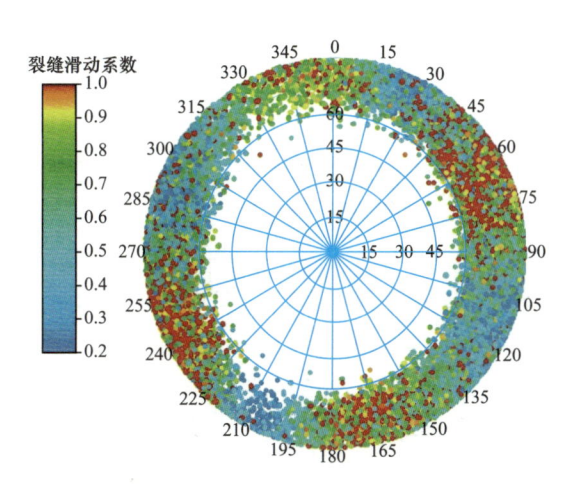

图10 裂缝在原地应力条件下的裂缝稳定性分析 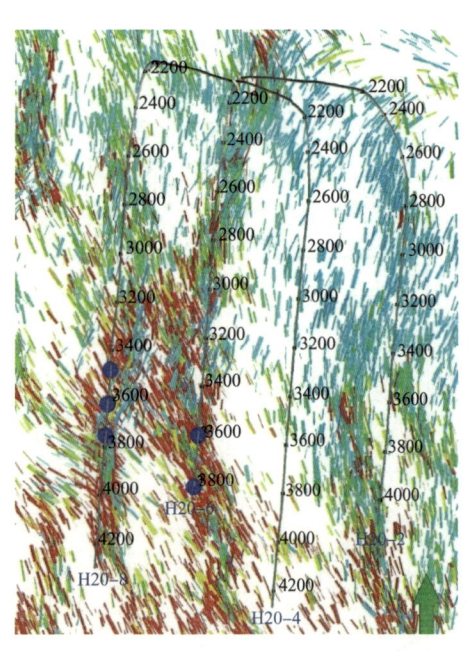　　图11 H20平台裂缝稳定性分布图（蓝色点为钻井液漏失点，色标见图10）

3　地质工程一体化场景下地质建模的应用

3.1　井位部署优化

开发初期井位部署多依赖地球物理资料，受限于复杂的构造地质特征以及资料的有限性，开发初期的井位部署方案需要随着开发的进行不断地优化，实现地下认识迭代、地质模型迭代、井位优化迭代。井位部署优化时要综合考虑储层品质、钻井品质和完井品质，滚动实施开发井钻探工程，提高部署设计实施的符合率[28]。比如，迭代更新的构造模型可以用来确定靶点海拔深度；蚂蚁体和裂缝模型可用于落实不稳定断层或裂缝带的位置，设计井轨迹时尽量避开；构造倾角和曲率可用来确定水平段构造变化，如上倾钻井尽量保证视倾角小于15°；已压裂井微地震事件可指示天然裂缝以及水力裂缝的扩展情况，作为井间距设计依据。

3.2　地质导向应用

川南页岩气区块多具备复杂的地表条件和地下构造特征，导致地震构造预测的不确定性较大，给钻井入靶和地质导向带来了困难。高质量的钻井和地质导向需要研究部门和工程作业部门的通力合作，图12给出了推荐的地质工程一体化地质导向工作流程。钻前阶段，研究部门根据地质模型开展地质设计，提出钻前风险提示，并针对性的制订地质导向策略，工程作业部门在钻前阶段可以收集到待钻井周边相关的模型和资料，开展工程设计和作业准备；实时钻井阶段，地质导向工作主要在作业部门展开，通过实时数据传输，结合导向剖面模型和真厚度域地层对比落实钻头的地层层位，同时考虑多种可能性以降低导向的不确定性，研究部门可将实时轨迹和随钻测井数据加载到三维模型中以跟踪钻进状况，并及时从地质角度提出地质导向建议；完钻后，作业部门将最终采集的数据和完钻导向模型提交给研究部门并总结经验教训，而研究部门需要根据完钻数据开展模型更新，在作业密集时可只更新构造模型，当构造模型更新时校深大于阈值（如15m）时，需对在原模型基础上制定的地质设计进行变更。

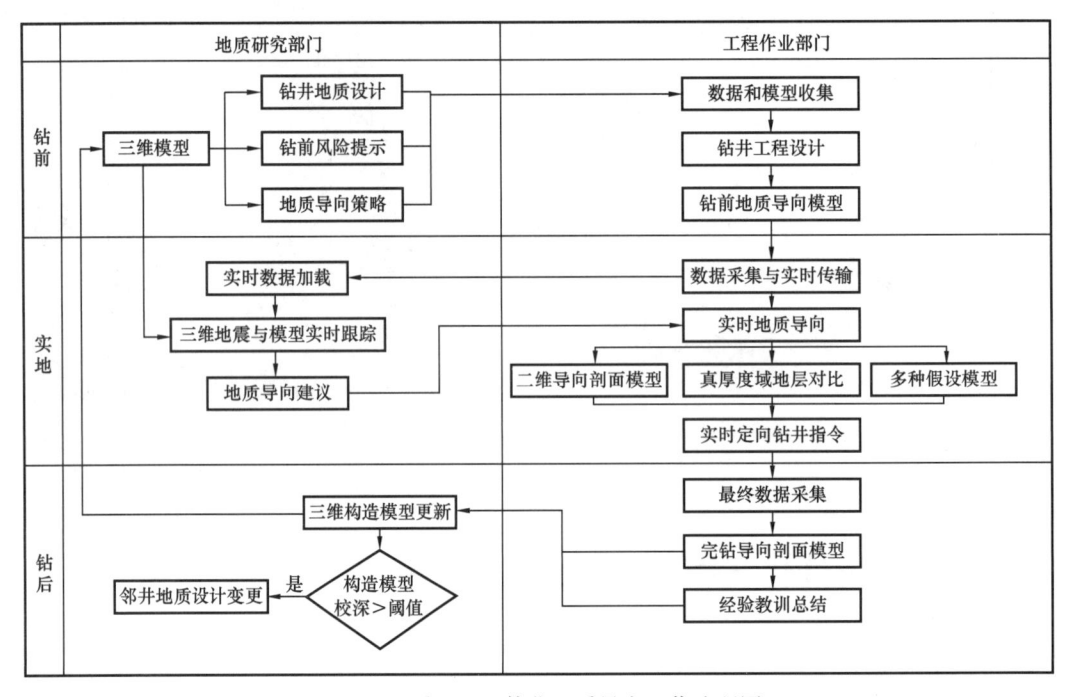

图 12 地质工程一体化地质导向工作流程图

通过此工作流程可见，在工程作业时，已钻井数据可以不断地迭代集成到新版本的地质模型中，而且在数据集成的过程中会通过地质建模流程来对新数据进行质量控制，从而"净化"掉低质量和错误的数据信息。另外，通过适时迭代建模，地质认识和工程经验也在建模过程中"固化"为定量的模型参数，因此不断迭代的地质模型成为钻井数据和经验知识的数字化汇聚中心。

3.3 地质模型的压裂工程应用

首先，模型可为测井系列不丰富的井提供过井及井周的模型属性，开展压裂设计。受不同井况的限制，页岩气水平井存在不同的测井系列，分别为特殊测井、常规测井及随钻测井。这就导致井间测井解释丰富性的差异，对于测井解释属性较少的井（如井筒条件差只有随钻测井曲线），可以通过三维模型来提供缺少的储层品质和完井品质属性，开展压裂分级及射孔簇设计[3]。

其次，模型为水力裂缝模拟提供基础。结合压裂施工参数、微地震监测数据、三维地质模型参数以及天然裂缝等信息可以开展水力裂缝的模拟。受页岩垂向应力非均质性的影响，在不同层段起裂会导致水力裂缝垂向延伸规律存在差异。以五峰组上部的观音桥段为例，该段钙质含量高、有机质含量低、应力高，虽然其厚度通常小于0.5m，但基于声波扫描测井解释的最小水平主应力要比龙马溪组高 20% 以上。水平段在观音桥段上部或下部起裂，水力裂缝的垂向展布会存在较大的差异。由图 13 可见，在观音桥段之上龙马溪组底部起裂的水力裂缝先在龙一$_1^1$到龙一$_1^3$内破裂延伸，然后向上和向下延伸改造龙一$_1^4$和五峰组，主力改造层段为龙一$_1^1$到龙一$_1^3$的主力产层；五峰组下段起裂的水力裂缝受上部高应力观音桥段和支撑剂沉降的影响，裂缝以在五峰组内破裂延伸为主，压力憋高突破上部应力隔挡后裂缝向上延伸，支撑剂分布仍在五峰组内为主，对上部裂缝的支撑效果较差。由此可见精确落实了水平段地层穿行情况的地质模型对压裂的重要性。

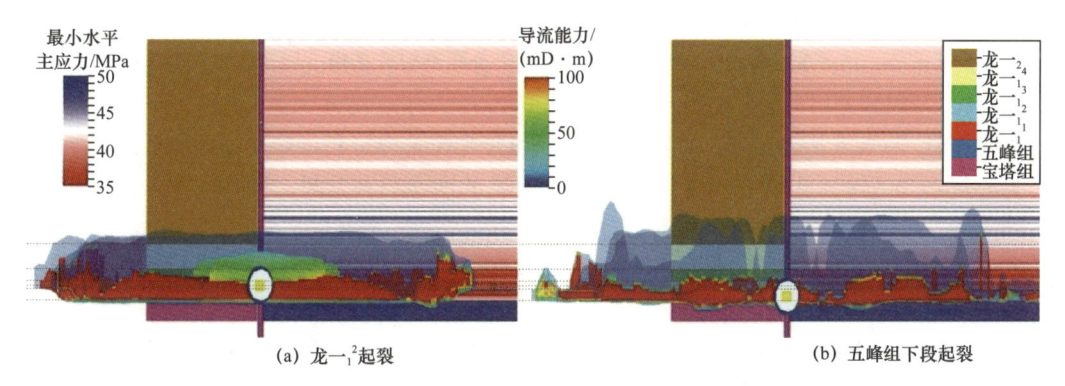

(a) 龙一$_1^2$起裂　　　　　　　　　　(b) 五峰组下段起裂

图 13　观音桥段上下不同起裂层段对水力裂缝扩展的影响模拟图

4　结语

地质工程一体化场景下，非常规储层的地质建模工作也需针对性地进行技术和流程的优化升级，从而适应其独特的数据基础和应用目的。三维构造和属性建模的重点和难点是水平井数据的应用，通过系统性地对水平井进行地层归位有助于建立可靠的构造和属性模型。天然裂缝建模依赖合理的地震预测，地球物理与地质学科的紧密结合有助于保证天然裂缝模型质量。在地质工程一体化场景下，地质建模需多学科交互、快速迭代以实现对钻、完井作业的及时支持。

从目前非常规地质建模的发展趋势来看，以下技术方面需在未来不断完善。

（1）碎屑岩储层非均质性的三维立体解剖。以致密油为代表的碎屑岩储层在水平井段同时存在垂向非均质性和平面非均质性的变化，如何更好地应用水平井资料解剖储层构型特征是该类型储层地质建模的重要挑战和必须解决的问题。

（2）基于地质模型的数据挖掘。非常规"井工厂"式的水平井开发方式带来了大量的水平井资料，在地质建模效率提升、基于地质模型挖掘工程和产能主控因素等方面，机器学习新技术将有助于实现地质建模技术的进一步升级。

致谢：感谢斯伦贝谢中国公司和中国石油浙江油田公司对本文发表的大力支持，本文所有建模过程均在斯伦贝谢 Petrel 平台开展。

参 考 文 献

[1] 邹才能.非常规油气地质[M].北京：地质出版社，2013.
[2] 邹才能，张国生，杨智，等.非常规油气概念、特征、潜力及技术——兼论非常规油气地质学[J].

石油勘探与开发，2013，40（4）：385-399.

［3］吴奇，梁兴，鲜成钢，等. 地质—工程一体化高效开发中国南方海相页岩气［J］. 中国石油勘探，2015，20（4）：1-23.

［4］Ma Y Z. Uncertainty analysis in reservoir characterization and management：How much should we know about what we don't know？［J］. AAPG Memoir 96，2011：1-15.

［5］Deutsch C V. Geostatistical reservoir modeling［M］. New York：Oxford University Press，2002.

［6］Yu X H，Ma Y Z，Gomez E，et al. Reservoir characterization and modeling：A look back to see the way forward［J］. AAPG Memoir 96，2011：289-309.

［7］乔辉，贾爱林，位云生. 页岩气水平井地质信息解析与三维构造建模［J］. 西南石油大学学报（自然科学版），2018，40（1）：78-88.

［8］Wang G H，Long S X，Ju Y W，et al. Application of horizontal wells in three-dimensional shale reservoir modeling：A case study of Longmaxi—Wufeng shale in Fuling gas field，Sichuan Basin［J］. AAPG Bulletin，2018，102（11）：2333-2354.

［9］邹才能，陶士振，白斌，等. 论非常规油气与常规油气的区别和联系［J］. 中国石油勘探，2015，20（1）：1-16.

［10］穆龙新，赵国良，田中元，等. 储层裂缝预测研究［M］. 北京：石油工业出版社，2009.

［11］Nelson R A. Geological Analysis of Naturally Fractured Reservoirs［M］. Second Edition. Houston：Gulf Professional Publishing，2001.

［12］Gale J F W，Laubach S E，Olson J E，et al. Natural fractures in shale：A review and new observations［J］. AAPG Bulletin，2014，98（11）：2165-2216.

［13］Gu H，Weng X，Lund J，et al. Hydraulic fracture crossing natural fracture at nonorthogonal angles：A criterion and its validation［J］. SPE Production & Operations，2012，February：20-26.

［14］鲜成钢. 页岩气地质工程一体化建模及数值模拟：现状、挑战和机遇［J］. 石油科技论坛，2018，37（5）：24-34.

［15］石学文，王利芝，赵圣贤，等. 页岩气储层建模及其钻完井工程应用［C］//CPS/SEG北京2018国际地球物理会议暨展览电子论文集. 石油地球物理勘探编辑部. 2018，1355-1358.

［16］Liang X，Wang L Z，Zhang J H，et al. An Integrated Approach to Ensure Horizontal Wells 100% in the Right Positions of the Sweet Section to Achieve Optimal Stimulation：A Shale Gas Field Study in the Sichuan Basin，China［C］//SPE-177474-MS，Abu Dhabi International Petroleum Exhibition and Conference，Abu Dhabi，UAE，9-12 November 2015.

［17］Xing Liang，Lizhi Wang，JiehuiZhang，Chenggang Xian，Gaocheng Wang，Chunduan Zhao，et al.100% in the Sweet Section：An Effective Geosteering Approach for Silurian Longmaxi Shale Play in Sichuan Basin［C］//SPE-176945-MS，SPE Asia Pacific Unconventional Resources Conference and Exhibition，Brisbane，Australia，9-11 November 2015.

［18］吴宗国，梁兴，董健毅，等. 三维地质导向在地质工程一体化实践中的应用［J］. 中国石油勘探，2017，22（1）：89-98.

［19］Ma Y Z，Gomez E. Sampling biases and mitigations in modeling shale reservoirs［J］. Journal of Natural Gas Science and Engineering，2019，71（102968）：1-11.

［20］Liang C，Jiang Z，Cao Y，et al. Deep-water depositional mechanisms and significance for unconventional hydrocarbon exploration : A case study from the lower Silurian Longmaxi shale in the southeastern Sichuan Basin [J]. AAPG Bulletin, 2016, 100 (5): 773-794.

[21] Helena Gamero-Diaz, Camron K. Miller, Richard Lewis. sCore: A mineralogy based classification scheme for organic mudstones [C] //SPE-166284-MS. SPE Annual Technical Conference and Exhibition, New Orleans, Louisiana, USA, 30 September-2 October 2013.

[22] Liang X, Xian C G, Shu H L, et al. Three-dimensional full-field and pad geomechanics modeling assists effective shale gas field development, Sichuan basin, China [C] //IPTC-18984-MS, International Petroleum Technology Conference, Bangkok, Thailand, 14-16 November 2016.

[23] 鲜成钢,张介辉,陈欣,等.地质力学在地质工程一体化中的应用[J].中国石油勘探,2017,22(1):75-88.

[24] 刘敬寿,丁文龙,肖子亢,等.储层裂缝综合表征与预测研究进展[J/OL].地球物理学进展,2019-03-05:1-25.

[25] Pedersen S I, Randen T, Sonneland L, et al. Automatic 3D fault interpretation by artificial ants [C] // 64th Meeting, EAGE Expanded Abstracts, G037, 2002.

[26] Xie J, Qiu K B, Zhong B, et al. Construction of a 3D geomechanical model for development of a shale gas reservoir in Sichuan basin [C] //SPE-187828-MS. SPE Russian Petroleum Technology Conference, Moscow, Russia, 16-18 October 2017.

[27] Jun Q, Xian C G, Liang X, et al. Characterizing and modeling multi-scale natural fractures in the ordovician-Silurian Wufeng-Longmaxi shale formation in South Sichuan basin [C] // URTEC-2691208-MS, SPE/AAPG/SEG Unconventional Resources Technology Conference, Austin, Texas, USA, 24-26 July 2017.

[28] 梁兴,王高成,张介辉,等.昭通国家级示范区页岩气一体化高效开发模式及实践启示[J].中国石油勘探,2017,22(1):29-37.

页岩气储层关键参数评价及进展

乔 辉，贾爱林，贾成业，位云生

（中国石油勘探开发研究院）

摘 要：目前，中国页岩气开发及储层评价工作正处于初步探索阶段。为有效利用地质及地球物理测井技术进行页岩储层评价，在前人研究的基础上，总结页岩气储层的基本特征，优选确定了TOC含量、矿物成分及矿物的脆性指数、储层物性及含气量。概括了页岩地层使用的测井系列、页岩气层的定性测井响应特征，详细论述了表征页岩储层特征的关键参数定量评价方法，从有机地球化学参数、页岩的矿物组分及脆性指数、页岩储层物性参数、储层含气性评价4个方面深入展开论述。最后，探讨了页岩储层综合评价研究进展及中国页岩气储层评价存在的问题，为建立适合中国页岩气储层的综合评价方法，优选优质页岩富集层段提供依据。

关键词：储层关键参数；定量评价；有机质；矿物成分；储层物性；含气性

随着全球能源需求的增长，页岩气等非常规资源开发发展迅速，已成为油气勘探开发的新领域。根据2013年美国能源信息署（EIA）的统计结果，全球页岩气可采资源量约为$220\times10^{12}m^3$，和常规天然气资源相当。近年来，北美地区在页岩气成藏理论、水平井钻井和压裂改造等方面取得了重要的理论与技术进步，促进了页岩气产业的快速发展[1-4]。美国页岩气每年的年产量从2003年的$265\times10^8m^3$快速上升至2015年的$4308\times10^8m^3$，占天然气总产量的比重超过50%。中国页岩气资源丰富，根据2015年中国国土资源部资源评价结果，中国页岩气的技术可采资源量为$21.8\times10^{12}m^3$。中国页岩气开发虽然起步较晚，尚处于初步开发阶段，但发展迅速。2015年页岩气产量为$45\times10^8m^3$，2016年全国页岩气产量达到$77\times10^8m^3$。随着页岩气勘探开发的推进，对于页岩地层的沉积作用、成岩作用机理及页岩气富集机理等方面的认识都有了长足的进步[5-7]。同时，随着页岩地层开发的不断深入，逐渐认识到页岩储层具有较强的非均质性，需要形成一系列适合中国页岩气储层评价的技术方法[8-10]。笔者通过总结近几年相关文献的主要观点，对页岩储层关键参数的定量评价方法进行简述，寄予为建立一套适合中国页岩气储层特征的综合评价方法，为优质页岩富集层段的评价与优选提供依据。

1 页岩储层基本特征

页岩储层的特殊性主要体现在富含有机质、富含黏土矿物、孔渗极低、孔喉结构以纳米级为主、矿物比表面积大、天然气赋存状态特殊等方面[11]（表1）。基于大量实验室测试数据统计分析表明，Barnett页岩的有机碳（TOC）一般为3%～13%，平均为3.9%[12]，中国南方龙马溪组富有机质页岩TOC含量较高，但各地区变化较大，一般为

0.07%～7.94%，平均为 2.54%[13]。与美国页岩的成熟度相比，中国页岩有机质成熟度普遍较高，镜质组反射率 R_o 在 2% 以上。页岩储层的另一重要特征是黏土矿物含量高，中国南方志留系富有机质页岩中黏土矿物含量通常在 50% 左右[13]。矿物成分主要为石英、长石和黏土矿物，其次为碳酸盐岩矿物，此外还包含少量的黄铁矿等，脆性矿物含量对页岩开发中的压裂效果有重要影响。页岩储层物性、孔隙结构主要利用氦气吸附、高压压汞、核磁共振等技术测得[14-17]，其普遍具有低孔隙度、特低渗透率致密的物性特征。中国志留系富有机质页岩储层的孔隙度范围为 0.77%～19.50%，平均为 5.05%，渗透率为 0.0013～0.058mD，平均 0.0102mD[13]。利用氩离子抛光技术和扫描电镜相结合观察储层的孔隙结构，富气页岩中含有大量微孔隙和微裂缝，是页岩气赋存的主要场所[14, 16]。页岩基质储层的孔隙结构很小，以微孔隙为主，后期的成岩作用及构造作用产生的次生孔隙及裂缝可在一定程度上改善储层的孔隙度。页岩气的赋存状态以吸附式和游离式为主，页岩气在微孔隙中主要以吸附态存在，而在较大的宏观孔隙和微裂缝中主要以游离态存在[16]。

表 1 中国和美国页岩气储层的基本特征（据文献 [13]，有修改）

页岩气区	埋深 /m	厚度 /m	有机碳质量分数 ω（TOC）/%	镜质组反射率 R_o/%	孔隙度 /%	含气量 /(m³/t)
黄金坝	2390～2516	32～40	0.6～6.5	2.8～3.0	1.0～7.0	1.35～3.48
焦石坝	2313～2595	38～42	3.5	2.2～3.1	1.2～7.2	0.44～5.19
Antrim	183～732	21～37	0.3～24	0.4～0.6	9.0	1.1～2.8
Ohio	610～1524	9～30	0～4.7	0.4～1.0	4.7	1.7～2.8
New Albany	183～1494	15～30	1～25	0.4～1.0	10～14	1.1～2.3
Barnett	1981～2591	15～61	4.5	0.5～2.0	4～5	8.5～9.9
Lewis	914～1829	61～91	0.45～2.5	1.6～1.88	3～5.3	0.4～1.3
Marcellus	1219～2591	15～61	3～12	0.4～1.3	10.0	1.7～2.8
Woodford	1829～3353	37～67	1～14	1.1～3.0	3～9	5.7～8.5

2 页岩储层关键参数优选

近年来，有不少学者对优质页岩储层主控因素进行了研究[18-25]，该研究为优选储层评价的关键参数奠定基础。有机碳含量（TOC）是页岩储层评价的一大重要指标，对长宁页岩气区块分析测试资料统计分析发现，页岩储层的有机质含量与页岩气总含气量、储层的孔隙度之间都存在较好的正相关关系（图 1a、b）。有机碳含量较高的页岩，有机质孔隙发育，其比表面大，为吸附态天然气的赋存提供了吸附剂，也为游离气的赋存提供了孔隙空间[14]。因此，有机碳含量为页岩储层评价的关键指标之一。

页岩储层中的矿物成分与含量对储层的孔隙度及含气性影响较大[20, 25]。长宁区块通过不同类型矿物含量与实测孔隙度的关系拟合发现，孔隙度随着石英含量增高而增大（图

1c），且TOC含量随着石英含量的增大而增大（图1d），石英为生物成因石英矿物，来源于较为丰富的硅质生物残体，间接增加了有机质的含量[5, 13]。丰富的有机质来源的石英和有机质伴生，发育丰富的微孔隙，具有较大的比表面积，增加了页岩中的可供页岩气吸附以及游离气赋存的空间，同时石英等脆性矿物越发育，越易形成天裂缝，微裂缝的存在可有效改善储层物性且有利于后期页岩气的压裂改造[25]。统计发现，碳酸盐矿物含量与孔隙度呈微弱的负相关关系（图1e），岩石薄片、扫描电镜可观察到碳酸盐胶结物，碳酸盐胶结物对页岩储层的孔隙度具有一定的消极影响。矿物成分及储层的脆性评价是储层评价的另一关键评价指标。

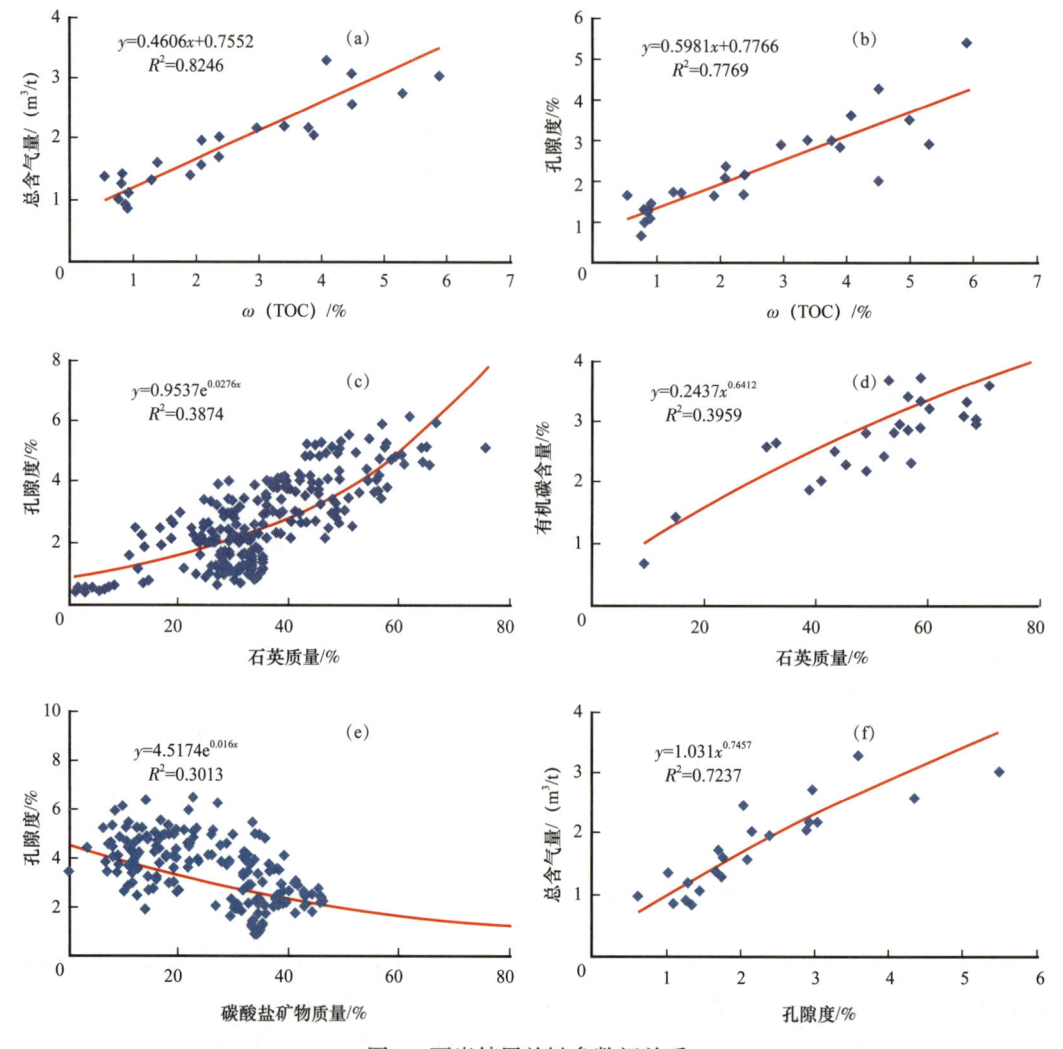

图1 页岩储层关键参数间关系

页岩储层的物性评价参数与常规储层类似，主要包括储层的孔隙度、渗透率、含气量、饱和度等，影响着页岩储层的质量[21-23]。页岩储层孔隙大小很大程度上决定着页岩的储气能力，长宁地区五峰—龙马溪组岩样的储层物性分析资料统计表明页岩储层的含气量与储层的孔隙度具有较好的正相关关系（图1f），储层孔隙度越大，储层的含气量越高。

页岩的含气量是储层评价的核心参数，是确定页岩气资源量，预测页岩气单井产气量最重要的储层影响因素之一，决定了页岩气区是否能够经济开发，对开采方案的编制均有重要意义，为页岩储层关键评价的重要指标[12]。

为此，在深入认识储层基本特征及储层质量影响因素的基础上，优选有机质含量、储层矿物成分与含量、储层脆性、储层物性及含气量作为最终页岩储层评价的关键指标参数。

3 页岩储层关键参数评价

3.1 页岩气测井系列

页岩气测井系列包括常规测井系列和特殊测井系列[20, 26-29]。常规测井系列主要包括自然伽马、井径、密度、补偿中子、声波时差及深浅电阻率测井。常规测井可以用于页岩储层的识别，但其在页岩储层矿物成分与含量的计算、裂缝的识别及定量表征以及岩石力学参数的计算等方面不适用。特殊测井系列主要包括元素俘获能谱（ECS）测井、声电成像测井、偶极声波测井及核磁共振等。其中，ECS元素测井可用于计算页岩储层的岩石矿物成分；声波测井，用于评价岩石力学性质；成像测井，用于识别裂缝；核磁共振测井，可用于评价页岩储层的孔隙度。

3.2 页岩气测井响应特征

页岩地层常表现出以下测井曲线特征[30]：（1）自然伽马值高，且有机质含量高的层段，总伽马与无铀伽马值差异幅度大；（2）电阻率测井曲线上显示电阻为中—低值，页理或微裂缝发育的层段，深浅电阻率有一定的幅度差；（3）三孔隙度中值，光电截面指数低值，在有机质含量高的层段密度明显降低。（4）井径一般呈现扩径现象。

含气页岩储层段测井曲线的响应特征主要表现出"四高两低"：高自然伽马、高电阻率、低密度、低光电截面指数、高声波时差、高中子孔隙度[27]（图2）。自然伽马值异常高的层段通常有机质含量高，对应的吸附气含量亦高；而深浅电阻率差异幅度大的层段及中—宏孔隙发育的层段，游离气的含量较高；通常情况下，高电阻率、低密度值、高声波时差值的层段对应的页岩储层段的含气饱和度也较高[31]。含气页岩储层在特殊测井曲线上也有明显的响应特征：电成像测井图上可见含气页岩层段的裂缝较发育；通常含气页岩层段的核磁共振图上可见其T_2谱随含气量增大而向增大方向移动[30]。

3.3 页岩储层关键参数定量评价

3.3.1 有机质评价

总有机碳含量（TOC）是评价含气页岩储层的一项重要参数。取心和分析化验资料仅能获取离散有限的数据点，而利用测井资料解释则可获取井段的连续数据。目前，国内外学者在利用测井解释确定TOC方面做了大量的工作[31-42]。Fertle[32]利用自然伽马能谱测井和岩心分析资料的线性回归方法确定TOC含量；Passey等[33]提出利用电阻率和孔隙度

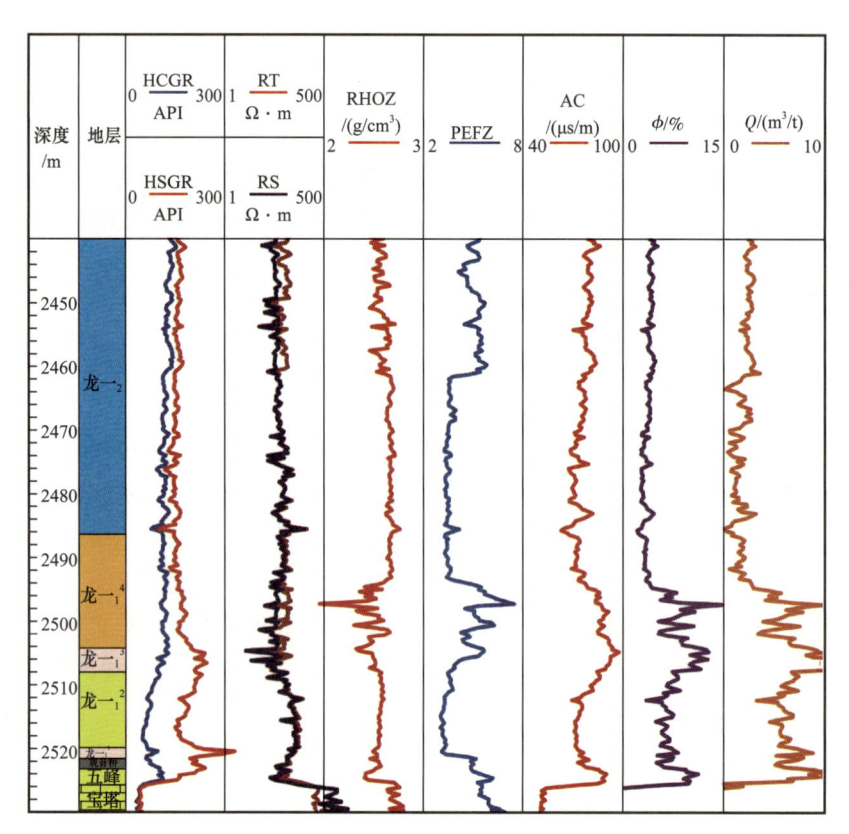

图 2　含气页岩层段测井响应特征

注：HCCR 为无铀伽马；HSGR 为总伽马；RT 与 RS 分别是深、浅阵列侧向电阻率；RHOZ 为密度；
PEFZ 为光电吸收截面指数；AC 为声波时差；ϕ 为孔隙度；Q 为总含气量

测井计算 TOC 的 $\Delta \lg R$ 法；朱光有等[34]对 $\Delta \lg R$ 法进行了改进；Carpentier 等[35]使用干酪根的含量计算页岩储层 TOC 含量；Schmoker 等[36]利用密度与 TOC 之间较好的相关性，采用岩心刻度密度测井曲线的方法计算页岩气储层 TOC 含量；Kwiatkowsky 等[37]利用了黄铁矿含量与 TOC 相关性；郭龙等[38]采用线性回归及神经网络预测等方法计算页岩储层的 TOC。

综上，总有机碳质量分数 TOC 的计算方法主要包括经典的 $\Delta \lg R$ 法及各种拟合法，如铀含量与岩心分析 TOC 拟合、密度测井与实验分析拟合 TOC、黄铁矿与实验分析拟合 TOC、利用干酪根含量计算 TOC、三孔隙度曲线计算法等。这些方法各有优缺点[30, 43]，重叠法受地层电阻率影响较大，因此对有机质成熟度高、含大量黄铁矿、电阻高的页岩储层并不适用；通常，U 元素法方法简单，但由于 U 与地层的 TOC 通常具有较好的相关性，因此实际应用效果较好，但若井眼扩径严重，则需对铀曲线进行校正之后方可使用。因此，在使用测井曲线计算 TOC 时，应该根据实际地层情况选择合适的计算方法，同时当地层差异较大时，对不同的层段可采用不同的方法分段计算、取值，从而提高 TOC 计算结果的准确度。

3.3.2　矿物成分及脆性评价

目前，传统的测井评价方法无法解决页岩储层岩性复杂、矿物类型多样的难题[43-46]。

元素俘获测井（ECS）可以测得硅、钙、硫、镁、铁等元素的质量分数，并选用相应的氧闭合分析模型确定出各种矿物组分[42]。若没有ECS资料，可依据X射线衍射测试岩样的矿物组成及含量，优选主要矿物建立岩石物理模型，采用最优化方法确定页岩储层各矿物组分的含量及储层的孔隙度[47]。

页岩层段岩石脆性的准确评价有助于压裂层段的优选和压裂参数的设计。目前常用的页岩储层脆性评价方法主要为声波法和矿物组分法[48-49]。声波法计算脆性指数是通过阵列声波测井获取纵横波参数，计算地层的杨氏模量与泊松比，并对两者进行归一化处理，求取平均值。矿物组分法可以计算脆性矿物的含量（一般为石英或石英与方解石占总矿物含量的百分比）。

两种方法都有一定的缺点，声波法需要采用阵列声波资料，受成本影响，在大量开发井中并不测横波；矿物组分法适用通常认为石英和方解石的脆性接近，但实际上两种矿物的岩石力学性质差异明显[43]。近年来有学者提出一种新的岩石脆性表征方法，即声波矿物组合法[49]，同时考虑脆性矿物的含量及对应声波参数对岩石脆性指数贡献大小，其计算公式如下：

$$BRIT = \frac{V_{石英} \times (YM_{石英}/PR_{石英})}{V_{石英} \times (YM_{石英}/PR_{石英}) + V_{方解石} \times (YM_{方解石}/PR_{方解石}) + V_{黏土} \times (YM_{黏土}/PR_{黏土})} \times 1$$

其中，$BRIT$ 为岩石脆性指数；V 为矿物体积；YM 为杨氏模量；PR 为泊松比。

3.3.3 储层物性评价

补偿声波测井、补偿中子测井和补偿密度测井是评价孔隙度常用的方法[51-55]。核磁共振测井也被用于计算有效孔隙度（不受总有机碳影响）[47, 52-53]。张晋言等[54]认为页岩地层对密度、孔隙度作有机质校正（类似于砂泥岩地层中对泥质含量校正的方法）后的孔隙度接近于地层的真实孔隙度；张作清等[51]使用ECS元素测井方法计算页岩储层的孔隙度；毛志强等[52]提出利用核磁共振资料结合声波时差测井可较准确地计算页岩储层的孔隙度；李军等[47]把页岩气储层的孔隙划分为有机孔隙、碎屑孔隙、黏土孔隙及微裂缝4类，并分别建立了这4类孔隙的定量计算模型。李亚男[53]使用中子—密度交会法，变骨架密度法及核磁测井计算孔隙度，且认为以纳米孔为主的页岩地层，由于T_2弛豫时间很短，无法准确测量，因此核磁共振测量孔隙度比实际地层总孔隙度要小。

页岩储层致密，基质渗透率为10^{-3}mD级别，气体的流动不符合达西定律，因此其岩心测得的孔渗相关性差。目前，常用自然伽马能谱测井、核磁共振测井及微电极测井计算页岩渗透率[30]。Shabro等[55]认为，页岩气储层的视渗透率受气体解析作用影响较小，提出建立一种微观意义上的气体流动模型，利用流体的分布情况评估页岩储层的视渗透率。李亚男[53]认为充气孔隙较大，气体在其中的流动符合达西渗流特征，渗透率与充气孔隙度的相关性较好，充气孔隙度越大，渗透率越大。

3.3.4 页岩含气量评价

在页岩储层中，天然气的赋存有3种状态：吸附气、游离气及少量的溶解气[56-59]。页岩储层总含气量的计算方法主要借鉴煤层气的解吸法，分别测量解吸气量、残余气量和损失气量，三者相加总和即得到页岩总含气量[27]，此外，页岩含气量也可利用其与影响它的参数之间的相关性（图2），拟合得到总含气量的计算公式。

目前，页岩气吸附气的含量主要采用等温吸附实验方法与测井解释结合进行计算。等温吸附法普遍采用经典的 Langmuir 模型[56]计算吸附气含量，计算过程中需要对页岩储层的温度、压力及有机碳含量等参数进行校正[30, 51]。张作清等[50]以等温吸附数据为基础，分析页岩吸附的影响因素，在 KIM 方程[57]原有形式的基础上，通过温度等系数修正页岩吸附气含量计算模型。部分学者利用吸附气含量与其影响参数的相关性，拟合得到吸附气的计算公式[40]。

页岩储层游离气含量的计算和常规储层的含气量计算方法类似，通常采用体积法计算。游离气含量计算准确与否的关键是页岩储层含水饱和度的计算精度。常规页岩储层含水饱和度多采用阿尔奇公式及其修正的公式计算。受页岩自身矿物成分与含量的影响，页岩岩电实验难以进行[27]，因此现场游离气含量一般采用解析法测得的总含气量减去等温吸附法计算的吸附气含量获得[40]，测井上常采用利用游离气含量与其影响因素的关系，通过多元参数拟合求取[51, 59]。

4 页岩储层综合评价及存在问题

4.1 页岩气储层评价进展

目前，页岩储层研究主要采用样品分析实验及测井方法对页岩储层 TOC、矿物组成、储集空间类型、储层孔隙度和渗透率特征、储层孔隙结构等及含气性等关键参数进行研究。在定性研究方面，主要包括采用实验测试数据的统计分析页岩有机质含量，采用激光拉曼光谱方法测定高演化程度烃源岩 R_o 值，基于全岩射线衍射方法分析页岩矿物成分与含量，利用氩离子抛光技术和扫描电镜相结合观察储层的孔隙形态并进行页岩孔隙分类，基于氮气吸附、高压压汞、核磁共振等技术对储层物性、孔隙结构进行研究[60-61]。在定量研究方面主要采用上述测井方法获取储层关键参数。

近年来，在储层评价中，国内外学者开始注意到页岩储层非均质性问题，并对页岩储层的非均质性进行评价。于炳松[11]通过分析富有机质页岩层段在纵横向上矿物组成、储气性能等方面的非均质性，寻找页岩气富集层段。郭英海等[62]运用实验技术手段，观测海相页岩样品在微观尺度下的非均质性，认为在微观尺度下矿物组分与含量，孔隙类型、孔径大小分布、形态等均具有较强的非均质性。随着页岩气开发的进行，对页岩储层非均质性的研究将会不断深入。

在储层研究的基础上，对储层进行分类评价是储层研究的一项重要工作。在常规储层评价中，储层分类评价主要利用岩心、测井、试油等资料对储层进行分级，并把分级评价指标落实到孔隙度、渗透率等储层参数上。影响页岩储层品质的因素较多，不仅包括地质上的孔隙度、渗透率以及 TOC 等参数，岩石脆性等参数也对其有重要影响。目前，不少学者对页岩储层分级评价进行了研究，大多是基于高分辨率仪器对孔隙大小、孔隙类型进行分类[14-15]。也有部分学者尝试利用影响页岩储层质量的关键因素对页岩储层进行定量分类评价[10, 63]。涂乙等[63]筛选出 10 个影响页岩储层的评价参数，采用灰色关联分析法分析各因子之间的相关性，计算影响页岩储层质量的各因子的权重和综合评价因子，对页岩储层进行了综合分类评价（表2）。

表2 页岩气储层质量综合评价分类[63]

页岩名称 \ 各参数权重	有机碳含量	成熟度	有效厚度	储量丰度	孔隙度	含气量	吸附气含量	压力	黏土矿含量	埋藏深度	综合评价因子	储层分类
Antrim	1.00	0.20	0.33	0.39	0.75	0.22	0.96	0.08	0.68	0.19	0.5295	Ⅱ
Ohio	0.30	0.34	0.31	0.24	0.39	0.28	0.88	0.40	0.79	0.94	0.5152	Ⅱ
New Albany	0.80	0.28	0.26	0.25	1.00	0.18	0.68	0.09	0.67	0.27	0.4932	Ⅲ
Barnett	0.30	0.53	0.37	1.00	0.42	1.00	0.27	0.70	0.75	1.00	0.6093	Ⅰ
Lewis	0.13	0.67	0.84	0.83	0.35	0.09	1.00	0.25	0.77	0.86	0.5578	Ⅱ
龙马溪组	0.17	0.81	0.72	0.43	0.27	0.28	0.55	1.00	1.00	0.00	0.4756	Ⅳ
筇竹寺组	0.13	1.00	1.00	0.23	0.27	0.07	0.68	0.90	0.57	0.04	0.4453	Ⅳ
权重系数	0.1560	0.0795	0.0885	0.0935	0.1163	0.0866	0.1025	0.0840	0.0857	0.1075		

4.2 中国页岩气储层评价存在的问题及发展趋势

（1）尽管页岩储层研究取得了较大的进步，但认识还不够全面。众所周知，页岩储层储集空间类型多样，但关于不同的储集空间类型对总孔隙的贡献、孔隙间的连通性及变化规律还认识不清。孔隙在成岩作用中的变化及及其与页岩气生成时间的匹配关系，孔隙度和渗透率测试方法差异对物性评价的影响等方面的研究尚不明朗，制约了我们对储层的认识，是下一步研究的方向。

（2）中国页岩气勘探开发处于初步开发阶段，早期主要关注富有机质页岩的生气能力，而对储层评价的基础理论与方法探讨不够完善，对优质页岩储层的主控因素缺乏系统全面的认识，直接影响到页岩气储层测井评价效果。因此，储层评价基础理论与方法的不断完善，将在今后页岩气的勘探开发中起重要作用。

（3）中国页岩气储层非均质性较强，各测井方法都有一定的适用条件，同一种页岩储层测井评价方法对不同地区、不同类型的页岩储层评价中的普适性受到制约。故应结合中国不同页岩气区块的实际储层特征建立与之相适用的测井评价体系。

（4）中国页岩气井多采用水平井开发，由于测井条件和地层环境的特殊性导致在测井曲线显示、数据处理及综合解释等方面与直井相比，有所不同，如何进行环境校正等对储层测井效果影响较大，是下一步研究的重点。

5 结束语

（1）页岩储层相对于常规油气储层的特殊性主要体现在富有机质、富黏土矿物、矿物成分复杂、孔隙度和渗透率极低、纳米级孔喉发育、天然气吸附赋存比例大等特征。影响页岩储层品质的参数较多，深入研究储层特征及主控因素，对优选页岩储层的关键参数评价具有重要意义。

（2）页岩储层的特殊性造成页岩储层测井系列、测井响应特征及储层定量评价方法不同于常规储层，决定了其测井评价重点在于计算总有机质含量、识别矿物成分与含量、计算岩石的脆性指数、储层的物性及含气量。

（3）需要探索一种更好地反映页岩储层地质及工程品质的页岩储层综合分类与评价方法，指导优质页岩富集层段的优选。

参 考 文 献

[1] Montgomery S L, Jarvie D M, Bowker K A, et al. Mississippian Barnett Shale, Fort Worth basin, north-central Texas: Gas-shale play with multitrillion cubic foot potential [J]. AAPG Bulletin, 2006, 89（2）: 155-175.

[2] Roussel N P, Shama M M. Optimizing fracture spacing and sequencing in horizontal-well fracturing [J]. SPE 127986, 2010.

[3] Mendoza E, Aular J, Sousa L. Optimizing horizontal-well hydraulic-fracture spacing in the Eagle ford [J]. SPE 143681, 2011.

[4] Yu W, Wu K, Zuo L, et al. Physical models for inter-well interference in shale reservoirs: Relative impacts of fracture hits and matrix permeability [C] //Society of Exploration Geophysicists, American Association of Petroleum Geologists, Society of Petroleum Engineers, 2016: 1535-1558.

[5] 梁超, 姜在兴, 杨镱婷, 等. 四川盆地五峰组—龙马溪组页岩岩相及储集空间特征 [J]. 石油勘探与开发, 2012, 39（6）: 691-698.

[6] 王秀平, 牟传龙, 王启宇, 等. 川南及邻区龙马溪组黑色岩系成岩作用 [J]. 石油学报, 2015, 36（9）: 1035-1047.

[7] 郭旭升, 胡东风, 文治东, 等. 四川盆地及周缘下古生界海相页岩气富集高产主控因素——以焦石坝地区五峰组—马溪组为例 [J]. 中国地质, 2014, 41（3）: 893-901.

[8] 刘乃震, 王国勇. 四川盆地威远区块页岩气甜点厘定与精准导向钻井 [J]. 石油勘探与开发, 2016, 43（6）: 1-8.

[9] 郭旭升, 李宇平, 刘若冰, 等. 四川盆地焦石坝地区龙马溪组页岩微观孔隙结构特征及其控制因素 [J]. 天然气工业, 2014, 34（6）: 9-16.

[10] 王伟明, 卢双舫, 田伟超, 等. 利用微观孔隙结构参数对辽河大民屯凹陷页岩储层分级评价 [J]. 中国石油大学学报: 自然科学版, 2016, 40（4）: 12-19.

[11] 于炳松. 页岩气储层的特殊性及其评价思路和内容 [J]. 地学前缘, 2012, 19（3）: 252-258.

[12] Loucks R G, Ruppel S C. Mississippian Barnett Shale: Lithofacies and depositional setting of a deep-water shale-gas succession in the Fort Worth Basin, Texas [J]. AAPG bulletin, 2007, 91（4）: 579-601.

[13] 王淑芳, 董大忠, 王玉满, 等. 中美海相页岩气地质特征对比研究 [J]. 天然气地球科学, 2015, 26（9）: 1666-1678.

[14] Ambrose R J, Hartman R C, Diaz Campos M, et al. New pore-scale considerations for shale gas in place calculations [C] //SPE Unconventional Gas Conference. SPE, 2010.

[15] Clarkson C R, Solano N, Bustin R M, et al. Pore structure characterization of North American shale gas reservoirs using USANS/SANS, gas adsorption, and mercury intrusion [J]. Fuel, 2013,

103: 606-616.

[16] Chalmers G R, Bustin R M, Power I M. Characterization of gas shale pore systems by porosimetry, pycnometry, surface area, and field emission scanning electron microscopy/transmission electron microscopy image analyses: Examples from the Barnett, Woodford, Haynesville, Marcellus, and Doig units [J]. AAPG bulletin, 2012, 96（6）: 1099-1119.

[17] 蒲泊伶, 董大忠, 牛嘉玉, 等. 页岩气储层研究新进展[J]. 地质科技情报, 2014, 33（2）: 98-104.

[18] 蒲泊伶, 董大忠, 耳闯, 等. 川南地区龙马溪组页岩有利储层发育特征及其影响因素[J]. 天然气工业, 2013, 33（12）: 41-47.

[19] 张晓明, 石万忠, 徐清海, 等. 四川盆地焦石坝地区页岩气储层特征及控制因素[J]. 石油学报, 2015, 36（8）: 926-939.

[20] Ross D J K, Bustin R M. The importance of shale composition and pore structure upon gas storage potential of shale gas reservoirs [J]. Marine and Petroleum Geology, 2009, 26（6）: 916-927.

[21] Jarvie D M, Hill R J, Ruble T E, et al. Unconventional shale-gas systems: The Mississippian Barnett Shale of north-central Texas as one model for thermogenic shale-gas assessment [J]. AAPG bulletin, 2007, 91（4）: 475-499.

[22] Wang F P, Reed R M. Pore networks and fluid flow in gas shales [C]//SPE annual technical conference and exhibition. Society of Petroleum Engineers, 2009.

[23] Bowker K A. Barnett shale gas production, Fort Worth Basin: Issues and discussion [J]. AAPG bulletin, 2007, 91（4）: 523-533.

[24] Caulton D R, Shepson P B, Santoro R L, et al. Toward a better understanding and quantification of methane emissions from shale gas development [J]. Proceedings of the National Academy of Sciences, 2014, 111（17）: 6237-6242.

[25] 何建华, 丁文龙, 王哲, 等. 页岩储层体积压裂缝网形成的主控因素及评价方法[J]. 地质科技情报, 2015, 34（4）: 108-118.

[26] 刘双莲, 陆黄生. 页岩气测井评价技术特点及评价方法探讨[J]. 测井技术, 2011, 35（2）: 112-116.

[27] 万金彬, 李庆华, 白松涛. 页岩气储层测井评价及进展[J]. 测井技术, 2012, 36（5）: 441-447.

[28] 王濡岳, 丁文龙, 王哲, 等. 页岩气储层地球物理测井评价研究现状[J]. 地球物理学进展, 2015, 30（1）: 228-241.

[29] Boyer C, Kieschnick J, Suarez-Rivera R, et al. Producing gas from its source [J]. Oilfield Review, 2006, 18（3）: 36-49.

[30] 杨小兵, 杨争发, 谢冰, 等. 页岩气储层测井解释评价技术[J]. 天然气工业, 2012, 32（9）: 33-36.

[31] 齐宝权, 杨小兵, 张树东, 等. 应用测井资料评价四川盆地南部页岩气储层[J]. 天然气工业, 2011, 31（4）: 44-47.

[32] Fertle H. Total organic carbon content determined from well logs [G]. SPE Formation Evaluation 15612, 1988: 407-419.

［33］Passcey Q R，Moretti F U，Stroud J D. A practical model for organic richness from porosity and resistivity logs［J］. AAPG Bulletion，1990，74（12）：1777-1794.

［34］朱光有，金强，张林晔. 用测井信息获取烃源岩的地球化学参数研究［J］. 测井技术，2003，27（2）：104-109.

［35］Carpentier B，Bessereau G. Wireline logging and source rocks estimation of organic carbon content by the CARBOLBG method［J］. The Log Analysts，1991，32（3）：279-297.

［36］Schmoker J W，Hester T C. Organic carbon in Bakken Formation，United States portion of Williston Basin［J］. AAPG Bulletion，1983，67（12）：2165-2174.

［37］Kwiatkowsky J，Galford J，Quircin J. Predicting pyrite and total organic carbon from well logs for enhancing shale reservoir interpretation［R］. SPE 161097，2012.

［38］郭龙，陈践发，苗忠英，等. 一种新的TOC含量拟合方法研究与应用［J］. 天然气地球科学，2009，20（6）：951-956.

［39］Robert R，Loucks R. Lmaging nanoscale pores in the Mississippian Barnett Shale of the Northern Fort Worth Basin［C］//AAPG Annual Convention Abstracts，California，2007，16：115.

［40］Autric A. Resistivity，radioactivity，and sonic transittime logs to evaluate the organic content of low permeability rocks［J］. The Log Analyst，1985，26（3）：36-45.

［41］Meyer B L，Nederlof M H. Identification of source rocks on wireline logs by density/resistivity and sonic transit time/resistivity crossplots［J］. AAPG Bulletin，1984，8（2）：121-129.

［42］Herron S L. A total organic log for source rock evaluation［J］. The Log Analyst，1987，28（6）：520-527.

［43］李霞，程相志，周灿灿，等. 页岩油气储层测井评价技术及应用［J］. 天然气地球科学，2015，26（5）：904-914.

［44］Grieser B，Bray J. Identification of production potential in unconventional reservoirs［C］//SPE 106623，2007：1-6.

［45］Rickman R，Mullen M J，Petre J E，et al. A practical use of shale petrophysics for stimulation design optimization：All shale plays are not clones of the barnett shale［C］//SPE，115258，2008：1-11.

［46］Britt L K，Schoeffler J. The geomechanics of a shale play：What makes a shale prospective［C］// SPE Eastern Regional Meeting. SPE，2009.

［47］李军，路菁，李争，等. 页岩气储层"四孔隙"模型建立及测井定量表征方法［J］. 石油与天然气地质，2014，35（2）：266-271.

［48］袁俊亮，邓金根，张定宇，等. 页岩气储层可压裂性评价技术［J］. 石油学报，2013，34（3）：523-527.

［49］Mullen M，Enderlin M. Fracability index-more than just calculating rock properties，SPE 159755［C］// Paper 159755-MS Presented at the SPE Annual Technical Conference and Exhibition. San Antonio，Texas. USA. 2012：8-10.

［50］刁海燕. 泥页岩储层岩石力学特性及脆性评价［J］. 岩石学报，2013，29（9）：3300-3306.

［51］张作清，郑炀，孙建孟. 页岩气评价"六性关系"研究［J］. 油气井测试，2013，22（1）：65-78.

［52］毛志强，张冲，肖亮. 一种基于核磁共振测井计算低孔低渗气层孔隙度的新方法［J］. 石油地球物

理勘探，2010，45（1）：105-109.

[53] 李亚男. 页岩气储层测井评价及其应用 [D]. 北京：中国矿业大学（北京），2014.

[54] 张晋言，孙建孟. 利用测井资料评价泥页岩油气"五性"指标 [J]. 测井技术，2012，36（2）：146-153.

[55] Shabro V, Tomes-Verdi C, Javadpour F. Numerical simulation of shale gas production：from pore-scale modeling of slip-flow, Knudsen Diffusion and Langmuir desorption to reservoir modeling of compressible fluid [C] //North American Unconventional Uas Conference and Exhibition, The Woodlands, Texas, USA. 2011：14-16，6.

[56] Lewis R, Ingraham D, Pearcy M, et al. New evaluation techniques for gas shale reservoirs [C] // Reservoir symposium. 2004：1-11.

[57] Kim A G. Estimating methane content of bituminous coalbeds from adsorption data [C] //U.S：Department. Of the Interior, Bureau of Mines, 1977：1-11.

[58] Zhou Q, Xiao X M, Tian H, et al. Modeling free gas content of the lower Paleozoic shales in the Weiyuan area of the Sichuan basin, China [J]. Marine and Petroleum Geology, 2015, 56：87-96.

[59] 李武广，钟兵，杨洪志，等. 页岩储层含气性评价及影响因素分析——以长宁—威远国家级试验区为例 [J]. 天然气地球科学，2014，25（10）：1653-1660.

[60] Slat R M, O'Brien N R. Pore types in Barnett and Woodford：Contributions to understanding gas storage and migration pathways in fine-grained rocks [J]. AAPG Bulletin, 2011, 95（11）：2017-2030.

[61] Loucks R G, Reed R M, Ruppel S C, et al. Morphology, genesis and distribution of nanometer-scale pores in siliceous mudstones of the Mississippian Barnett shale [J]. Journal of Sedimentary Research, 2009, 79（12）：848-861.

[62] 郭英海，赵迪斐. 微观尺度海相页岩储层微观非均质性研究 [J]. 中国矿业大学学报，2015，44（2）：300-307.

[63] 涂乙，邹海燕，孟海平，等. 页岩气评价标准与储层分类 [J]. 石油与天然气地质，2014，35（1）：153-158.

基于叠后地震数据的裂缝预测与建模
——以太阳—大寨地区浅层页岩气储层为例

王建君[1]，李井亮[2]，李　林[1]，马光春[2]，杜　悦[1]，

姜逸明[2]，刘　晓[2]，于银华[2]

（1. 中国石油浙江油田分公司；
2. 北京蓝海智信能源技术有限公司）

摘　要：太阳—大寨地区在钻井过程中常见钻井液漏失、压裂施工压力高等现象，鉴于叠后三维地震资料几何属性定量预测裂缝的可靠性和精度存在尺度及破碎程度的量化分级问题，提出了基于地震几何属性的裂缝地震相识别和裂缝确定性提取及建模方法；综合应用倾角、曲率和非连续性等多属性，以贝叶斯概率模型为基础，通过无监督聚类分析获得最佳的聚类效果和聚类数；垂向采用逐个时间切片扫描法建立裂缝的空间体系，并对追踪得到的所有线状结构进行清理去噪，简化复杂几何结构，对清理后的裂缝进行网格化重构，计算了裂缝的几何（拓扑）参数，建立了高精度离散裂缝模型。结果表明，应用该方法在太阳—大寨浅层页岩气区块得到成功应用，准确预测了水平井钻进过程中的断裂和裂缝发育位置，断层（裂缝）预测准确率达到92%，有效规避了钻井液漏失，并对压裂设计提供了有力支撑。

关键词：无监督聚类分析；裂缝地震相；离散裂缝网络；裂缝网格重构；浅层页岩气；太阳—大寨地区

太阳—大寨区块构造上处于川南低陡褶带叙永复向斜中，研究区内五峰组底界埋深500～1500m，核心建产区埋深500～2000m，五峰组—龙一$_1$亚段优质页岩发育，优质页岩段具有厚度相对较薄（3～5m）、有机质丰度及热演化程度高（TOC质量分数：2.58%～3.21%，R_o：1.99%～3.08%）、含气性好（3.30～5.51m^3/t）、储集性能好（孔隙度：3.98%～5.41%）、脆性矿物含量高（51%～75%）和目的层超压（压力系数1.25～1.62）等特征，是有利的页岩气开发层系。区内地层倾角变化大（2°～40°），太阳背斜核部一般小于10°，断层以三、四级断裂为主，小断层和离散裂缝异常发育。水平井钻井过程中经常钻遇到0.5～10m断距的微断裂，常规地震解释较难有效识别和拾取。相比于深层页岩储层，浅层页岩储层对裂缝更加敏感，在钻井过程中由于地层埋深较浅，这些微小的断裂或裂缝会造成较为严重的钻井液漏失，影响钻井质量和钻完井安全。因而对于微小断层或离散裂缝的准确描述及级别划分可以有效规避钻井风险，提高钻完井质量，其意义重大。

对于次断层级别裂缝（sub-seismic fault）的研究由来已久。Marfurt等[1]提出空间或

时间滑动的多时窗方法估算地震倾角和方位角，为断裂异常部位的识别提供了更稳健的方法基础，同时为构造导向滤波、相干振幅梯度等裂缝描述算法提供了更好的基础数据。Pedersen等[2]、Randen等[3]和Van等[4]均提出蚂蚁追踪算法，在倾角、方差等属性基础上计算蚂蚁体，该方法在国际上得到广泛应用。孙乐等[5]在乌夏地区综合应用相干体、方差体等技术手段检测地震反射不连续性，应用蚂蚁追踪技术进行断裂系统追踪，取得了很好的效果。程超等[6]在任丘潜山雾迷山组油藏断裂系统分析中应用蚂蚁追踪取得了良好的效果，认为该技术可以用在本区剩余油预测中，并具有较好的推广价值。随着蚂蚁追踪技术在复杂断裂识别，碳酸盐岩裂缝研究领域的进一步应用[7-8]，使得裂缝叠后预测技术在充分利用蚂蚁追踪属性方面得到了长足发展，多属性融合裂缝各向异性研究及其对油田开发的影响研究日益深入[9-10]，叠前地震资料蚂蚁追踪技术也得到了充分利用[11]。Hale[12]在研究断面提取和断距估算时提出最大似然属性（Likelyhood），在增强断裂地震成像效果的基础上提高了断层可识别的效果；马德波等[13]利用最大似然属性进行哈拉哈塘地区热瓦普区块奥陶系走滑断裂识别，取得良好的应用效果。王浩等[14]将最大似然法预测裂缝与相干法预测裂缝进行对比，揭示最大似然法可展示裂缝发育细节，表征中小尺度裂缝，与实钻结果具有更高的吻合度。

叠后裂缝预测技术不断发展，但三维裂缝定量提取技术进展却较为缓慢。本文将介绍基于贝叶斯无监督聚类的离散裂缝地震相和基于CT扫描的三维裂缝提取技术。基于贝叶斯无监督聚类的方法通过对观察的数据进行分类，使得类内差异最小化，类间的差异最大化，这种基于模型的聚类（Model-based Cluster Analysis）与传统的聚类方法如K-mean等相比，具有评估模型优劣的贝叶斯标准；因此它不仅可以给出模型参数化的选择，而且还可以客观地给出最优的聚类数。采用逐层扫描裂缝提取技术，在曲率增强属性体上沿时间（或深度）切片自动追踪、提取裂缝；对提取的裂缝做清理处理，去除非裂缝属性的噪声影响，由于每根裂缝具有拓扑参数，可通过连通性分析，构建单井上裂缝网络模型，描述与井相连的裂缝特征，井与井之间的裂缝分布、连接路径，有效指导页岩气井的钻井和压裂返排。

1 地震几何属性计算及地质解释

对滤波后的地震数据体进行地震几何属性的计算，包括地震倾角、不连续性和曲率属性。倾角曲率属性描述断裂—微断裂信息。地震倾角属性（图1a）描述地层的地貌和断裂系统的大尺度的不连续性。非连续性属性（图1b）描述的是大尺度的断裂产生的不连续性。和地震倾角属性相比，不连续性属性更清晰地定义了断裂系统，但是不能显示构造地貌特征。在倾角平面图上，断层和裂缝往往表现为长条形的线状特征，该特征被保留的是它可被观察到的长度，但形状却无法保留在这一属性中。为了更好地描述断层和裂缝在平面上的线型特征，计算了曲率属性。Roberts定义了用于测定地层曲率的几种曲率属性，其中最大曲率（图1c）对微断裂—裂缝系统的线状特征十分敏感，可以用于识别出由于局部构造引起的小断层和裂缝。

(a) 地震倾角属性

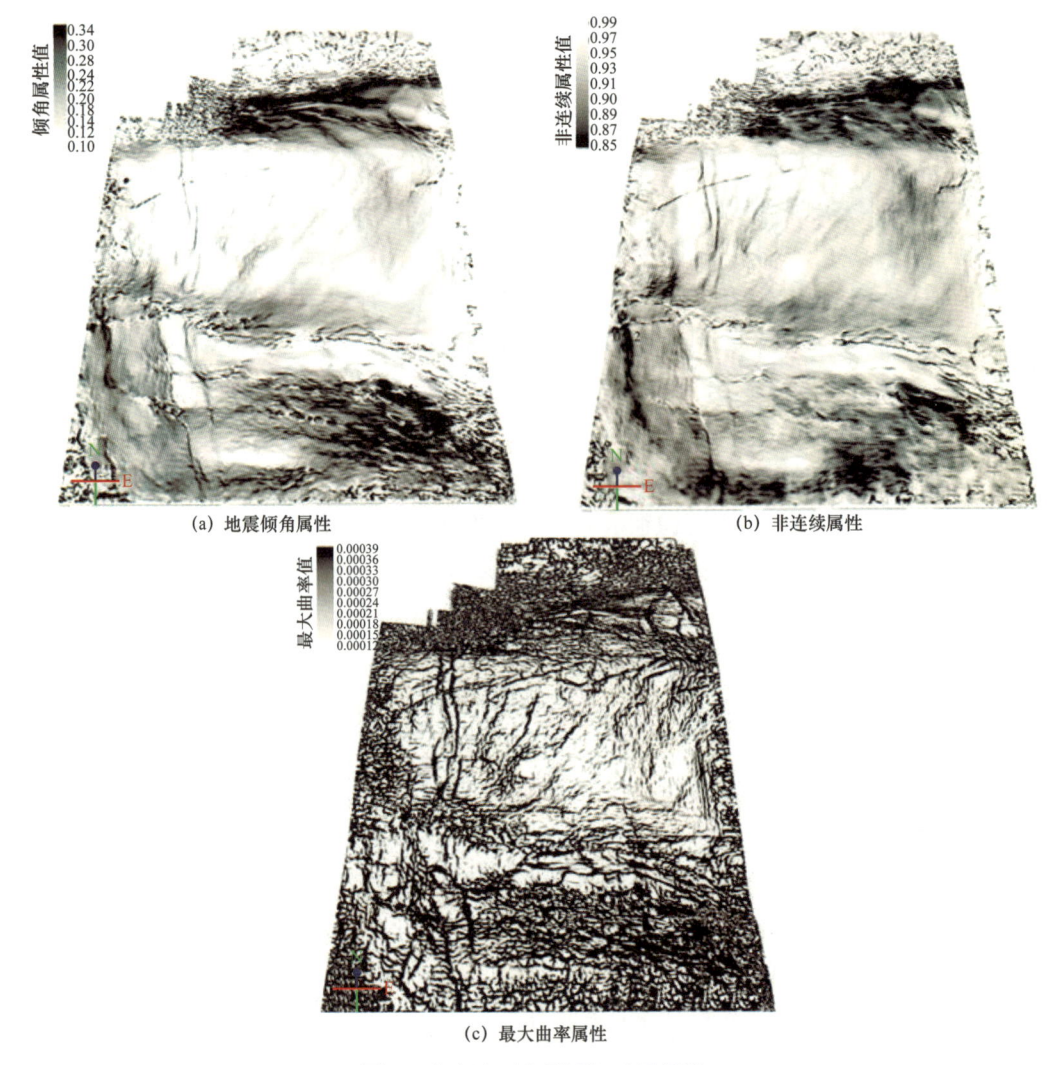

(b) 非连续属性

(c) 最大曲率属性

图 1 大寨地区龙马溪组地震属性

2 裂缝地震相

2.1 多属性无监督聚类分析

选用几何属性中的最大曲率属性、地震倾角属性、非连续属性作为多属性裂缝地震相分相的输入数据，通过无监督贝叶斯聚类分析的方法对协方差矩阵的特征值进行分解得到与裂缝相关的裂缝地震相，并且每一种裂缝地震相具有清晰的物理和地质意义[15]。

假设每个单因素属性为[16]：$y_1, y_2, \cdots, y_n, y_i \in R^d$，即有 d 个指标（可以认为 d 为所选取主成分的个数），n 个观测，$y_i = (y_{i1}, y_{i2}, \cdots, y_{id})$。一般情况下所有数据组成一个 $n \times d$ 矩阵

$$Y=\begin{pmatrix} y_{11}y_{12}\cdots y_{1d} \\ y_{21}y_{22}\cdots y_{2d} \\ \cdots \\ y_{n1}y_{n2}\cdots y_{nd} \end{pmatrix}=\begin{pmatrix} y'_1 \\ y'_2 \\ \cdots \\ y'_n \end{pmatrix} \qquad (1)$$

式中，y_1，y_2，$\cdots$，y_n 就是由前 d 个主成分计算的样本值。

一般情况下，最大可能的聚类数 M 由先验知识确定，M 应该尽可能小，以减少计算的复杂性。假定所考虑前数据可以分成 K（$2 \leq K \leq M$）类，第 j 个类的数据都是服从均值参数为 μ_j、协方差阵为 $\sum_j$ 的多元正态分布，其密度函数为：

$$\phi_j(y|\mu_j,\sum\nolimits_j)(2\pi)^{-d/2}|\sum\nolimits_j|^{-1/2}=\exp\left\{-\frac{1}{2}(y-\mu_j)'\sum\nolimits_j^{-1}(y-\mu_j)\right\} \qquad (2)$$

设 P_j 为第 j 个类在总样本中的混合比例，$0<P_j<1$，$j=1$，2，$\cdots$，K，且 $\sum_{j=1}^{k}=1$。设参数 $\theta=(P_1, P_2, \cdots P_{K-1}, ; \mu_1, \mu_2\cdots, \mu_k; \sum_1, \sum_2, \cdots, \sum_k)$，则所考虑问题的对数似然函数为：

$$l(\theta|y_1,y_2,\ldots,y_n)=\sum_{i=1}^{n}\log\left[\sum_{j=1}^{k}P_j\phi_j(y_i|\mu_j),\sum\nolimits_j\right] \qquad (3)$$

由于协方差阵 $\sum_j$ 可以分解为 $\sum_j=\lambda_j D_j A_j D_j'$（频谱解），式（3）中 $\lambda_j=|\sum_j|^{1/d}$ 称为模型体积；D_j 是由协方差阵 $\sum_j$ 的正交标准特征向量所组成的矩阵，故称为模型的方向；A_j 是一个对角阵，其对角线上元素是与由协方差阵 $\sum_j$ 的特征根从左上角按照从大到小的顺序排列到右下角的对角阵相差一个常数的 λ_j 倍，所以 $|A_j|=1$，称 A_j 为模型的形状。由于模型中这三个参数的变化，可以演变出如下 14 个模型（表 1）。相分析采用了 14 个模型，选择最优的模型实际上就是求出了对应模型的参数和对应的贝叶斯信息准则（Bayesian Information Criterion，BIC），并比较 BIC 的大小，找出最大的 BIC，其对应的模型最优。

表 1 14 种模型的一些特征

编号	模型	参数数目	M 步的求法	总体归类
1	$\lambda DAD'$	$\alpha+\beta$	显示	一般情况（椭球）
2	$\lambda_j DAD'$	$\alpha+\beta+(K-1)$	迭代	
3	$\lambda DA_j D'$	$\alpha+\beta+(K-1)(d-1)$	迭代	
4	$\lambda_j DA_j D'$	$\alpha+\beta+(K-1)d$	迭代	
5	$\lambda D_j AD'$	$\alpha+K\beta-(K-1)d$	显示	
6	$\lambda D_j A D_j'$	$\alpha+K\beta-(k-1)(d-1)$	迭代	
7	$\lambda D_j A_j D_j'$	$\alpha+K\beta-(k-1)$	显示	
8	$\lambda_j D_j A_j D_j'$	$\alpha+K\beta$	显示	

续表

编号	模型	参数数目	M步的求法	总体归类
9	λB	$\alpha+d$	显示	加权
10	$\lambda_j B$	$\alpha+d+(K-1)$	迭代	
11	λB_j	$\alpha+Kd-(K-1)$	显示	
12	$\lambda_j B_j$	$\alpha+Kd$	显示	
13	λI	$\alpha+1$	显示	球体
14	$\lambda_j I$	$\alpha+d$	显示	

说明：1. 模型 9-12 中的 B 或 B_j 为对角阵，模型 13，14 中的 I 为单位阵
2. 模型中参数 $\alpha=Kd+(K-1)$，$\beta=d(d+1)/2$

2.2 太阳—大寨浅层页岩气裂缝地震相

应用太阳—大寨地震数据进行无监督裂缝地震相聚类分级，最终将裂缝地震相分为 6 级。1、2 相线性特征明显，倾角、曲率最大，相干最小，对应断层级别，所占比例为 5.64%；3、4 相在断层周围成片状分布，对应次一级别的裂缝，所占比例为 27.05%；5 相大面积分布，相所占比例为 40.39%，是地震尺度下能够检测到的较小级别的裂缝；6 相是倾角、曲率最小，相干最大，裂缝发育程度最低，所占比例为 26.93%（图 2、图 3）。

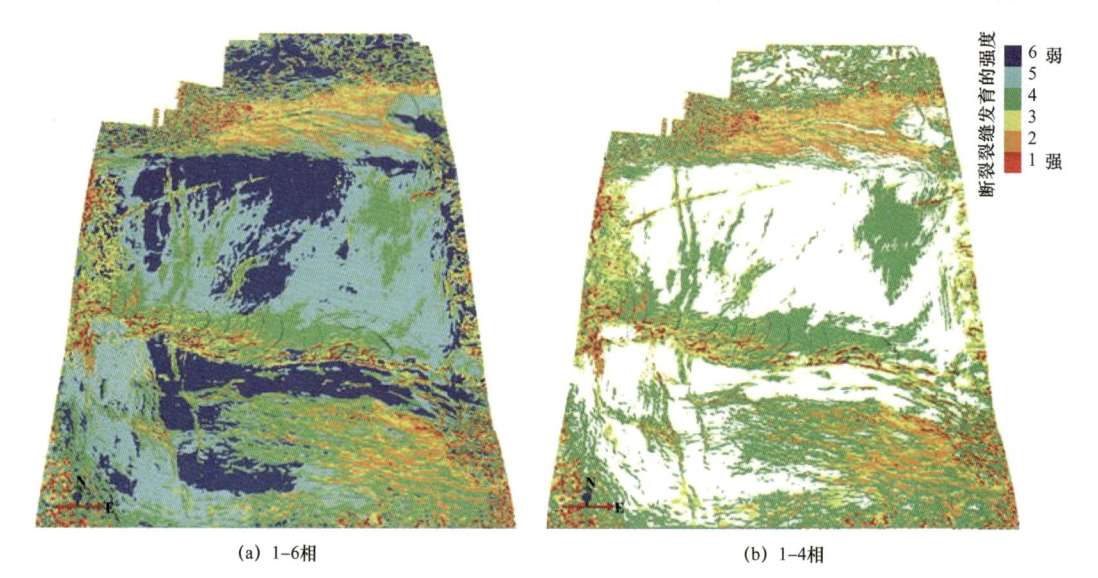

(a) 1-6相　　　　(b) 1-4相

图 2　大寨地区龙马溪组裂缝地震相分布图

整合各种相关的地质属性来描述断裂与裂缝的空间分布，裂缝地震相能够增强构造线性的特征变化，进一步将地质特征与这些线性变化特征联系起来，对断裂—裂缝系统的空间发育程度进行了很好的量化分级。

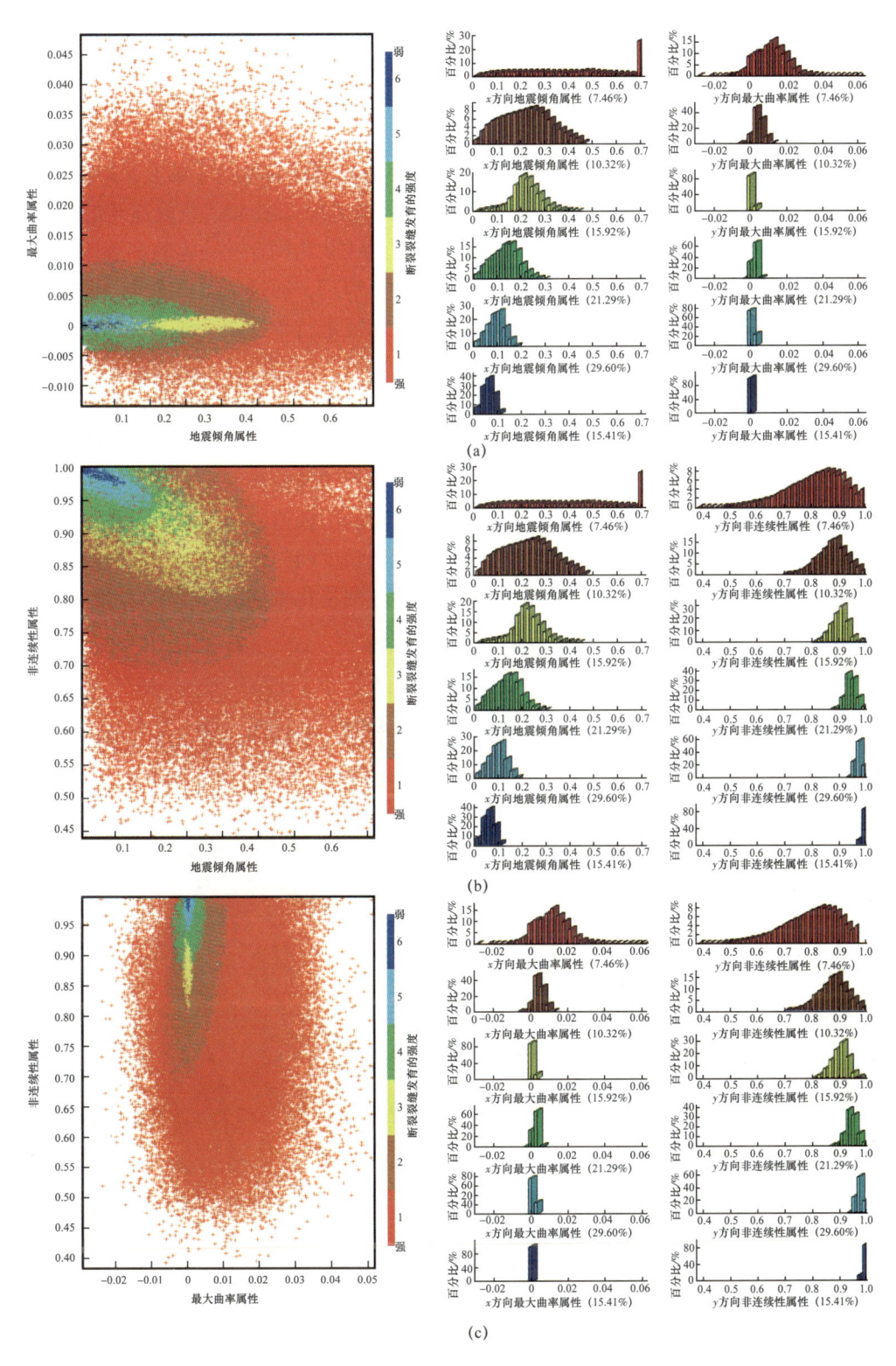

图 3 大寨地区龙马溪组地震倾角与最大曲率属性交会图（a）；地震倾角与非连续性属性（b）；最大曲率与非连续性属性（c）

3 离散裂缝网络建模

3.1 建模方法

对目的层进行沿层自动追踪，在曲率增强裂缝追踪通道属性的基础上，采用了CT扫描的方式，将数据按照等间隔进行沿时间切片的自动追踪（图4），获取连续的断裂、裂缝线状结构，受噪声及其他非裂缝信息的影响追踪的线状结构存在一定的多解性，通过自动清理算法，将非裂缝信息去除。第一，对长度进行清理，通过定义长度的门槛值获取研究区域的裂缝线状特征；第二，通过对拾取的线状特征进行单元式管理，沿着任意裂缝轮廓线的线状单元的长度与总长度比小于定义的段长度百分比门槛值时进行删除；当共享公共点的两个连续线段单元的夹角大于定义的段角度门槛值时，将沿着该线的点删除，将两个线段单元的端点进行连接。第三，对并行的两根线状特征进行清理，当短裂缝轮廓线与长裂缝轮廓线的比值小于并行长度百分比门槛值时删除短裂缝轮廓线，同时兼顾两根并行裂缝线状特征的角度门槛值，它们并行的角度大于定义的门槛值，认为两者不平行、否则视为平行进行删除。第四，定义两根裂缝线状特征重叠长度的门槛值，大于定义的长度进行删除。清理后的裂缝线状特征通过目的层位的剪切、网格化重构的方式建立离散裂缝网络（DFN）模型。

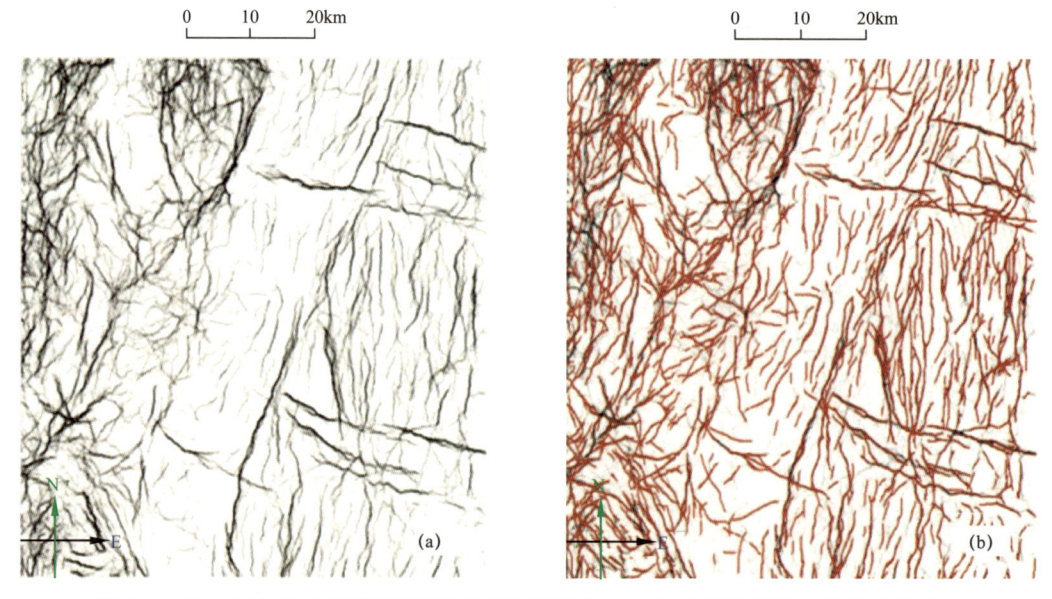

图4 太阳—大寨地区龙马溪组增强曲率属性切片（a）及裂缝沿层自动追踪结果（b）

3.2 浅层页岩气离散裂缝网络

通过对太阳—大寨浅层页岩气离散裂缝网络模型的构建，统计本区域的裂缝主要以东西向（80°～100°）、北西（110°～130°）、南东向为主（图5）；裂缝长度集中在40～400m（图6），这部分占到全区裂缝的76%。通过对完钻水平井钻遇离散裂缝组数进行统计表明，在水平段200～400m的间隔内，水平井会钻遇一组离散裂缝。

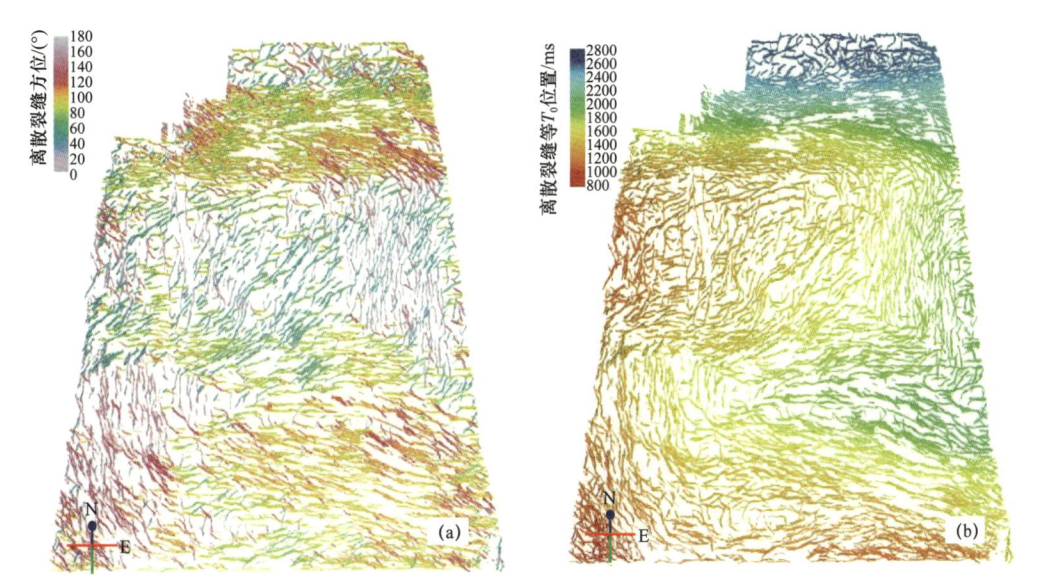

图 5 太阳—大寨地区龙马溪组地层离散裂缝网络模型方位分布图（a）及等 T_0 分布图（b）

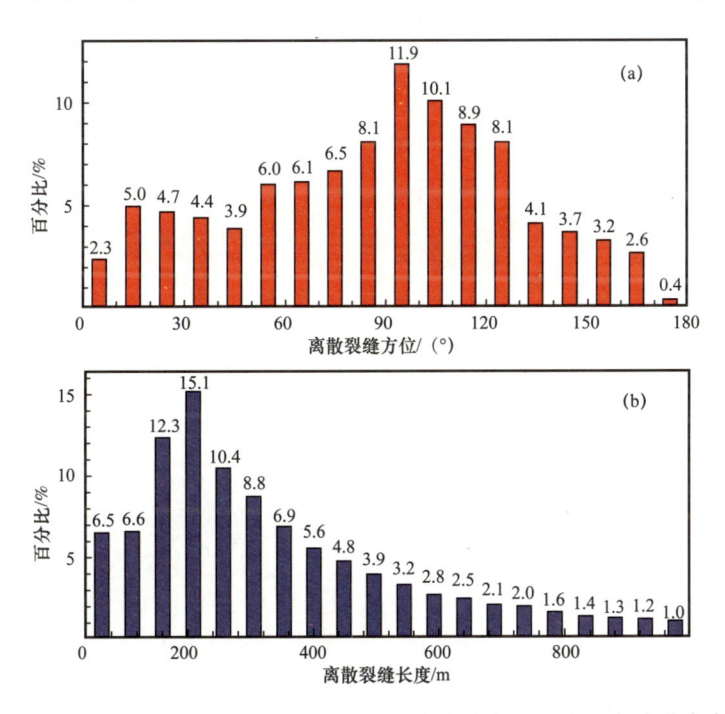

图 6 太阳—大寨地区龙马溪组离散裂缝网络模型方位分布图（a）及长度分布直方图（b）

3.3 单井离散裂缝提取

将裂缝线状结构从地震属性中确定性的拾取出来，构建水平井DFN模型。根据搜索深度段（图7中轨迹黄色段）、垂向延展深度（h）、搜索半径（R）搜索种子裂缝，以建立种子裂缝为起始点，通过连通性判断，将所有与种子裂缝直接或间接连接的裂缝重新组合，建立井上的DFN（图7）。

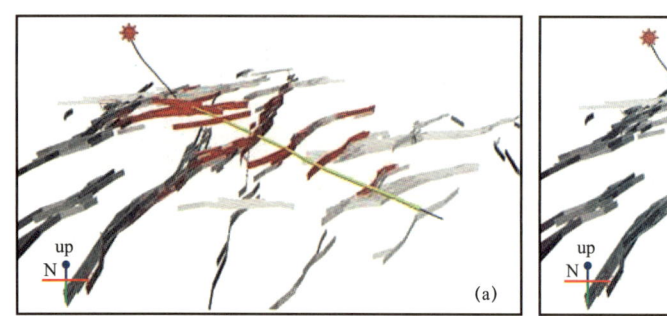

图7 太阳—大寨地区龙马溪组种子裂缝（蓝色）(a)及裂缝搜索连通关系（红色）(b)示意图

通过建立单井的DFN模型进行连通性分析。沿井轨迹搜索构建单井离散裂缝网络模型（图8），通过不同岩颜色表征单井离散裂缝分布，同时对两口单井中公共部分的裂缝用特殊颜色标明（黄色），该组裂缝即为沟通两口单井的裂缝通道，可以读取连通位置的深度及裂缝方位等信息，结合岩石力学、油藏等多学科进行钻井液漏失预测、压裂施工、压后分析等多学科综合应用[17-18]。

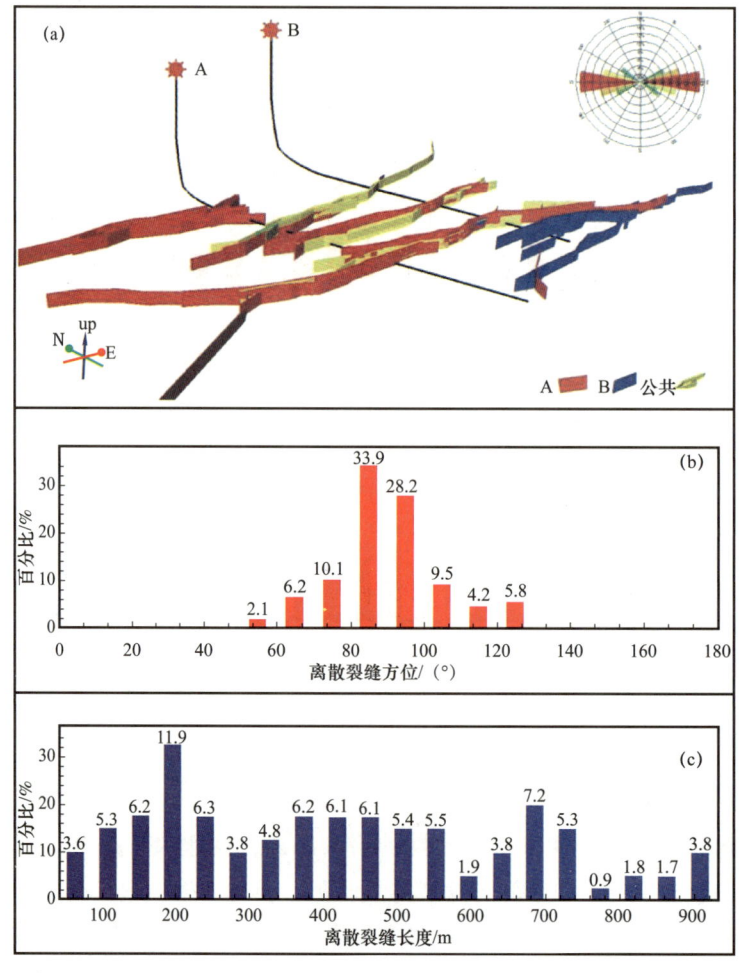

图8 太阳—大寨地区龙马溪组沿水平井轨迹搜索离散裂缝分布(a)、方位(b)、长度(c)统计图

4 多尺度裂缝模型的动态校正

DFN模型描述了大尺度的裂缝分布是介于钻井尺度（如岩心，成像测井等）与地震解释断层间的裂缝模型。由于测井储层预测类型与不同储层的发育情况对地震裂缝发育起着标定性作用[19-20]，所以测井综合研究缩小了井和地震数据之间的尺度差距，而研究表明DFN是钻井过程中最容易引发钻井液漏失的裂缝[21]。通过在开发井钻井过程中的钻井液漏失位置、钻遇断层位置、生产井套变位置的数据来标定离散裂缝的可靠性，是非常有效的质量控制方法。通过成像测井钻井诱导缝的方向标定泥浆漏失套变位置裂缝的方位信息，通过对多口已完钻井对裂缝信息的质量控制，进一步证明在叠后地震几何属性的基础上确定性提取离散裂缝模型的准确性、可靠性对浅层页岩气钻井地质导向、泥浆漏失预测有着重要的预测意义，同时提高了优质储层钻遇率。

水平井D与水平井E钻井过程中共计发生7处钻井液漏失（图9中黄的圆球的位置），漏失总量为80.2m³，漏失主要发生在龙一$_1^1$小层及五峰组。通过对水平井沿井筒轨迹搜索离散裂缝（D井绿色、E井粉色）可以看到钻井液漏失发生的位置都在离散裂缝发育区，从而证明了裂缝预测模型的准确性。

图9 太阳地区龙马溪组钻井液漏失与离散裂缝发育位置叠合图

钻井地质导向过程中通过临井对比和曲线拟合获取的断层信息，可以看到地质导向显示水平井C的5组断层的断距为0.5～2.6m（图10）。钻遇的5组断层信息在离散裂缝网络上都有裂缝信息被提取出来，再一次表明依靠地震几何属性确定性提取的裂缝可信度高，裂缝发育位置准确。

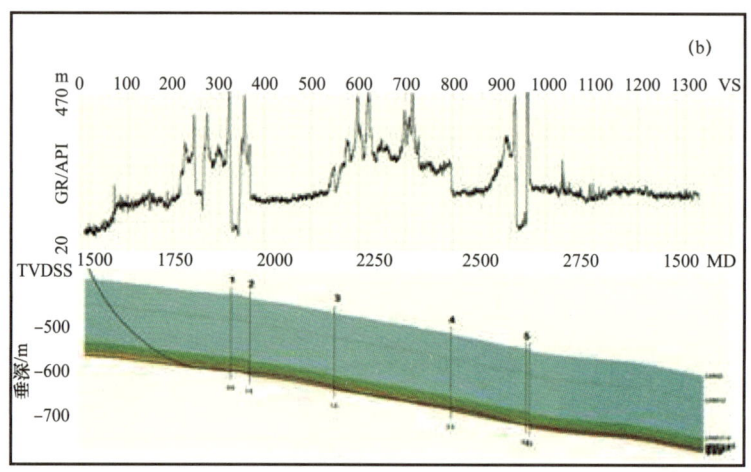

图 10 太阳地区龙马溪组水平井地质导向小断层（a）与离散裂缝发育位置（b）对比图

5 离散裂缝模型在水平井钻井中的应用

5.1 钻井液漏失预警

（1）钻前优化

通过总结已完钻井的地质导向裂缝与钻井液漏失位置分布规律，离散裂缝变得具有预测性；将总结出的钻遇裂缝发育特征、钻井液漏失的发育特征结合，设计井轨迹进行钻前井轨迹优化。

（2）钻中预警

在即将到达裂缝发育位置时的提前预警，降低因小断层引起的钻井出层，能够提前做好钻具优化及降低钻井液密度等措施，避免漏失发生。

以前文钻遇钻井液漏失的 D 井与 E 井井组为例，在钻 F 井时，根据 D 井与 E 井钻井液漏失位置及离散裂缝的连通关系、多井钻井液漏失规律的统计结果，把离散裂缝分为

3组，第1组为平行于最大水平主应力方向，最容易活化引发钻井液漏失离散裂缝（粉色）；第2组为与最大水平主应力方向夹角小于20°，为次一级引发钻井液漏失的裂缝；第3组为与最大水平主应力方向大于20°，是最不容易引发钻井液漏失的裂缝组。通过钻前裂缝分组对钻井液漏失风险位置提示。（图11中红色圈位置）F井钻井过程中采取了优化钻具、降低钻井液密度等措施，预测的位置并未发生钻井液漏失，但是地质导向证实了预测的1、2、3位置处钻遇了小断层，位置4证实地层倾角有变化，证实了预测的4个位置天然裂缝的发育。

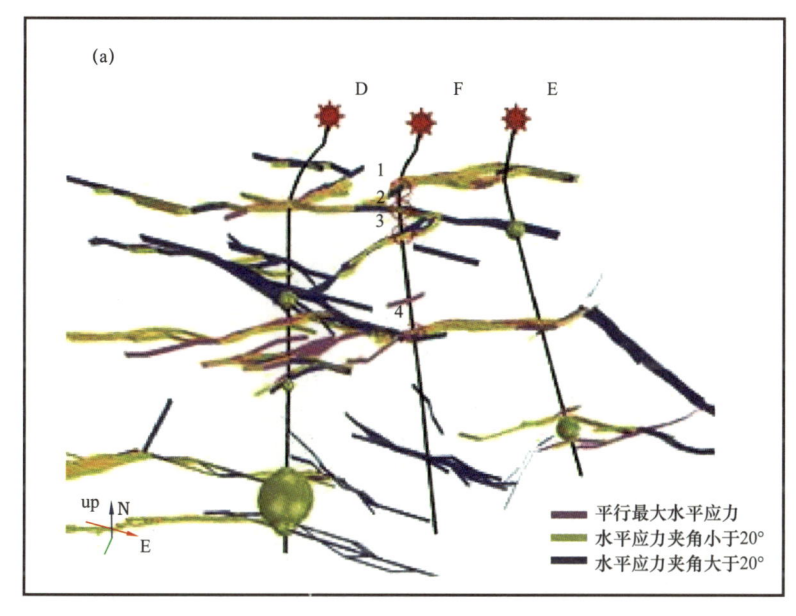

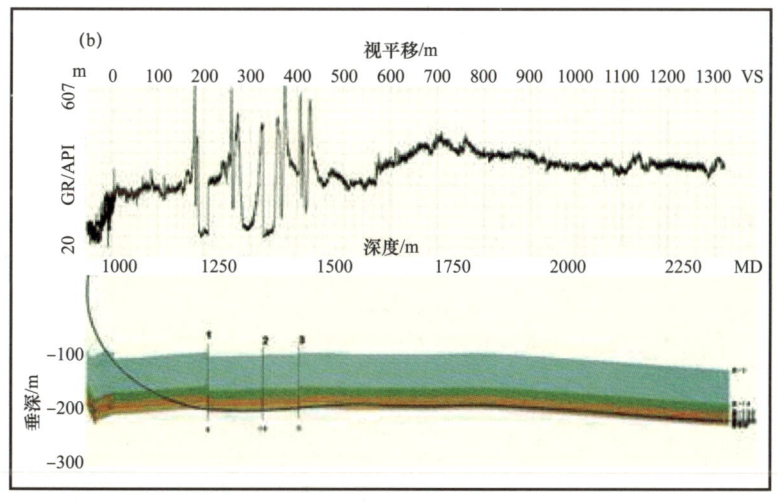

图11 太阳地区龙马溪组钻井液漏失预测（a）与实钻地质导向位置（b）对比图

龙马溪页岩离散裂缝的活化是引起钻井过程中钻井液漏失的主要原因。通过统计多口井多处钻井液漏失的位置与之裂缝发育的位置匹配，可以看到钻井液漏失发育的位置都有离散裂缝发育且发育的级别较高，尤其是在浅层页岩气中的裂缝在钻井过程中更容易被活化，容易引发漏失风险[22-23]。统计结果显示本区龙马溪组完钻井钻井液漏失24处，累计钻井液漏失353m³，钻井液漏失位置与裂缝发育位置相对应，漏失位置的水平井轨迹与离散裂缝近垂直，离散裂缝接近平行于最大水平主应力方向（表2）。

表2 太阳—大寨地区龙马溪组钻井液漏失位置裂缝信息统计表

漏失次数	漏失位置/m	裂缝地震相	井轨迹与裂缝夹角/(°)	裂缝与最大水平主应力夹角/(°)	漏失量/m³
1	3915	3、4	89	0	10
2	3097	3、4	90	5	9
3	3187	3、4	90	5	35
4	3781	3、4	40	40	6
5	3793	3、4	40	40	14
6	1508	3、4	80	5	6
7	1563	3、4	85	10	7
8	1581	3、4	85	10	6
9	1591	3、4	85	10	3
10	2216	3、4	85	15	26
11	2361	3、4	89	15	10
12	2024	3、4	85	10	7
13	1228	3、4	80	25	13
14	1938	3、4	89	15	12
15	1907	4、5	85	20	10
16	1400	3、4	85	10	47
17	1907	3、4	55	0	10
18	2195	3、4	70	15	4
19	2465	3、4	90	45	10
20	2515	3、4	85	40	27
21	1228	3、4	85	15	10
22	1904	3、4	90	10	10
23	3530	3、4	45	40	36
24	2356	3、4	45	40	25
平均			77	18	15

综合本地区的岩石物理力学参数、地应力分布、离散裂缝网络模型，计算了岩体的破裂压力，并用钻井数据校正了防止钻井液漏失的最大钻井液密度分布（图12），结果显示，完整岩体防止钻井液漏失的最小钻井液密度都在 2.5g/cm³ 以上，而裂缝断裂带处防止钻井液漏失的最小钻井液密度约 1.7~2.0g/cm³，且随深度增加而增大。该防止钻井液漏失的最小钻井液密度图在钻井工程中和观测到的现象有较好的对应关系。

准确的离散裂缝发育位置是预测钻井液密度窗口的重要输入数据，确保安全钻井液密度窗口分布的合理、准确性，能够真正意义的指导钻井。

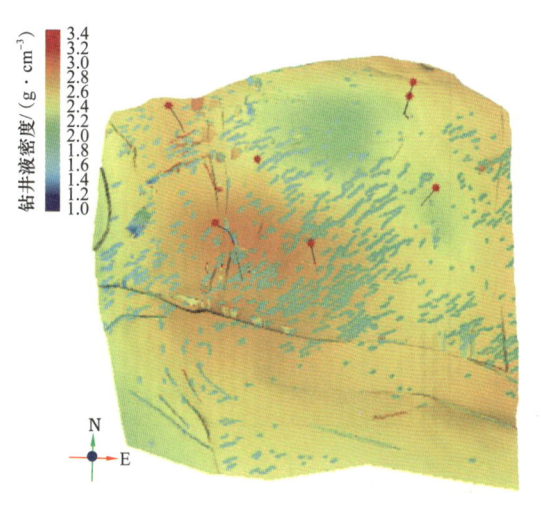

图12 大寨地区龙马溪组钻井液漏失窗口图

5.2 指导压裂施工

裂缝的发育程度与位置与压裂完井有很强的相关性[24]，浅层页岩气研究区目的层储层薄、储层物性非均质性不强，通过对压裂施工参数与离散裂缝发育位置、裂缝地震相的级别的对比研究，结果表明在离散裂缝不发育或发育较弱的区域，施工压力高（图13），同时加砂强度低（图14），反之施工压力低、加砂强度高[25-26]。施工过程中造成大量的压裂液滤失的情况少，说明了天然裂缝的发育对浅层页岩气储层压裂改造是有利的。通过裂缝地震相划分的断裂—裂缝的强弱与加砂规模、施工压力呈现很好的线性关系。裂缝地震相3、4相代表裂缝系统发育、断裂—裂缝的级别较高，表现为加砂强度高、施工压力低的特点。裂缝地震相5、6相代表裂缝系统欠发育或不发育、断裂—裂缝级别低，表现为加砂强度低、施工压力高的特点。

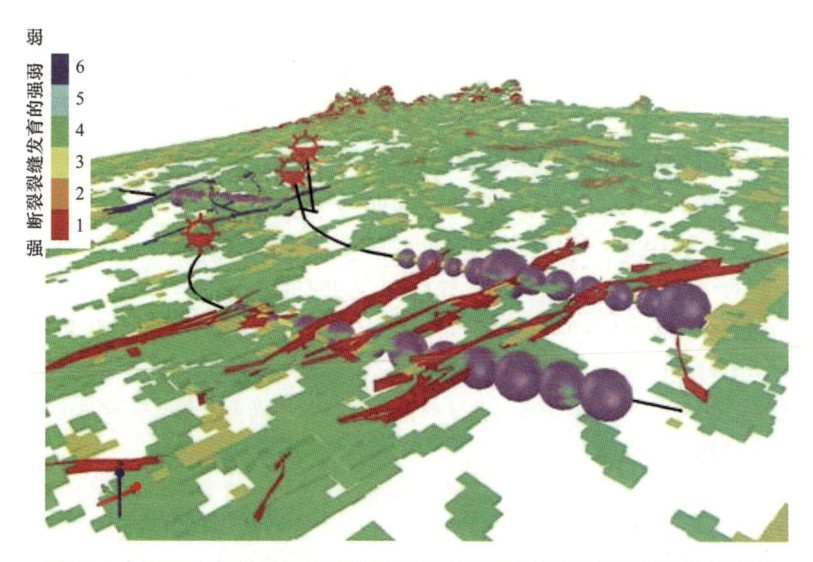

图13 太阳地区龙马溪组压裂施工压力与离散裂缝及裂缝地震相对比图

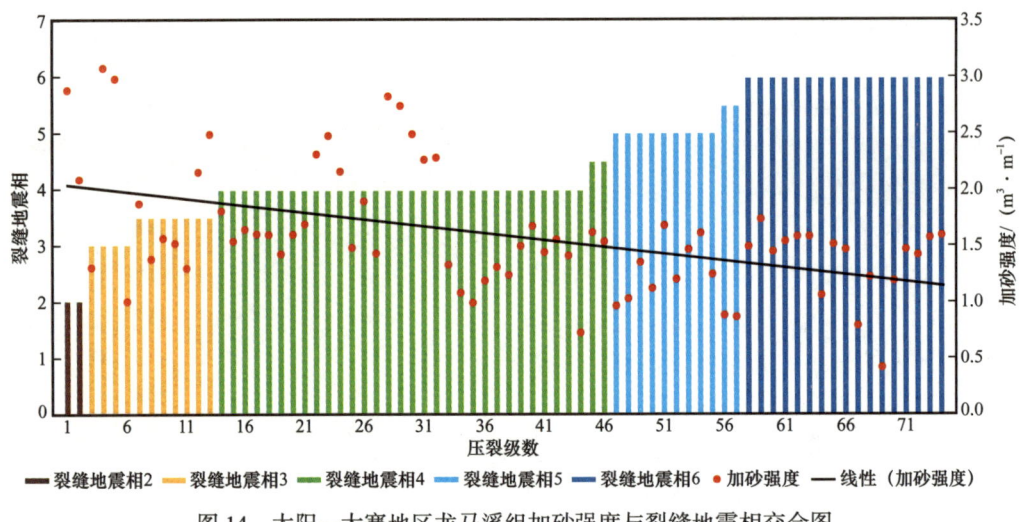

图14 太阳—大寨地区龙马溪组加砂强度与裂缝地震相交会图

6 结论

（1）基于贝叶斯概率模型的无监督聚类分析得到的裂缝地震相模型能定量描述有效裂缝系统的空间分布各向异性。

（2）基于叠后地震数据确定性提取裂缝线状结构能准确定位裂缝的空间位置。太阳—大寨浅层页岩气17口水平井钻遇裂缝预测准确度达92%，大大降低了钻井风险。

（3）浅层页岩气开发中天然裂缝的准确预测在提高水平井有效储层钻遇率及压裂施工改造等方面起到关键作用。减少了水平井钻井过程由于小断层和裂缝造成钻井出层问题，在压裂施工过程中裂缝相级别越高系统压力越低、同时施工压力跨度越大、加砂强度越高。

参 考 文 献

［1］Marfurt K J，Kirlin R L，Farmer S L，et al. 3-D seismic attributes using a semblance-based coherency algorithm［J］. Geophysics，1998，63（4）：1150-1165.

［2］Pedersen S I，Randen T，Sonneland L，et al. Automatic 3D fault interpretation by artificial ants［C］// 2002，64th Meeting，EAEG Expanded Abstracts，G037.

［3］Randen T，Monsen M，Signer C，et al. Three dimensional texture attributes for seismic data analysis［C］// 2000，SEG Annual International Meeting Expanded Abstracts 19，668.

［4］Van B P，Pepper R. Seismic signal processing method and apparatus for generating a cube of variance values：United States Patent 615155［P］.2000-11-21.

［5］孙乐，王志章，李汉林，等.基于蚂蚁算法的断裂追踪技术在乌夏地区的应用［J］.断块油气田，2014，21（6）：716-721.

［6］程超，杨洪伟，周大勇，等.蚂蚁追踪技术在任丘潜山油藏的应用［J］.西南石油大学学报（自然科版），2010，32（2）：48-52.

［7］王军，李艳东，甘利灯.基于蚂蚁体各向异性的裂缝表征方法［J］.石油地球物理勘探,2013,48（5）：

763-769.

[8] 李楠,王龙颖,黄胜兵,等.利用高清蚂蚁体精细解释复杂断裂带[J].石油地球物理勘探,2019,54(1):182-190.

[9] 罗群,王井伶,罗家国,等."非常规油气缝—孔耦合富烃假说"概述[J].岩性油气藏,2019,31(4):1-12.

[10] 霍丽娜,张建军,郑良合,等.多属性断层解释技术在煤层气储层解释中的应用[J].石油地球物理勘探,2014,49(增刊1):221-227.

[11] 王蓓,刘向君,司马立强,等.磨溪龙王庙组碳酸盐岩储层多尺度离散裂缝建模技术及其应用[J].岩性油气藏,2019,31(2):124-133.

[12] Hale D.Fault surfaces and fault throws from 3Dseismic images[C]//Seg Technical Program Expanded Abstracts,2012:205-220.DOI:10.1190/segam2012-0734.1.

[13] 马德波,赵一民,张银涛,等.最大似然属性在断裂识别中的应用:以哈拉哈塘地区热瓦普区块奥陶系走滑断裂的识别为例[J].天然气地球科学,2018,29(6):817-825.

[14] 王浩,陈志强,刘远洋.川东北元坝地区长兴组岩相及裂缝表征预测[J].天然气勘探与开发.2018,41(3):30-36.

[15] 孙萌思,刘池洋,杨阳.塔中地区鹰山组碳酸盐岩储层地震正演模拟[J].地质论评,2017,63(4):49-50.

[16] Banfield J D,Raftery A E.Model-based Gaussian and non-Gaussian clustering[J].Biometrics,1993,49:803-821.

[17] 曾凌翔.一种页岩气水平井均匀压裂改造工艺技术的应用与分析[J].天然气勘探与开发,2018,41(3):95-100.

[18] 蒋廷学,卞晓冰,王海涛,等.深层页岩气水平井体积压裂技术.天然气工业,2017,37(1):90-96.

[19] 闫建平,崔志鹏,耿斌,等.四川盆地龙马溪组与大安寨段泥页岩差异性分析.岩性油气藏,2016,28(4):16-23.

[20] 闫建平,温丹妮,李尊芝,等.基于核磁共振测井的低渗透砂岩孔隙结构定量评价方法:以东营凹陷南斜坡沙四段为例[J].地球物理学报,2016,59(4):1543-1552.

[21] 张少龙,闫建平,唐洪明,等.致密碎屑岩气藏可压裂性测井评价方法及应用——以松辽盆地王府断陷登娄库组为例[J].岩性油气藏,2018,30(3):133-142.

[22] 董大忠,施振生,管全中,等.四川盆地五峰组—龙马溪组页岩气勘探进展、挑战与前景[J].天然气工业,2018,38(4):67-76.

[23] 曾义金.页岩气开发的地质与工程一体化技术[J].石油钻探技术,2014,42(1):1-6.

[24] 糜利栋,姜汉桥,李俊键.页岩气离散裂缝网络模型数值模拟方法研究[J].天然气地球科学,2014,25(11):1795-1803

[25] 王文东,苏玉亮,慕立俊,等.致密油藏直井体积压裂储层改造体积的影响因素[J].中国石油大学学报(自然科学版),2013,37(3):93-97.

[26] 冯国强,赵立强,卞晓冰,等.深层页岩气水平井多尺度裂缝压裂技术[J].石油钻探技术,2017,45(6):77-82.

蜀南地区五峰—龙马溪组页岩微观孔隙结构及分形特征

朱汉卿，贾爱林，位云生，贾成业，袁 贺

（中国石油勘探开发研究院）

摘 要：南方上奥陶统五峰组—下志留统龙马溪组海相页岩是中国页岩气主力开发层位，页岩微观孔隙结构特征的研究对于页岩含气性和开发储量的评价有重要意义。本文采用场发射扫描电镜和低温氮气吸附实验方法对蜀南地区长宁区块五峰—龙马溪组页岩微观孔隙结构进行了定性评价和定量表征。实验结果表明，蜀南地区五峰—龙马溪组页岩以有机质孔隙为主，局部可见粒间孔和粒内孔发育。氮气吸附回滞环属于H4型，对应纳米级孔隙类型为狭缝型；五峰—龙马溪组页岩平均比表面积17.35m^2/g，平均孔体积16.70mm^3/g，平均孔径9.82nm；页岩纳米级孔隙表面具有分形特征，分形维数平均值为2.681；有机碳含量的增加使得纳米级孔隙数量增多，页岩分形维数增大，孔隙表面粗糙程度增大，页岩比表面积增大，页岩吸附能力增强。

关键词：五峰—龙马溪组页岩；氮气吸附实验；孔隙结构；分形维数

页岩气作为一种非常规资源，近年来在中国发展迅速。中国已经先后建成涪陵、长宁—威远和昭通3个国家级示范区[1]，发展前景广阔。与常规天然气藏不同，页岩气主要由游离气和吸附气两部分组成，游离气主要赋存在粒间孔和微裂缝中，吸附气主要赋存在有机质表面[2]，其中，吸附气占页岩气总含气量的20%~80%[3-4]。页岩储层极其致密，发育丰富的纳米级孔隙，这些纳米级孔隙是吸附气和游离气主要的储存场所，页岩纳米级微观储层孔隙结构的研究对页岩含气性和开发储量的评价有重要意义。

近年来，对于页岩储层微观孔隙结构的研究逐渐从定性描述向定量表征发展[5]。在催化领域应用较广泛的气体吸附法越来越多地用来表征致密页岩储层的孔隙结构[6-8]。气体吸附实验是将气体作为探针，通过吸脱附过程来探测多孔材料的孔隙结构特征，气体吸附实验可以得到多孔材料的孔隙结构参数（如比表面积、孔体积、孔径分布、分形维数），从而实现微观孔隙结构的定量表征，然而，前人很少从分形的角度分析页岩的孔隙结构特征。

本文选取长宁示范区评价井岩心样品，运用场发射扫描电镜和低温氮气吸附实验，对蜀南地区上奥陶统五峰组—下志留统龙马溪组页岩微观孔隙类型和孔隙结构特征进行研究，并基于低温氮气吸附数据求取页岩孔隙分形维数，从分形的角度分析页岩孔隙结构，以期深化对示范区页岩储层微观孔隙结构的认识。

1 样品采集与分析方法

本次实验页岩样品采集自蜀南地区长宁国家级页岩气示范区一口评价井的黑色页岩，取样层位为上奥陶统五峰组（2块次）和下志留统龙马溪组（7块次），样品深度范围2200～2400m。实验共涉及TOC含量测试、页岩成熟度测试、岩石密度测试等页岩基本地球化学指标的测试以及X衍射实验、氩离子抛光场发射扫描电镜实验和低温氮气吸附实验。TOC含量测试仪器为CS230碳硫分析仪，测试方法参照国家标准GB/T 19145—2003；镜质组反射率测试仪器为QDI-302型显微分光度计，测试方法参照行业标准SY/T 5124—2012；岩石密度测试方法参照国家标准GB/T 6949—2010；全岩矿物含量以及黏土矿物组成定量测试分析使用X射线衍射仪，测试方法参照行业标准SY/T 5163—2010。以上实验在中国石油西南油气田公司勘探开发研究院分析实验中心完成，实验温度15～25℃，实验湿度＜70%。

氩离子抛光场发射扫描电镜实验在北京理化中心完成，实验仪器为Helios NanoLab 600型双束电镜，其二次电子成像技术较为成熟，首先使用高速氩离子束对制备的页岩样品表面进行轰击，然后使用扫描电镜对页岩的矿物组成、孔隙类型、孔隙结构特征进行定性观察，放大倍数为4000～60000。相对于机械研磨，氩离子抛光、切割制样更能还原真实的内部结构，可以直观有效地判断页岩中发育的孔隙类型。

低温氮气吸附实验在中国石油天然气集团公司非常规油气重点实验室完成，实验仪器为麦克ASAP 2420比表面仪。实验前首先对80目页岩粉样在110℃温度下进行8小时脱气预处理，以此去除吸附剂表面的物理吸附物质，然后以99.99%纯度的氮气为吸附质，实验温度为–195.6℃，在相对压力为0.04～0.99范围内进行吸附—吸附实验，测试不同相对压力下的氮气吸附量，根据吸附等温线形态定性评价页岩纳米孔隙形态，根据吸附数据定量评价页岩孔隙结构特征以及分形特征，包括比表面积、孔体积、平均孔径以及分形维数等特征参数。

采用BET方程计算页岩样品比表面积。当相对压力介于0.05～0.35时，氮气吸附量与P/P_0符合多层吸附BET方程：

$$\frac{P}{V(P_0-P)} = \frac{1}{V_m C} + \frac{C-1}{V_m C}\frac{P}{P_0} \tag{1}$$

式中，V为实验吸附量，cm^3/g；V_m为单分子层的饱和吸附量，cm^3/g；P为实验压力，MPa；P_0为实验温度下氮气的饱和蒸气压，0.1MPa；C为与样品吸附能力有关的常数。

在得到实验吸附量V后，将$P/[V(P_0-P)]$对P/P_0（$0.05<P/P_0<0.35$）作图，得到直线斜率和截距值，从而计算得到单分子层饱和吸附量V_m，最后根据式（2）计算得到页岩的比表面积：

$$A_s = \left(\frac{V_m N a_m}{22400}\right) \times 10^{-18} \tag{2}$$

式中，A_s为比表面积，m^2/g；N为Avogadro常数，6.022×10^{23}；a_m是一个氮气分子在试样表面所占的面积，$0.162nm^2$。

采用BJH方程计算页岩样品的孔体积。当$P/P_0>0.4$时，发生毛细管凝聚现象，真实孔径半径由开尔文孔径和吸附层厚度组成，即：

$$r_p = r_k + t \tag{3}$$

$$r_k = -\frac{2\sigma V_L}{RT\ln(P/P_0)} \tag{4}$$

$$t = 0.326 \times \left[\frac{5}{\lg(P_0/P)}\right]^{1/3} \tag{5}$$

式中，r_p为孔隙半径，nm；r_k为开尔文半径，nm；t为吸附层厚度，nm；σ为液氮的表面张力，8.9 mN/m；V_L为液氮的摩尔体积，34.64 cm³/mol；R为气体常数，8.314J/（K·mol）；T为绝对温度，77.4K。

平均孔径采用式（6）计算，计算公式如下：

$$r_a = \frac{\sum r_i V_i}{\sum V_i} \tag{6}$$

式中，r_i为根据BJH孔径分布得到的分段孔径，nm；V_i为分段孔径对应的孔体积，mm³/g。

2 实验结果与讨论

2.1 页岩样品基本特征

蜀南地区五峰—龙马溪组页岩样品基本参数见表1。其中，镜质组反射率为1.91%～2.61%，平均值为3.32%，页岩热演化程度达到了过成熟阶段。页岩样品的有机碳含量分布范围为0.26%～4.77%，平均值为2.28%，龙马溪组上段有机碳含量普遍较低（N6～N9），五峰组和龙马溪组下段有机碳含量均>2%（N1～N5），是有利的开发层段。页岩密度介于2.48～2.65g/cm³，平均值为2.56g/cm³。且随着有机质含量的增加，岩石密度线性降低（图1）。页岩基质密度约为2.71g/cm³，而有机质密度为1.04g/cm³，页岩中有机质含量的变化对页岩整体密度的影响较大，有机质含量越高，页岩密度越低。

X射线衍射仪测试结果显示（表2），五峰—龙马溪组页岩的主要矿物成分为石英（40.0%～64.4%）、黏土矿物（12.3%～30.8%）和碳酸盐矿物（8.3%～33.2%），从矿物组分三角

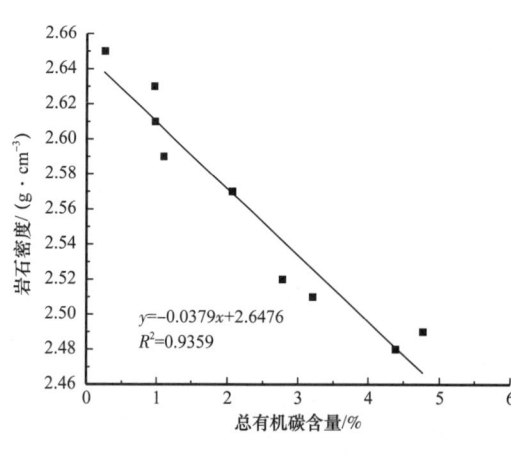

图1 有机碳含量与岩石密度交会图

图（图2）中可以看出，大多数样品石英含量＞50%，为硅质页岩；另外还有少量长石（0～15.6%）和黄铁矿（0～5.1%）。黏土矿物中，主体为伊利石（48%～94%）和绿泥石（6%～44%），部分样品含少量伊/蒙混层，不含蒙皂石和高岭石。

表1 蜀南五峰—龙马溪组页岩样品基本参数

样品编号	样品层位	样品深度/m	有机质含量/%	镜质组反射率/%	岩石密度/(g·cm⁻³)
N1	五峰组	2394.8	2.07	2.04	2.57
N2	五峰组	2391.4	4.39	2.21	2.48
N3	龙马溪组	2385.8	3.21	1.94	2.51
N4	龙马溪组	2379.0	4.77	2.61	2.49
N5	龙马溪组	2372.7	2.78	2.05	2.52
N6	龙马溪组	2355.0	0.97	2.18	2.61
N7	龙马溪组	2338.2	0.96	1.91	2.63
N8	龙马溪组	2298.6	0.26	2.24	2.65
N9	龙马溪组	2238.8	1.10	2.43	2.59

表2 蜀南地区五峰—龙马溪组页岩样品矿物组分

样品编号	石英	长石	方解石	白云石	黄铁矿	黏土矿物	I	I/S	C
N1	56.0	3.6	9.3	9.3	3.7	18.1	60		40
N2	64.4		13.5	5.7	4.1	12.3	76		24
N3	62.4		9.8	7.9	1.4	18.5	94		6
N4	46.1		10.3	7.7	5.1	30.8	65	12	23
N5	57.1	5.3	8.3		4.2	25.1	55	5	40
N6	48.8	15.6	7.1	7.3		21.2	74		26
N7	52.6	11.0	10.2			26.2	74		26
N8	54.2		19.5		6.3	20.0	48	12	40
N9	40.0	4.0	33.2			22.8	56		44

注：I：伊利石；I/S：伊/蒙混层；C：绿泥石

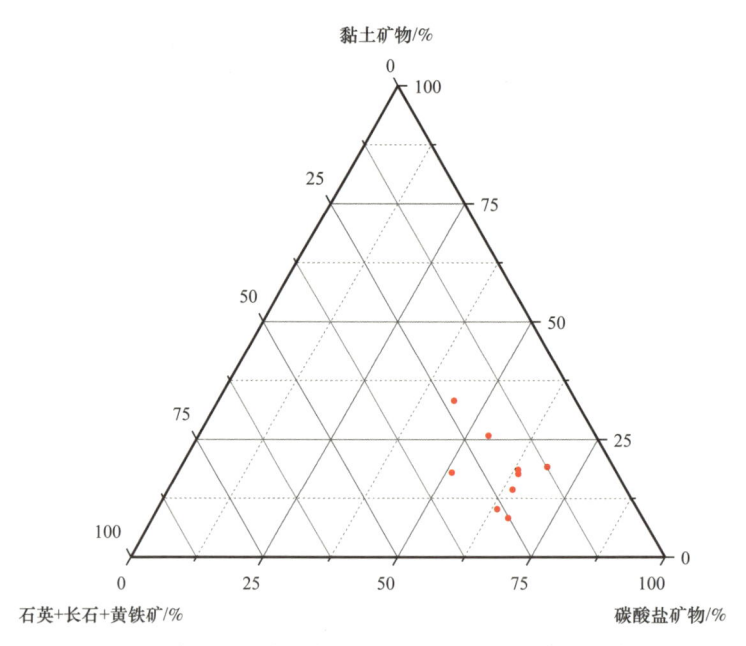

图 2 蜀南地区五峰—龙马溪组页岩样品矿物组成三角图

根据扫描电镜观测结果,将蜀南地区五峰—龙马溪组页岩的孔隙类型分为有机质孔隙和无机孔隙[9]。

研究区五峰—龙马溪组富有机质页岩中有机质孔大量发育(图3),孔径<200nm,孔隙形态主要狭缝型和不规则状。有机质孔隙在页岩中分布相对集中,数量受控于有机质颗粒大小的控制[10],是一种发育在有机质颗粒内的粒内孔隙,连通性较好,是页岩气主要的吸附和储存空间。有机质孔是在有机质生烃过程中形成的孔隙[11],研究区页岩处于过成熟生气阶段,有利于有机质孔隙的发育。

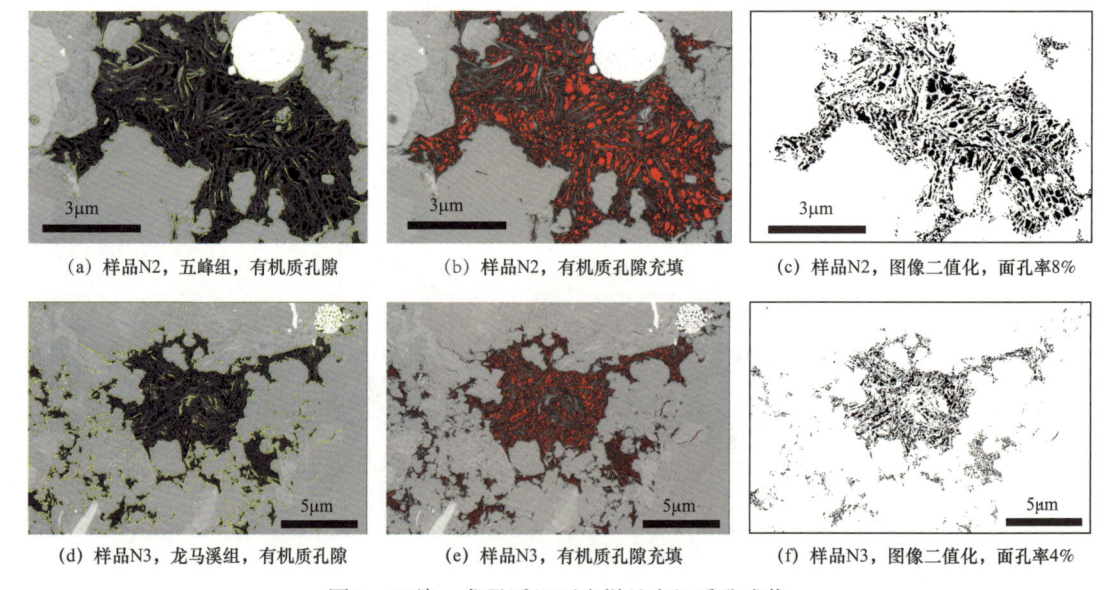

(a) 样品N2,五峰组,有机质孔隙   (b) 样品N2,有机质孔隙充填   (c) 样品N2,图像二值化,面孔率8%

(d) 样品N3,龙马溪组,有机质孔隙   (e) 样品N3,有机质孔隙充填   (f) 样品N3,图像二值化,面孔率4%

图 3 五峰—龙马溪组页岩样品有机质孔成像

2.2 页岩孔隙类型

研究区页岩无机孔隙主要发育在龙马溪组上部低有机质含量的样品中，包括粒内孔和粒间孔两种类型。粒内孔主要有碳酸盐颗粒内溶孔、黏土矿物粒内孔以及黄铁矿粒内孔三种类型。碳酸盐颗粒内溶蚀孔隙较为常见（图4a），是由于地层流体对碳酸盐矿物溶蚀作用形成的，其形态多成椭圆形，数量较多，但空隙间的连通性较差，孔径多<100nm；黏土矿物粒内孔在研究区较为少见（图4b），原因是黏土矿物这类塑性矿物受压实作用影响较大，粒内孔隙消失殆尽，这类孔隙不是气体储存的主要空间；黄铁矿粒内孔主要发育在草莓状黄铁矿内（图4c），如果黄铁矿晶体内充填有机质，则会形成有机质孔；粒间孔隙主要发育在脆性矿物的边缘（图4b），由于脆性矿物的抗压实作用形成，孔隙形态一般呈狭缝型，孔径较大，连通性较好。

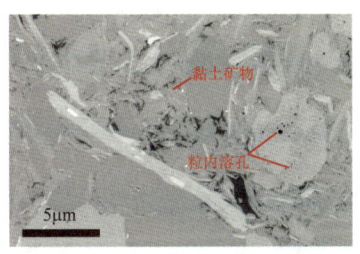

(a) 样品N8，碳酸盐矿物粒内溶孔　　(b) 样品N6，有机质孔，粒内溶孔，黏土矿物孔隙，粒间孔　　(c) 样品N4，有机质孔，黄铁矿粒内孔，粒内溶孔

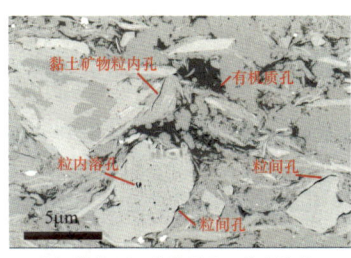

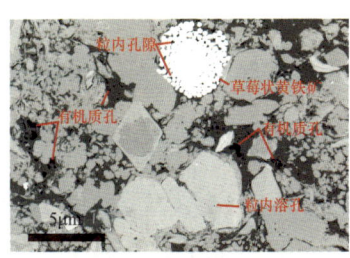

图4　五峰—龙马溪组页岩样品无机孔成像

2.3 孔隙结构特征

2.3.1 低温氮气吸附等温线

由77.4K低温氮气吸附实验得到9个页岩样品的等温吸附线（图5），依据国际纯粹与应用化学联合会（IUPAC）2015年对等温吸附线的最新分类[12]，蜀南地区页岩样品的等温吸附线属于Ⅳ(a)型等温线。等温线整体呈反"S"形，在相对压力$P/P_0>0.45$后，吸附进入毛细管凝聚阶段，吸附等温线和脱附等温线不重合，形成回滞环，回滞环的形态反映了吸附剂的孔隙结构特征[13]。根据IUPAC对回滞环的分类[12]，页岩样品的吸附回滞环近似H4型，反映了纳米级孔隙形态呈狭缝型，其回滞环较小，在P/P_0低压段有明显的吸附量，与微孔充填有关。样品N3在低压段出现吸附等温线和脱附等温线不重合的现象，塑性孔隙的膨胀或者氮气分子在极小孔隙（与分子大小相当）孔隙中的不可逆吸附都会导致低压区的不闭合现象[14]。

2.3.2 页岩孔隙结构参数特征

研究区页岩样品孔隙结构参数见表3。实验结果表明，9个页岩样品BET比表面积介于$9.16\sim26.81m^2/g$，平均值为$17.35m^2/g$；BJH孔体积介于$10.69\sim22.31mm^3/g$，平均值为$16.70mm^3/g$；平均孔径介于$7.49\sim12.10nm$，平均值为$9.82nm$。其中，五峰组和龙马溪组下部样品（N1～N5）比表面积介于$16.89\sim26.81m^2/g$，平均值为$22.12m^2/g$，孔体积介于$17.52\sim22.31mm^3/g$，平均$19.27mm^3/g$；龙马溪组上部（N6～N9）比表面积介于

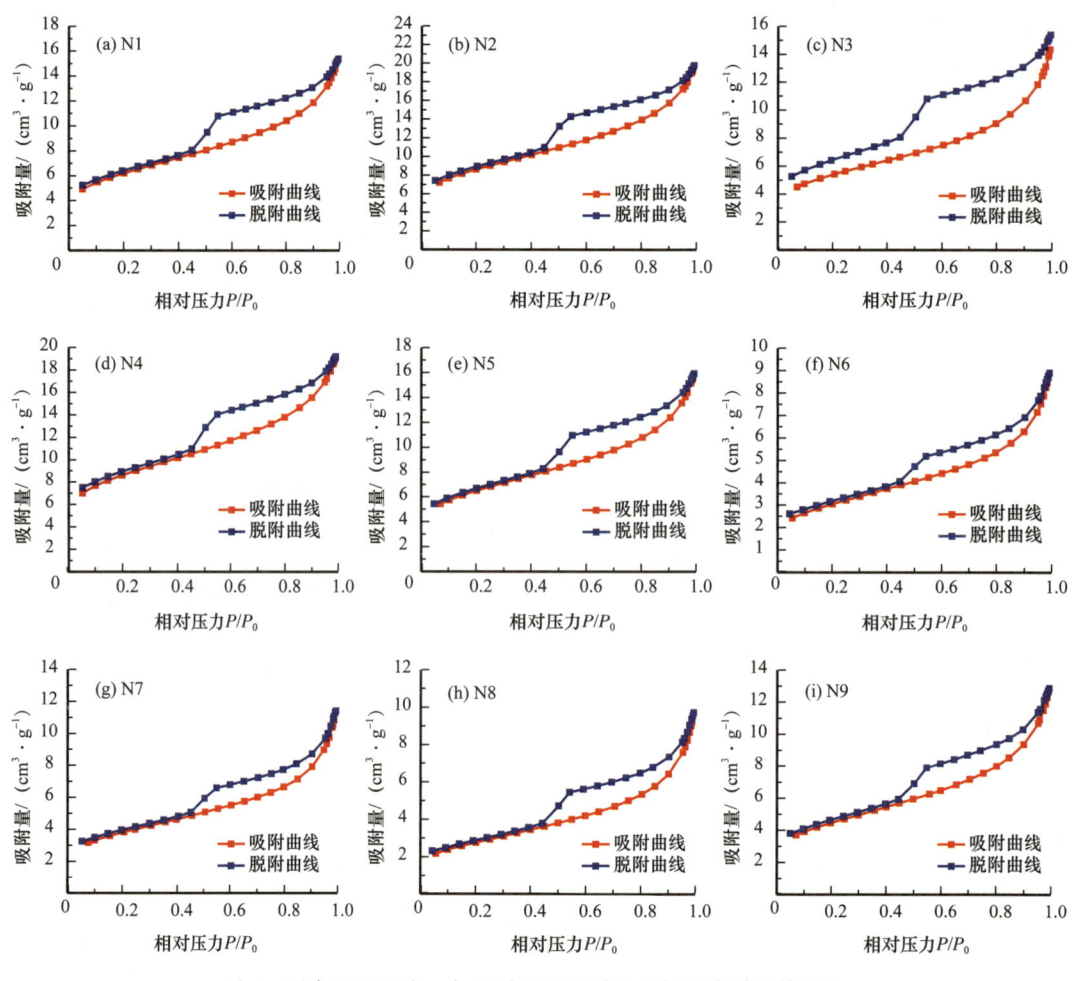

图5 蜀南地区五峰—龙马溪组页岩样品氮气吸附脱附等温线

9.16~14.36m²/g，平均值为11.39m²/g，孔体积介于10.69~15.73mm³/g，平均值为13.48mm³/g，下部页岩比表面积和孔体积明显大于上部页岩。从孔径分布直方图上来看（图6），9个页岩样品孔径分布呈现相似的特征，其纳米级孔径主体分布在15nm以下，且孔径分布呈双峰特征，主峰分布在1.7~2.1nm以及8~15nm两个位置。

表3 页岩样品孔隙结构参数

孔隙结构参数	N1	N2	N3	N4	N5	N6	N7	N8	N9	均值
BET比表面积/(m²·g⁻¹)	19.70	26.80	16.89	26.81	20.40	9.84	12.20	9.16	14.36	17.35
BJH孔体积/(mm³·g⁻¹)	18.20	22.31	17.52	20.26	18.06	10.69	14.26	13.23	15.73	16.70
平均孔径/nm	9.20	8.24	9.22	7.49	9.12	11.04	12.07	12.10	9.89	9.82
分形维数	2.688	2.715	2.702	2.714	2.697	2.659	2.669	2.623	2.663	2.681

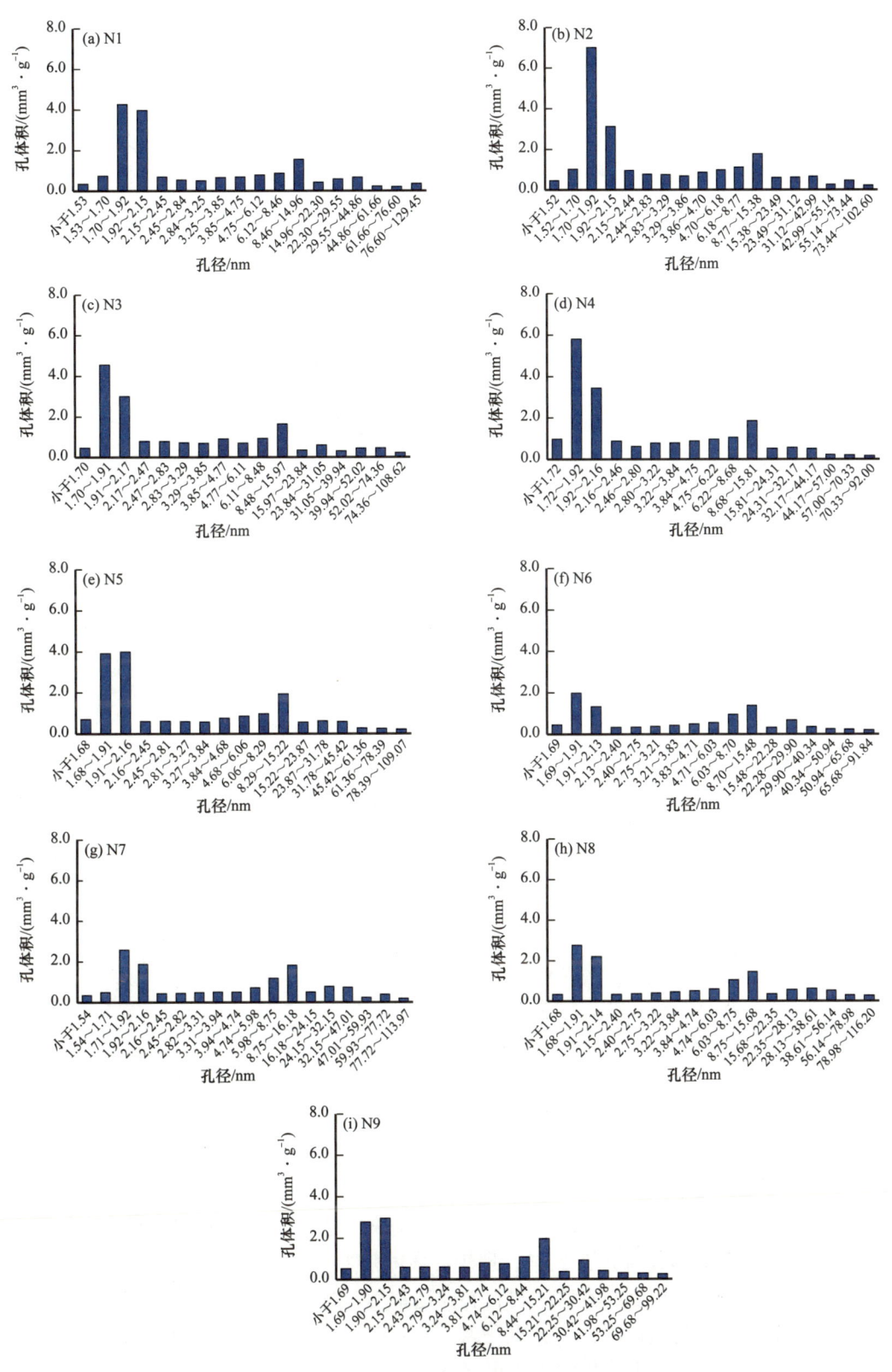

图 6 五峰—龙马溪组页岩孔径分布直方图

从比表面积、孔体积以及平均孔径三者的关系来看（图7a、b），随着页岩比表面积的增大，其孔体积增大，平均孔径减小，线性相关系数分别高达0.91和0.87。随着页岩孔径的减小，发育在有机质中的纳米级孔隙数量增多，从而可供甲烷吸附的比表面积增大，孔体积亦增大。从总有机碳含量与比表面积的关系来看（图7c），随着有机质含量的增大，页岩比表面积线性增大（R^2=0.89），说明有机质孔是页岩比表面积主要的贡献者，随着有机碳含量的增多，页岩中的有机质孔数量增多，从而可供甲烷吸附的比表面积增大，有机碳含量是控制页岩纳米级孔隙发育的主控因素。

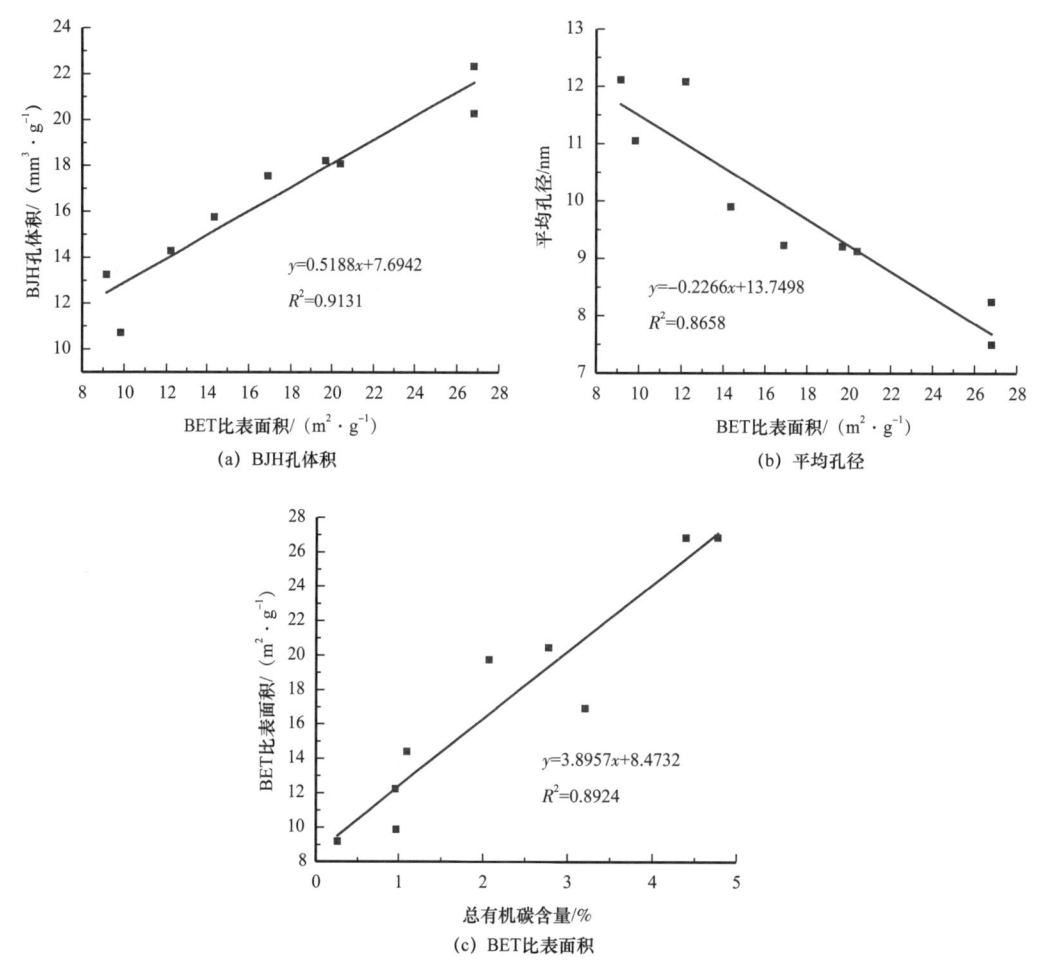

图7 页岩比表面积与孔体积、平均孔径以及有机质含量的关系

2.4 页岩纳米孔隙分形特征

以分形几何理论为基础的分形维数可以真实地表征多孔介质表面的粗糙程度及不规则程度，通常情况下，其数值介于2～3。理想的表面是光滑的，其分形维数为2，然而由于原子堆积排列的错位等原因，真实的材料表面通常是凹凸不平的，分形维数越大，表明该材料的粗糙程度及非均质性越强。页岩孔隙表面粗糙，表面的不规则性能够创造出比理论值更大的真实比表面，从而为甲烷提供更多的吸附位，增大了页岩的吸附能力。通过高压

压汞、气体吸附、核磁共振等实验数据均可以计算岩石孔隙结构的分形维数。本文根据实验测得的低温氮气吸附数据，运用 Frenkel–Halsey–Hill（FHH）方程进行分形维数的计算，FHH 吸附式如下：

$$\frac{V}{V_m} = C\left[\mathrm{RT}\ln\left(P_0/P\right)\right]^{\alpha} \qquad (7)$$

式中，C 为特征常数，α 为与分形维数和吸附机制相关的参数。将式（7）进行对数处理，得到对数形式的方程：

$$\ln V = A + \alpha \ln\left[\ln\left(P_0/P\right)\right] \qquad (8)$$

以 $\ln V$ 对 $\ln[\ln(P_0/P)]$ 作曲线，找出回归系数最好的一段，得到曲线斜率 α，如图 8 所示。由图可知，9 条曲线直线拟合效果都非常好，其相关系数均大于 0.99，说明页岩孔隙表面具有分形特征，由此可求得拟合曲线的斜率 α。前人研究表明[15]，在考虑表面张力效应的情况下，分形维数与 α 有如下对应关系：

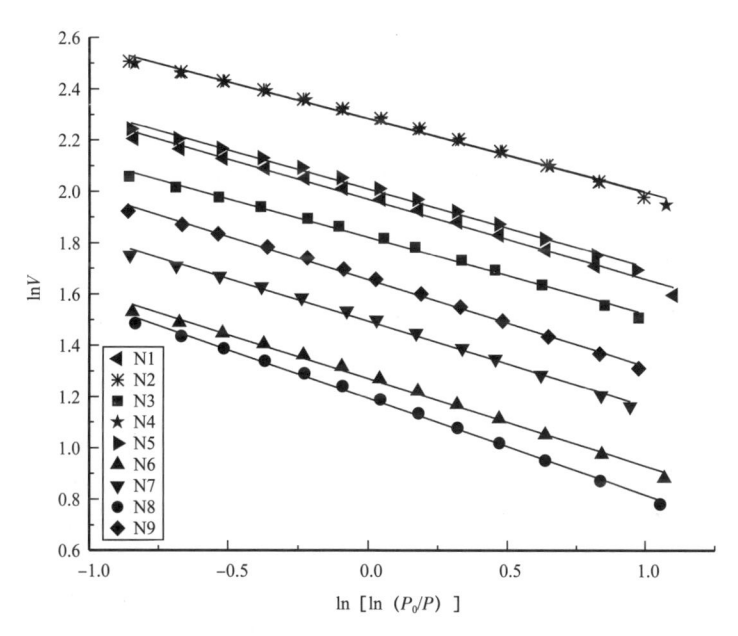

$$D = 3 + \alpha \qquad (9)$$

图 8　低温氮气吸附等温线 $\ln[\ln(P_0/P)]$ 与 $\ln V$ 交会图

根据 FHH 方程计算得到的页岩样品分形维数为 2.623～2.715（表 3），平均值为 2.681，表明页岩孔隙表面具有较强的粗糙度和非均质性。其中，五峰组和龙一段分形维数均值为 2.70，龙二段分形维数均值为 2.65，说明五峰组和龙马溪组下部富有机质页岩孔隙表面粗糙度和非均质性强于龙马溪组上部页岩。从分形维数与总有机碳含量的关系也可以看出（图 9），随着页岩总有机碳含量的增大，页岩分形维数呈指数增大。有机质含量的增大一方面使得页岩样品中纳米级孔隙增多，另一方面使得孔隙表面的粗糙程度增大，从而为甲烷提供了更多的吸附空间，增强了页岩的吸附性能。

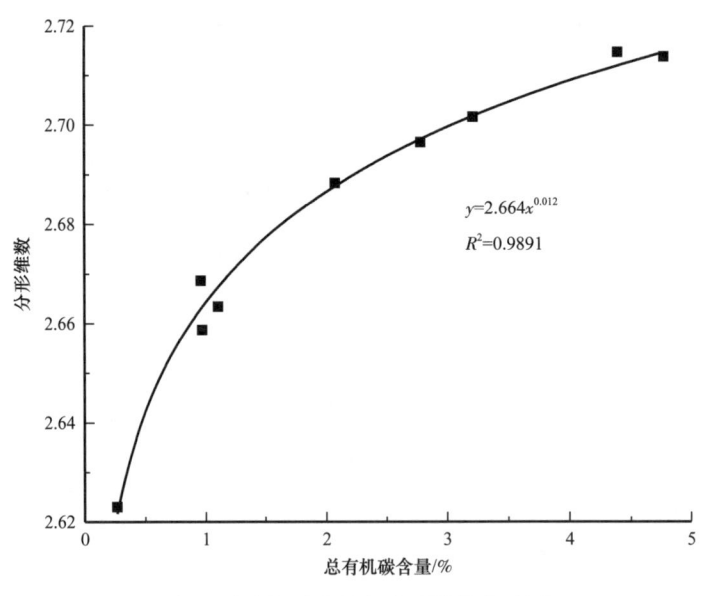

图 9　总有机碳含量与分形维数交会图

3　结论

（1）蜀南地区五峰—龙马溪组页岩孔隙分为有机质孔和无机孔两种类型，其中五峰—龙马溪组下部富有机质页岩主要发育有机质孔，发育位置相对集中，呈狭缝型和不规则状，连通性较好，数量受控于有机质颗粒的大小；龙马溪组上部低有机质页岩主要发育各种类型无机孔，包括粒内孔和粒间孔两类。粒内孔以碳酸盐颗粒内溶蚀孔隙最为常见，孔隙呈圆形，连通性较差；粒间孔主要发育在脆性矿物边缘，孔径较大，呈狭缝型。

（2）低温氮气吸附实验表明，吸附回滞环属于H4型，纳米级孔隙类型为狭缝型；五峰—龙马溪组页岩平均比表面积17.35m²/g，平均孔体积16.70mm³/g，平均孔径9.82nm，五峰组和龙马溪组下部页岩比表面积和孔体积明显大于龙马溪组上部页岩；页岩孔径分布呈双峰特征，主峰分布在1.7～2.1nm以及8～15nm两个位置；页岩孔隙结构参数之间有较好的相关性，随着平均孔径的减小，页岩比表面积和孔体积增大；有机碳含量控制了页岩纳米级孔隙的发育，随着页岩有机碳含量的增大，页岩有机质孔数量增多，页岩比表面积增大。

（3）页岩纳米级孔隙表面具有分形特征，分形维数平均值为2.681，表明孔隙表面较粗糙，从而提供了比理想表面更大的比表面积；有机碳含量的增加使得页岩分形维数增大，孔隙表面粗糙程度增大，页岩比表面积增大，页岩吸附能力增强。

参 考 文 献

［1］贾爱林，位云生，金亦秋．中国海相页岩气开发评价关键技术进展［J］．石油勘探与开发，2016，43（6）：1-7.

［2］Loucks S R G, Reed R M, Ruppel S C, et al. Morphology, genesis, and distribution of nanometer-scale pores in siliceous mudstones of the Mississippian Barnett Shale [J]. Journal of sedimentary Research, 2009, 79: 848-861.

[3] Curtis J B. Fractured shale-gas system. AAPG Bulletin [J]. 2002, 86 (11): 1921-1938.

[4] 张寒，朱炎铭，夏筱红，等．页岩中有机质与黏土矿物对甲烷吸附能力的探讨［J］．煤炭学报，2013，38（5）：812-816.

[5] 姜振学，唐相路，李卓，等．川东南地区龙马溪组页岩孔隙结构全孔径表征及其对含气性的控制［J］．地学前缘，2016，23（2）：126-134.

[6] 杨峰，宁正福，张世栋，等．基于氮气吸附实验的页岩孔隙结构表征［J］．天然气工业，2013，33（4）：135-140.

[7] 邓恩德，金军，王冉，等．黔北地区龙潭组海陆过渡相页岩微观孔隙特征及其储气性［J］．科学技术与工程，2017，17（24）：190-195.

[8] 武瑾，梁峰，苔文，等．渝东北地区龙马溪组页岩储层微观孔隙结构特征［J］．成都理工大学学报（自然科学版），2016，43（3）：308-319.

[9] Loucks R G, Reed R M, Ruppel S C, et al. Spectrum of pore types and networks in mudrocks and a descriptive classification for matrix-related mudrock pores [J]. AAPG Bulletin, 2012, 96: 1071-1098.

[10] 侯宇光，何生，易积正，等．页岩孔隙结构对甲烷吸附能力的影响［J］．石油勘探与开发，2014，41（2）：248-256.

[11] Paul C, Hackley. Geological and geochemical characterization of the Lower Cretaceous Pearsall Formation, Maverick Basin, south Texas: Afuture shale gas resource [J]. AAPG Bulletin, 2012, 96 (8): 1449-1482.

[12] Matthias T, Katsumi K, Alexander V, et al. Physisorption of gases, with special reference to the evaluation of surface area and pore size distribution (IUPAC Technical Report) [J]. Pure Appl. Chem, 2015, 87: 1051-1069.

[13] 近藤精一，石川达雄，安部郁夫．吸附科学［M］．第二版．李国希，译．北京：化学工业出版社，2007.

[14] Gregg S J, Sing K S W. Adsorption, surface area and porosity [M]. 2 nd. New York: Academic Press, 1982.

[15] Qi H, Ma J, Wong P. Adsorption isotherms of fractal surfaces [J]. Colloids Surface A: Physicochemical and Engineering Aspects, 2002, 206: 401-407.

基于氩气吸附的页岩纳米级孔隙结构特征

朱汉卿，贾爱林，位云生，贾成业，金亦秋，袁 贺

（中国石油勘探开发研究院）

摘 要：以川南地区龙马溪组页岩为研究对象，应用场发射扫描电镜（FE-SEM）定性描述页岩镜下孔隙形态及其类型，创新使用低温氩气（Ar）吸附实验定量测量页岩样品的比表面积、孔体积以及孔径分布，实现了页岩小于100nm（纳米级）孔隙的连续测量，并根据Frenkel-Halsey-Hill（FHH）模型研究了页岩孔隙结构的分形特征，探讨有机质对页岩孔隙结构及分形特征的影响。结果表明：川南地区龙马溪组页岩储层主要发育有机质孔、粒间孔及粒内孔三大类，主体为有机质孔。Ar吸附等温线表明，纳米级孔隙以狭缝型为主，孔径主体分布在10nm以下的微孔和介孔中，呈"三峰"特征，微孔主要集中在0.6~0.9nm以及1.8~2.0nm，介孔主要集中在4.0~5.0nm。纳米级孔隙分形维数2.55~2.64，表现出较强的非均质性。有机碳含量控制了页岩纳米级孔隙的发育，TOC的增大使得页岩中微孔含量及其所占比例增大，分形维数增大，孔隙结构趋于复杂，有利于页岩储层吸附能力的增强。

关键词：龙马溪组；页岩；氩气吸附；孔隙结构；分形维数

近年来，中国四川盆地及其周缘下志留统富有机质页岩的勘探开发取得了诸多进展[1-3]。页岩作为一种非常规储层，针对其储层特征的研究越来越受到国内外学者的重视[4-6]。页岩微观孔隙结构特征的表征是页岩储层特征的一个重要方面，直接影响页岩储层的储集性能和吸附性能[7-9]。从材料学的角度来看，页岩是一种孔隙结构非常复杂的多孔介质，孔径分布广泛，本文根据国际纯粹与应用化学联合会（IUPAC）对孔径的分类[10]，将直径<2nm的孔隙定义为微孔，直径介于2~50nm的孔隙定义为介孔，直径>50nm的孔隙定义为宏孔，将直径<100nm的孔隙统称为纳米级孔隙。目前针对页岩储层微观孔隙结构的研究方法主要分为图像观测法和流体渗透法两类[11]，其中，图像观测法主要是通过氩离子抛光技术，在扫描电镜下观察微观孔隙类型及分布特征[12]，从而建立对储层结构的直观认识。流体渗透法主要是通过注入流体来间接探测页岩的微观孔隙结构，以高压压汞法和气体吸附法为代表[13]。高压压汞法测量进度取决于实验的最大压力，且高压会对页岩储层产生破坏，影响测量的准确性。气体吸附法是一种常用的表征吸附材料孔隙结构的实验方法，近年来在页岩储层表征领域得到了广泛的应用[14-16]。最常用的吸附剂是氮气。77K下的氮气吸附作为多孔材料孔径分析的标准实验方法被广泛适用，但是其本身也存在一定的局限性[10]，首先氮气分子的四极距性质使其在吸附时会与吸附剂表面官能团发生相互作用，从而影响氮分子在吸附剂表面的取向，实验在极低相对压力条

件下很难达到平衡，无法得到准确的吸附数据，进而无法准确表征介质中微孔的分布；其次，氮气分子是棒状分子，吸附时的分子界面不确定，为测量带来了不确定性。与氮气分子相比，氩气为球形单原子分子，且四极距为零，不会与表面官能团发生相互作用，吸附时的分子截面积稳定，为含微孔和介孔页岩的孔径分析提供更为准确的分析结果，氩气吸附是国际纯粹与应用化学联合会推荐的表征含微孔和介孔介质孔径分布的实验方法[10]，它可为纳米级孔隙结构特征研究提供依据。

1 实验样品

本次实验共计12块页岩样品，取自川南地区的一口评价井，取样层位为下志留统龙马溪组（O_3l），取样深度为2355.0～2393.7m，取样间隔最小0.56m，最大6.8m。页岩样品中的总有机碳（TOC）含量分析使用LECO CS-230碳硫分析仪；矿物组分分析采用TTR Ⅲ多功能X射线衍射仪。页岩样品的地球化学以及矿物成分数据见表1。结果表明，实验页岩样品总有机碳含量为0.82%～4.37%，平均值为2.45%。矿物组分上，页岩主要由石英、碳酸盐矿物以及黏土矿物组成，其中，石英含量为11.1%～48.0%，平均值为30.36%；碳酸盐矿物由方解石和白云石组成，方解石含量为5.5%～29.7%，平均值为13.64%，白云石含量为3.6%～20.3%，平均值为10.48%；黏土矿物含量为15.0%～46.3%，平均值为35.89%，其中，伊利石占黏土矿物的46%～71%（平均52.33%），伊/蒙混层占20%～49%（平均33.92%），绿泥石占2%～22%（平均11.75%），高岭石占1%～4%（平均2%）；另外，页岩中还含有少量钾长石、钠长石以及黄铁矿，含量分别为0.5%～4.5%（平均1.43%）、1.9%～9.9%（平均5.04%）、0.5%～10%（平均3.17%）。

表1 N1井页岩样品TOC及矿物组成数据

样品号	样品深度/m	ω（TOC）/%	矿物含量/%						
			石英	钾长石	斜长石	方解石	白云石	黄铁矿	黏土矿物
1	2393.7	2.55	11.9	0.7	2.2	29.7	20.3	1.6	33.6
2	2392.7	3.07	33.9	1.2	3.9	8.3	10.5	2.2	40.0
3	2391.4	2.61	11.1	0.6	2.9	18.9	15.5	4.7	46.3
4	2390.3	2.05	32.3	1.5	9.9	12.1	7.7	1.2	35.3
5	2389.8	2.54	33.0	0.5	1.9	22.1	17.5	10.0	15.0
6	2385.8	2.78	48.0	0.7	2.0	8.8	14.7	3.4	22.4
7	2379.0	4.37	28.7	0.8	4.0	11.1	15.9	5.9	33.6
8	2372.7	2.39	35.7	0.9	4.4	5.5	4.9	3.1	45.5
9	2368.5	2.61	29.5	1.5	3.8	10.0	6.0	3.1	46.1

续表

样品号	样品深度/m	ω(TOC)/%	矿物含量/%						
			石英	钾长石	斜长石	方解石	白云石	黄铁矿	黏土矿物
10	2363.1	1.17	34.4	1.2	8.7	12.4	3.6	1.2	38.5
11	2359.3	0.87	29.1	3.0	9.3	12.7	5.1	1.1	39.7
12	2355.0	0.82	36.7	4.5	7.5	12.1	4.0	0.5	34.7

低温氩气实验采用美国 Quantachrome 公司生产的 Autosorb IQ 比表面积及孔径分析仪进行，采用体积测量法计算吸附量。实验前需要对页岩样品进行预处理，包括磨样称重和脱气处理，在对样品进行110℃真空脱气处理8小时后，以氩气为吸附质，实验温度为87.5K，在相对压力 $5.0\times10^{-7}\sim0.99$ 的范围内进行吸附—脱附实验，得到吸附—脱附曲线。比表面积、孔体积以及孔径分布等孔隙结构参数由非定域泛函理论（NLDFT）模型计算得到。

2 实验结果及分析

2.1 页岩微观孔隙类型及形态特征

经过氩离子抛光过的页岩样品在扫描电镜下可以观察到丰富的纳米级以及微米级孔隙发育，根据场发射扫描电镜的图像分析，可将川南地区龙马溪组页岩的孔隙类型分为三类：有机质孔、粒间孔及粒内孔。

有机质孔是一种分布在干酪根内的粒内孔，是残留在基质孔隙中的原油裂解生气过程中形成的[17]。有机质孔隙构成了富有机质页岩主要的孔隙连通网络[18]，且为甲烷提供了主要的吸附和储存空间。研究区龙马溪组下部页岩中有机质孔广泛发育（图1），形态呈蜂窝状、椭圆形，孔隙边缘较为光滑。有机质孔隙的形态说明其受后期压实作用的影响较小。

粒间孔在研究区主要呈现两种类型，一种发育在脆性矿物和塑性矿物之间（图1e），脆性矿物的存在使得黏土矿物发生弯曲，同时阻止了黏土矿物的进一步压实，从而形成了狭缝型粒间孔；另一种发育在碳酸盐矿物边缘，局部溶蚀作用导致了这种孔隙的形成，通常呈现不规则形图（图1f）。

粒内孔在研究区主要呈现三种类型，一类是碳酸盐颗粒内部由于溶蚀作用形成的粒内溶孔（图1b、d、e），这类孔隙通常呈圆形，孔径为纳米级，这类孔隙的连通性通常较差，对甲烷气体在页岩中的渗流贡献较小[19]；另一种与黄铁矿晶体有关（图1d），这是一种常见的粒内孔类型，形状多为不规则形，如果黄铁矿晶体内被有机质充填，则会形成有机质孔；第三种与黏土矿物有关，这类孔隙通常呈狭缝形（图1c、e），这类孔隙主要是由于蒙皂石在成岩过程中向伊/蒙混层和伊利石转化，黏土矿物体积缩小造成的。这类孔隙不仅可以作为气体渗流的通道，而且黏土矿物本身也具有一定的吸附性能[20]，黏土矿物内的孔隙可以为甲烷提供吸附空间。

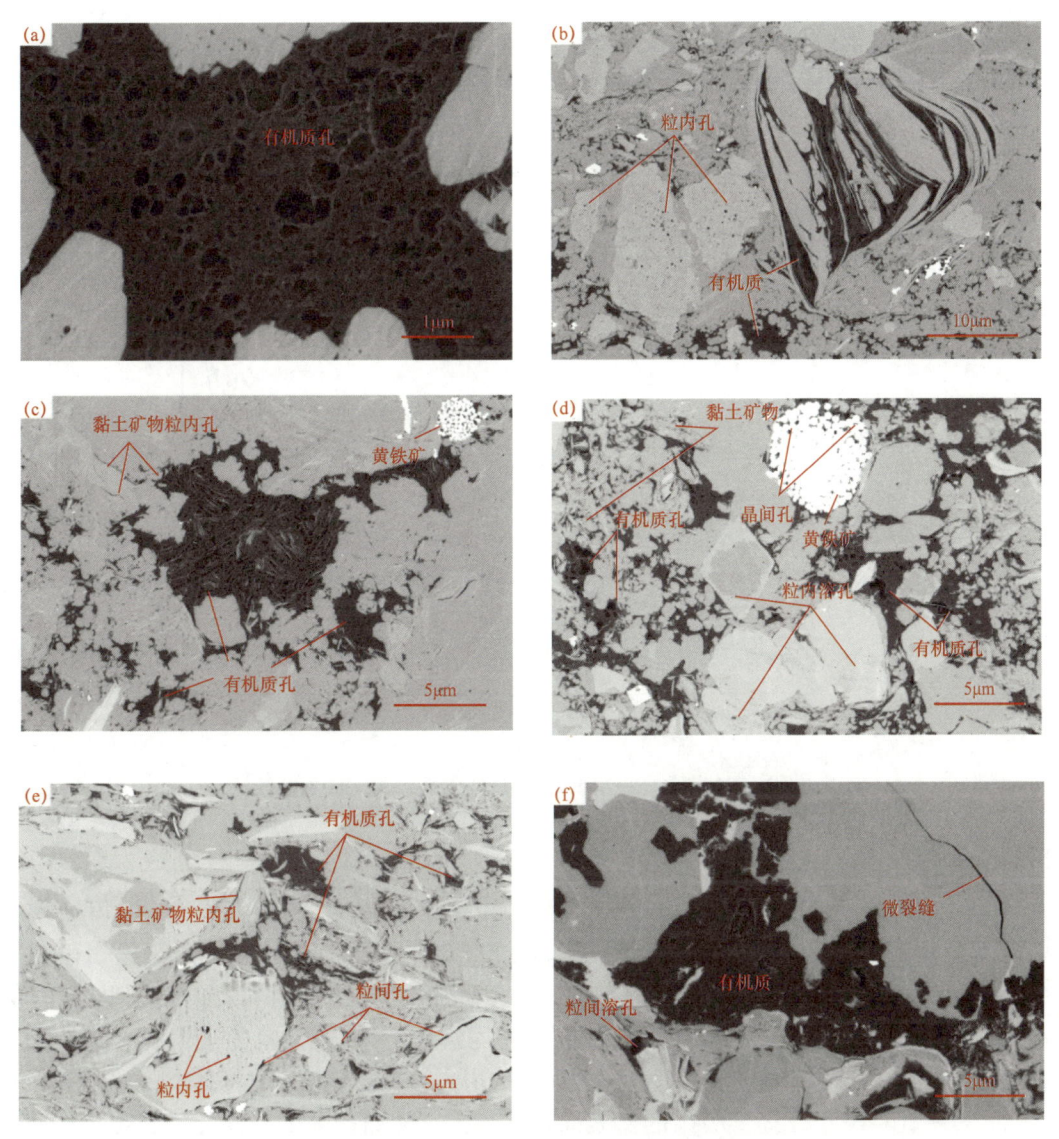

图 1 N1 井龙马溪组页岩场发射扫描电镜图像

（a）5 号样，2389.8m 有机质孔密集发育；（b）5 号样，2389.8m 视域内发育有机质孔以及碳酸盐矿物内溶蚀孔；（c）4 号样，2390.3m 有机质孔发育，零星发育粒内溶蚀孔以及黏土矿物粒内孔；（d）7 号样，2379.0m 孔隙类型多样，有机质孔、草莓状黄铁矿晶内孔、粒内溶蚀孔发育；（e）9 号样，2368.5m 有机质孔、碳酸盐矿物粒内溶孔、粒间孔、黏土矿物粒内孔等发育；（f）11 号样，2359.3m 发育有机质孔以及粒间溶孔，发育一条较大微裂缝

2.2 基于氩气吸附的页岩孔隙结构特征

2.2.1 吸附—脱附等温线

根据 87.5K 下的 Ar 吸附—脱附实验，得到页岩样品的 Ar 吸附—脱附曲线（图 2）。2015 年 IUPAC 更新了等温吸附的分类[10]，本次实验的吸附—脱附曲线与该分类的Ⅳ（a）型相近，其特点是在相对压力 P/P_0 小于 0.4 时，曲线类型与 I 型曲线类似，气体主要发生微孔充填和单层吸附，当 $P/P_0>0.4$ 时，吸附曲线出现明显的上凹状，说明在孔道中发生

了凝聚作用，且该阶段脱附曲线不可逆，形成回滞环，根据IUPAC的分类，该回滞环属于H3型，说明研究区页岩孔隙主体属于狭缝型孔隙。当相对压力P/P_0接近于1时等温线急剧上升，吸附没有达到饱和，说明页岩中存在宏孔。

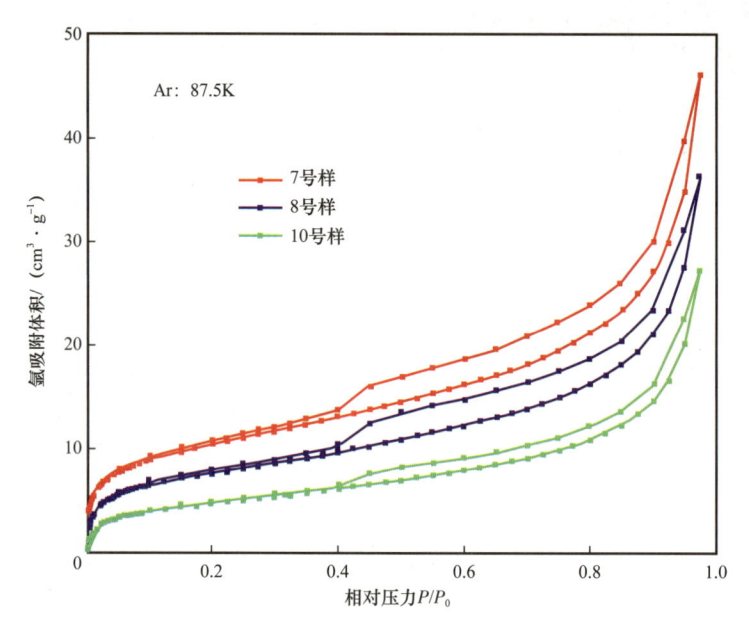

图 2　N1井龙马溪组页岩样品Ar气体吸附—脱附曲线

2.2.2　孔隙结构参数特征

利用非定域泛函理论（NLDFT）方法对氩气吸附数据进行处理，得到川南地区龙马溪组富有机质页岩孔隙结构参数（表2）。实验结果表明，页岩样品比表面积为16.85～63.74m²/g，平均值为33.06m²/g；总孔体积为5.05～9.14mL/100g，平均值为6.72mL/100g；平均孔径为13.08～21.75nm，平均值为16.50nm，且比表面积、孔体积和平均孔径之间存在一定的关系，随着比表面积的增大，页岩储层孔体积增大，平均孔径减小，这表明孔径越小，提供吸附的比表面积就越大，从而孔隙的吸附能就越强；按照孔径尺寸的分类统计，微孔比表面积占总比表面积的32.35%～68.48%，平均值为52.25%，介孔比表面积占总比表面积的28.24%～58.44%，平均值42.13%，宏孔比表面积占总比表面的3.28%～9.21%，平均值5.61%，微孔+介孔提供了页岩90%以上的比表面积，是页岩气吸附的主要场所；微孔体积占总孔隙体积的2.30%～8.10%，平均值为5.16%，介孔体积占总孔隙体积的48.65%～65.42%，平均值55.92%，宏孔体积占总孔隙体积的29.74%～46.76%，平均值为38.92%，介孔+宏孔提供了页岩90%以上的孔体积，是页岩储存的主要场所，且由于吸附实验的测量尺度问题，这一比例还不包括100nm以上的宏孔，如果考虑100nm以上的宏孔体积，这一比例将更大。

从孔径分布曲线可以看出（图3），川南地区龙马溪组页岩样品孔径主要分布在小于10nm的微孔和介孔中。从峰的位置来看，微孔主要集中在0.6～1.0nm以及1.8～2nm位置，介孔主要集中在4～5nm的位置，这与氮气吸附实验使用传统的BJH方法得到的介孔分布存在差异[19]。H3型回滞环的等温线使用BJH方法计算得到的孔径分布通常是假峰[21]。

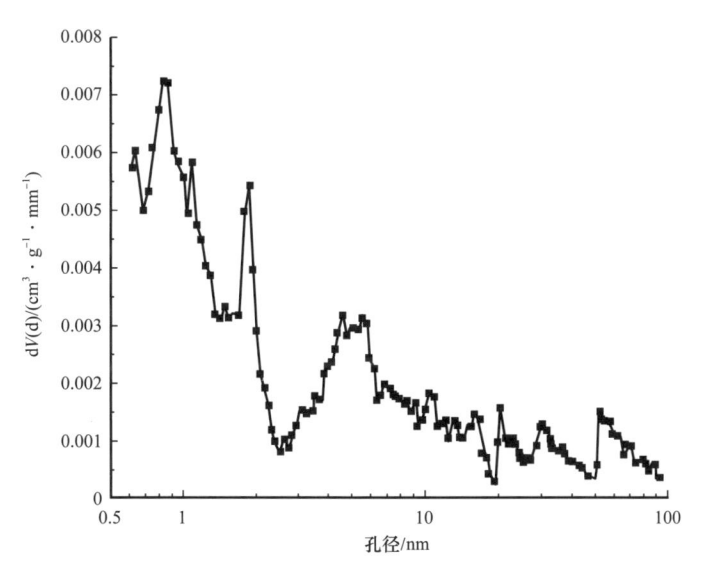

图 3 基于 Ar 吸附的 N1 井龙马溪组 7 号页岩样品孔径分布

表 2 页岩样品氩气吸附实验孔隙结构参数

样品编号	总比表面积/($m^2·g^{-1}$)	总孔体积/($mL·100g^{-1}$)	平均孔径 nm	微孔（<2nm） 比表面/($m^2·g^{-1}$)	微孔 孔体积/($mL·100g^{-1}$)	介孔（2~50nm） 比表面/($m^2·g^{-1}$)	介孔 孔体积/($mL·100g^{-1}$)	宏孔（50~100nm） 比表面/($m^2·g^{-1}$)	宏孔 孔体积/($mL·100g^{-1}$)
1	35.20	7.23	18.98	16.76	0.35	15.94	4.73	2.50	2.15
2	42.81	6.74	13.09	27.17	0.49	13.89	3.90	1.75	2.35
3	37.18	7.54	15.69	20.95	0.41	14.42	4.12	1.81	3.01
4	23.07	5.07	15.45	11.71	0.25	10.31	3.06	1.05	1.76
5	30.48	5.44	15.13	19.04	0.35	10.17	2.99	1.27	2.10
6	32.56	7.40	17.29	17.02	0.34	13.46	3.60	2.08	3.46
7	63.74	9.14	13.08	43.65	0.74	18.00	4.94	2.09	3.46
8	39.88	7.11	13.82	23.62	0.46	14.68	4.02	1.58	2.63
9	36.34	8.25	16.7	19.26	0.40	14.98	4.37	2.10	3.48
10	20.99	5.59	17.68	9.13	0.21	10.49	3.11	1.37	2.27
11	17.59	6.10	21.75	5.69	0.14	10.28	3.29	1.62	2.67
12	16.85	5.05	19.36	6.33	0.15	9.27	2.83	1.25	2.07
平均	33.06	6.72	16.50	18.36	0.36	12.99	3.75	1.71	2.62

2.2.3 孔隙分形特征

分形理论被广泛用于描述不规则物体的形貌特征[22]。分形维数通常 D 介于 2～3，其中 2 代表是一个光滑表面，3 代表孔隙表面非常粗糙，非均质性强。页岩孔隙表面具有分型特征[23-25]，本文根据 Ar 吸附数据，运用 Frenkel–Halsey–Hill（FHH）模型进行分形维数的计算，FHH 吸附式如下：

$$\frac{V}{V_m} = C\left[RT\ln\left(\frac{P_0}{P}\right)\right]^{\alpha} \quad (1)$$

其中，V 为在压力 P 时的吸附量，m³/t；V_m 为单层吸附量，m³/t；R 为气体常数，J/(mol·K)；T 为绝对温度，K；P_0 为 Ar 在温度 T 时的饱和蒸气压，MPa；c 为特征常数，α 为与分形维数和吸附机制相关的参数。将式（1）进行对数处理，得到对数形式的方程：

$$\ln V = cons\tan t + \alpha \ln\left[\ln\left(\frac{P_0}{P}\right)\right] \quad (2)$$

对等温吸附数据进行处理，使用吸附曲线，取相对压力 P/P_0 大于 0.4 的点，以 $\ln V$ 对 $\ln[\ln(P_0/P)]$ 作曲线图（图 4），从而根据线性回归得到曲线的斜率 α，再根据 $D=3+\alpha$ 得到分形维数 D，数据见表 3，线性相关系数均大于 0.99，分形维数为 2.55～2.64，均值 2.6，说明页岩纳米级孔隙具有较强的非均质性。

表 3 N1 井龙马溪组页岩样品分形维数（R^2= 相关系数）

样品标号	拟合方程	R^2	分形维数 D
1	y=-0.4172x+2.2321	0.9955	2.58
2	y=-0.3646x+2.2433	0.9997	2.64
3	y=-0.3901x+2.1956	0.9979	2.61
4	y=-0.4004x+1.8084	0.9996	2.60
5	y=-0.3869x+1.8835	0.9983	2.61
6	y=-0.3956x+2.1102	0.9932	2.60
7	y=-0.3574x+0.9967	0.9967	2.64
8	y=-0.3687x+2.2494	0.9983	2.63
9	y=-0.3994x+2.2159	0.9962	2.60
10	y=-0.4166x+1.7842	0.9983	2.58
11	y=-0.4472x+1.6866	0.9970	2.55
12	y=-0.4254x+1.4659	0.9933	2.57

(a) 7号样　$y=-0.3574x+2.5315$　$R^2=0.9968$

(b) 8号样　$y=-0.3687x+2.2494$　$R^2=0.9983$

(c) 10号样　$y=-0.4166x+1.7842$　$R^2=0.9968$

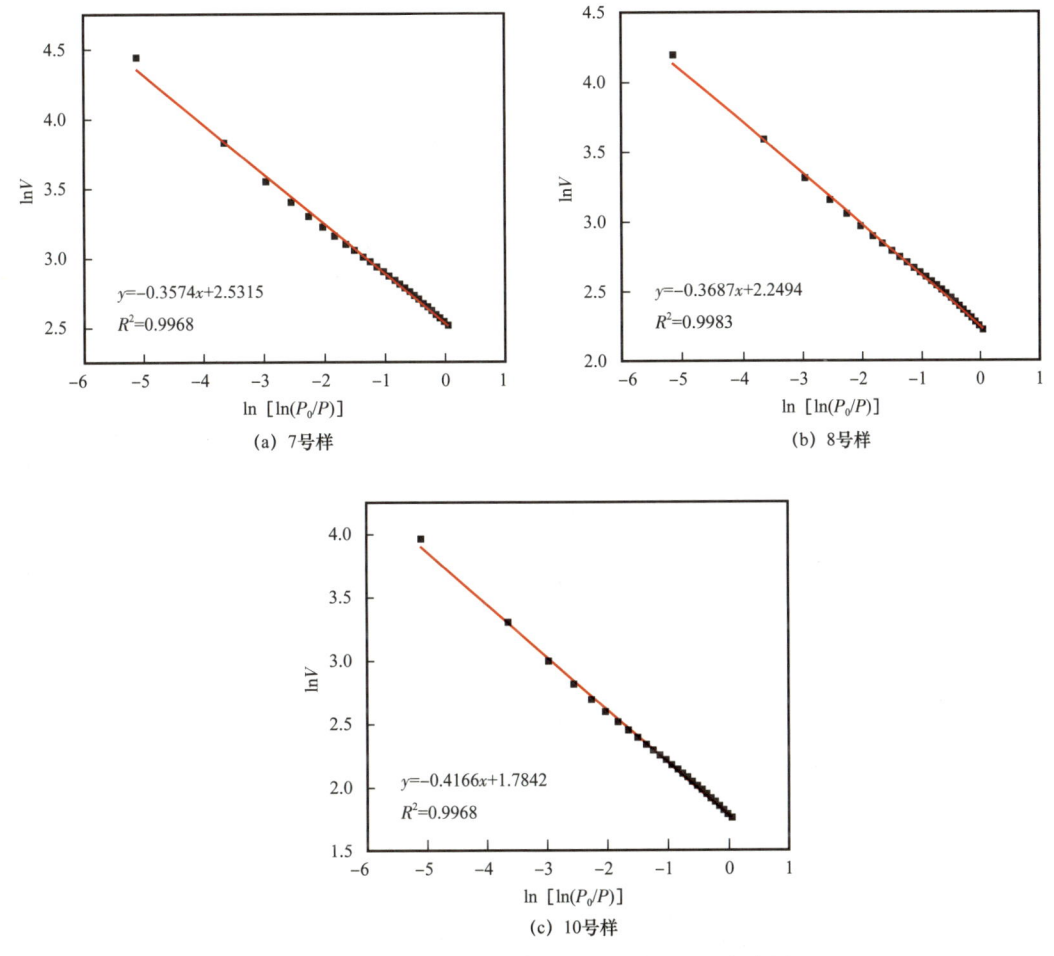

图 4　Ar 吸附等温线 $\ln[\ln(P_0/P)]$ 和 $\ln V$ 交会图

2.3　有机碳含量、孔隙结构参数和分形维数的相关关系

大量研究表明，有机碳含量是影响页岩纳米级孔隙发育的主控因素[16, 19]。从氩离子抛光扫描电镜可以看出，页岩中大部分孔隙为有机质孔，仅发育少量粒间孔和粒内孔，有机碳含量的多少决定了页岩中孔隙的多少，从而决定了页岩孔隙结构特征。

研究区龙马溪组页岩有机碳含量与微孔及介孔比表面积存在较好的正相关关系（图 5a、b），相关系数分别为 0.89 和 0.68，而与宏孔不存在明显的相关性（图 5c），从有机碳含量与各类孔隙占比的关系来看，有机碳含量越高，页岩中微孔占比越高（图 5d），而介孔和宏孔占比越低（图 5e、f），说明有机碳含量的增大有利于微孔的发育。前人研究认为页岩中微孔的含量与页岩的吸附气量呈明显的正相关关系[26-27]。微孔不仅可以提供巨大的比表面积，为甲烷气体提供了大量的吸附位置，而且随着孔径的减小，孔隙中的吸附势增大，吸附能力增强。

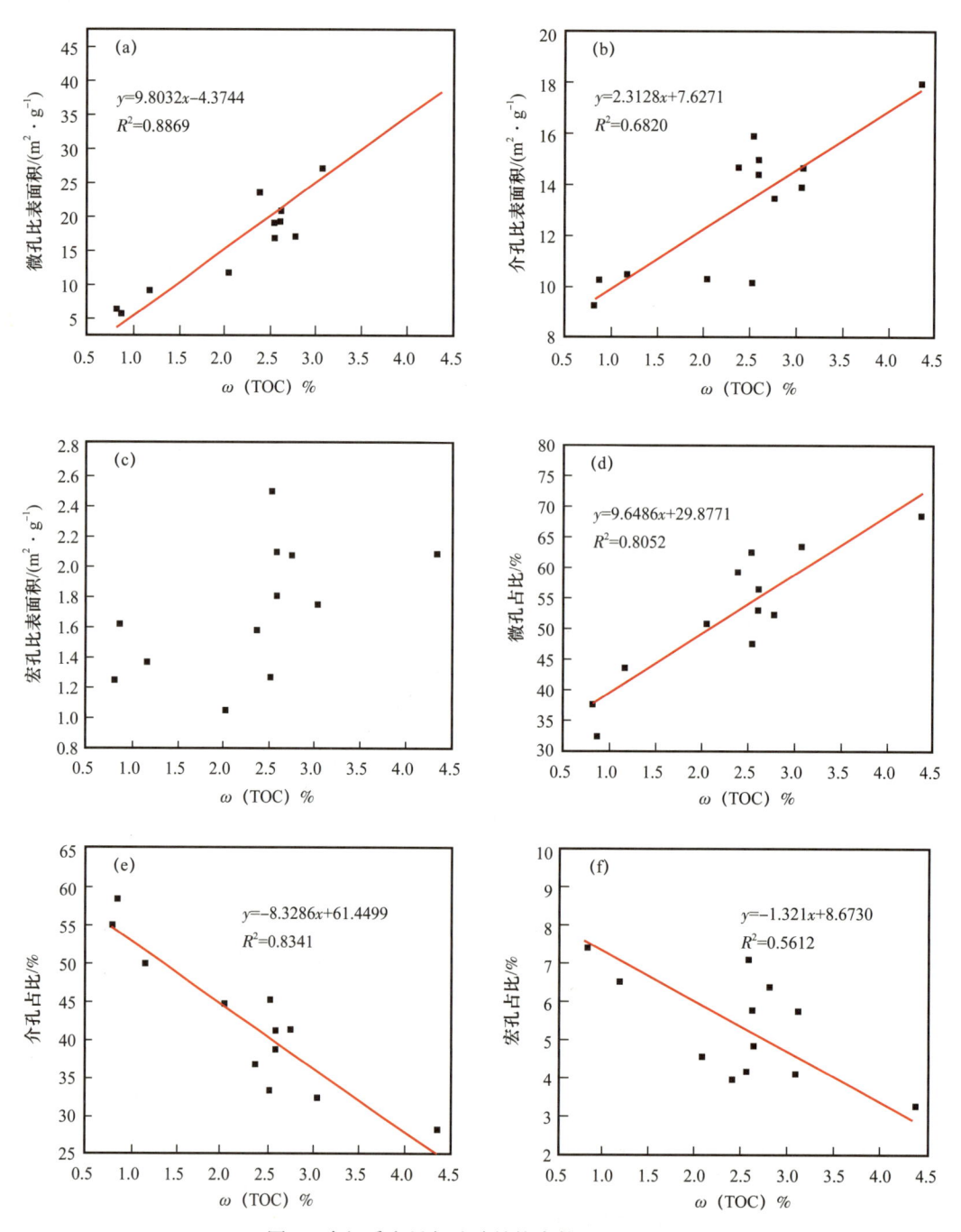

图 5　有机质含量与孔隙结构参数的相关关系

从图 6 可以看出，随着有机质含量以及页岩孔隙中微孔占比的增大，分形维数增大（图 6a、b），相关系数分别为 0.71 和 0.93，有机碳含量的增大使得页岩中微孔的含量增多，微孔的增多使得页岩孔隙结构复杂化，从而导致了页岩孔隙分形维数的增大，即孔隙结构非均质程度增大；分形维数与平均孔径呈明显的负相关关系（图 6c），平均孔径大的页岩其孔隙非均质性小，孔径的减小增大了页岩孔隙结构的非均质性。

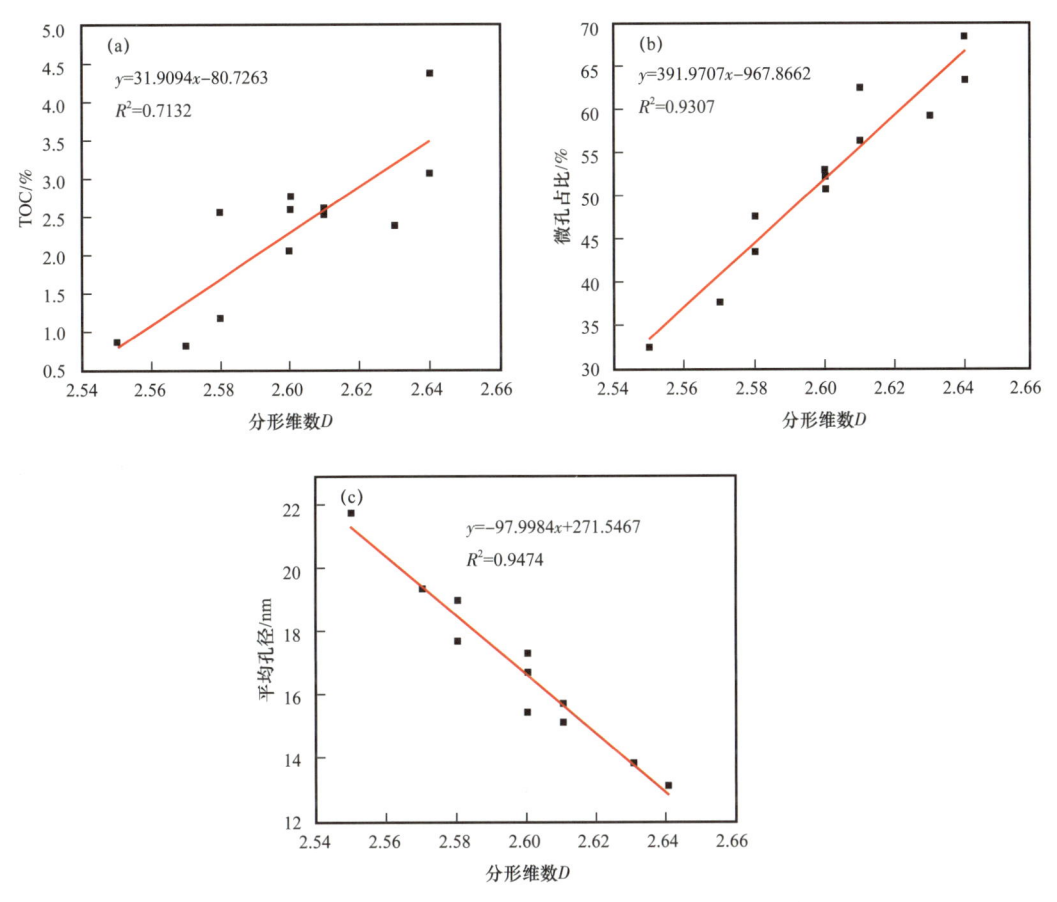

图 6 分形维数与 TOC、微孔占比以及平均孔径的相关关系

3 结论

（1）川南地区下志留统龙马溪组页岩孔隙类型可以分为有机质孔，粒间孔和粒内孔三大类，其中有机质孔占主体，为页岩气的储集和吸附提供了空间。

（2）首次使用低温氩气（Ar）吸附实验定量测量页岩样品的比表面积、孔体积以及孔径分布，实现了页岩小于 100nm（纳米级）孔隙的连续测量，页岩孔隙呈狭缝型，平均比表面积为 33.06m^2/g，平均孔体积为 6.72mL/100g，孔径主体分布在 10nm 以下的微孔和介孔中，微孔和介孔提供了页岩 90% 以上的比表面，介孔和宏孔提供了页岩 90% 以上的孔体积。

（3）基于氩气吸附数据的 FHH 模型表明，页岩纳米级孔隙具有分形特征，研究区龙马溪组页岩样品纳米级吸附孔隙分形维数为 2.55～2.64，表明页岩孔隙具有复杂的孔隙结构和非均质性。

（4）总有机碳含量控制页岩纳米级孔隙结构特征，随着 TOC 的增大，页岩中微孔含量以及微孔所占比例增大，平均孔径减小，页岩孔隙分形维数增大，孔隙结构趋于复杂。

参 考 文 献

［1］董大忠，邹才能，杨桦，等.中国页岩气勘探开发进展与发展前景［J］.石油学报，2012，33（S1）：107-114.

［2］郭彤楼，张汉荣.四川盆地焦石坝页岩气田形成与富集高产模式［J］.石油勘探与开发，2014，41（1）：28-36.

［3］贾爱林，位云生，金亦秋.中国海相页岩气开发评价关键技术进展［J］.石油勘探与开发，2016，43（6）：1-7.

［4］蒋裕强，董大忠，漆麟，等.页岩气储层的基本特征及其评价［J］.天然气工业，2010，30（10）：7-12.

［5］刘树根，王世玉，孙玮，等.四川盆地及其周缘五峰组—龙马溪组黑色页岩特征［J］.成都理工大学学报（自然科学版），2013，40（6）：621-638.

［6］Kitty L M，Mark R，David N A，et al. Organic matter-hosted pore system，Marcellus Formation（Devonian），Pennsylvania［J］. AAPG Bulletin，2013，97：177-200.

［7］Curtis M E，Ambrose R J，Sondergeld C H，et al. Microstructural investigation of gas shales in two and three dimensions using nanometer-scale resolution imaging［J］. AAPG Bulletin，2012，96：665-677.

［8］何建华，丁文龙，付景龙，等.页岩微观孔隙成因类型研究［J］.岩性油气藏，2014，26（5）：30-35.

［9］魏祥峰，刘若冰，张廷山，等.页岩气储层微观孔隙结构特征及发育控制因素：以川南—黔北XX地区龙马溪组为例［J］.天然气地球科学，2013，24（5）：1048-1059.

［10］Matthias T，Katsumi K，Alexander V，et al. Physisorption of gases，with special reference to the evaluation of surface area and pore size distribution（IUPAC Technical Report）［J］. Pure Appl. Chem，2015，87（9/10）：1051-1069.

［11］顾忠安，郑荣才，王亮，等.渝东涪陵地区大安寨段页岩储层特征研究［J］.岩性油气藏，2014，26（2）：67-73.

［12］Loucks R G，Reed R M，Ruppel S C，et al. Morphology，genesis，and distribution of nanometer-scale pores in siliceous mudstones of the Mississippian Barnett Shale［J］. Journal of sedimentary Research，2009，79：848-861.

［13］孙文峰，李玮，董智煜，等.页岩孔隙结构表征方法新探索［J］.岩性油气藏，2017，29（2）：125-130.

［14］Clarkson C R，Solano N，Bustin R M，et al. Pore structure characterization of North American shale gas reservoirs using USANS/SANS，gas adsorption，and mercury intrusion［J］. Fuel，2013，103：606-616.

［15］杨峰，宁正福，张世栋，等.基于氮气吸附实验的页岩孔隙结构表征［J］.天然气工业，2013，33（4）：135-140.

［16］李贤庆，王哲，郭曼，等.黔北地区下古生界页岩气储层孔隙结构特征［J］.中国矿业大学学报，2016，45（6）：1172-1183.

［17］Loucks R G，Reed R M，Ruppel S C，et al. Spectrum of pore types and networks in mudrocks and a descriptive classification for matrix-related mudrock pores. AAPG Bulletin，2012，96：1071-1098.

［18］Paul H. Geological and geochemical characterization of the Lower Cretaceous Pearsall Formation，

Maverick Basin, south Texas: A future shale gas resource [J]. AAPG Bulletin, 2012, 96 (8): 1449-1482.

[19] 武瑾, 梁峰, 吝文, 等. 渝东北地区龙马溪组页岩储层微观孔隙结构特征 [J]. 成都理工大学学报 (自然科学版), 2016, 43 (3): 308-319.

[20] 吉利明, 邱军利, 夏燕青, 等. 常见黏土矿物电镜扫描微孔特征与甲烷吸附性 [J]. 石油学报, 2012, 33 (2): 249-256.

[21] Pieter B, Kevin S, Helge S, et al. On the use and abuse of N_2 physisorption for the characterization of the pore structure of shales [J]. The Clay Minerals Society Workshop Lectures Series, 2016, 21 (12): 151-161.

[22] Hansen J P, Skjeltorp A T. Fractal pore space and rock permeability implications [J]. Physical Review B: Condensed Matter Physics, 1988, 38 (4): 2635.

[23] Yang F, Ning Z F, Liu H Q. Fractal characteristics of shales from a shale gas reservoir in the Sichuan Basin, China [J]. Fuel, 2014, 115: 378-384.

[24] 杨峰, 宁正福, 王庆, 等. 页岩纳米孔隙分形特征 [J]. 天然气地球科学, 2014, 25 (4): 618-623.

[25] 徐勇, 吕成福, 陈俊国, 等. 川东南龙马溪组页岩孔隙分形特征 [J]. 岩性油气藏, 2015, 27 (4): 32-39.

[26] Ross D J K, Bustin R M. The importance of shale composition and pore structure upon gas storage of shale gas reservoirs [J]. Marine and Petroleum Geology, 2009, 26 (6): 916-927.

[27] Chalmers G R L, Bustin R M. The organic matter distribution and methane capacity of the Lower Cretaceous strata of northeastern British Columbia [J]. Canada International Journal of Coal Geology, 2007, 70 (1): 223-239.

三、气藏工程类

页岩气压裂水平井控压生产动态预测模型及其应用

贾爱林[1]，位云生[1]，刘 成[2]，王军磊[1]，齐亚东[1]，贾成业[1]，李 波[3]

（1.中国石油勘探开发研究院；2.中国石油浙江油田公司；3.华能国际电力开发公司）

摘 要：为理清页岩气压裂水平井采取控压生产制度延缓裂缝闭合与提高单井最终可采量（EUR）的内在联系，明确控压生产的气藏工程意义，以有限导流裂缝为基本流动单元，引入变应力敏感系数，建立地层—裂缝耦合的压裂水平井生产动态预测模型并求解，并与经典模型、Saphir 软件数值模型的计算结果进行对比，在此基础上，采用所建立的模型分析了应力敏感性特征参数、控压生产制度等因素对压裂水平井生产动态的影响，进而对某两口实例井在不同控压制度下的生产动态进行拟合和预测。研究结果表明：（1）所建立的模型与经典模型、Saphir 软件数值模型计算的结果基本吻合；（2）应力敏感性使得裂缝导流能力随着气井生产而递减，导致气井累计产气量减低，且应力敏感性越强，累计产气量减低的幅度越大；（3）与放压方式相比，采用控压方式生产的页岩气压裂水平井的初期产气量及累计产气量虽偏低，但最终累计产气量却更高，可见页岩气压裂水平井长期控压生产更具合理性。结论认为，所取得的研究成果可以为页岩气井控压生产制度的推广提供理论支撑。

关键词：压裂水平井；变应力敏感系数；非线性流动；井底压力；控压生产；累计产气量

通常认为是储层应力敏感导致介质传导能力下降而造成的[1-3]。根据有效应力与原地应力、孔隙压力之间的函数关系[4]可知，随着页岩气井的生产，地层压力逐渐衰减，地层/裂缝受到的有效应力逐渐增加，由于裂缝中支撑剂流失、嵌入和被压碎，页岩气流动的通道发生变形，导致其渗流能力降低。页岩气的开发实践表明，控压限产的生产方式可以有效提高单井的最终可采气量（以下简称为 EUR）[5-7]。在北美地区，相比于 Barnett、Marcellus 等页岩气藏，Haynesville 页岩由于地层压力较高，储层应力敏感性特征显著，在气藏的实际开发过程中广泛采用控压限产的生产方式，实践证明该方式较放压生产方式，单井 EUR 可普遍提高 28%[8]；国内长宁—威远、昭通等页岩气示范区的地质条件与 Haynesville 页岩气藏相似，经过几年的现场试验后也逐步在推广控压限产的生产方式[9-10]。

与放压生产相比，控压限产有利于降低压裂液返排率，更重要的作用在于保持人工裂缝的长期开启，缓解渗流场的应力敏感效应[4, 11]。裂缝系统的应力敏感效应直接受有效应力场控制，当原地应力场的变化幅度很小时，利用线性关系，有效应力可近似转换为孔隙压力，将压力场的模拟结果与应力敏感曲线相结合可用于气井生产动态的分析[12-14]，该方法虽简洁有效，但有效应力估计值偏低，导致单井 EUR 预测值偏高，同时由于无法模拟井底压力对有效应力场的直接影响，使得放压和控压两种生产方式所引起的产量差异不明显。尤其对于页岩等非常规储层，压力场变化会引起原地应力场发生显著变化，国内外学者较多利用流固耦合数值模型[4, 15-16]、应力解析模型[17-18]来精确表征生产过程中压力场—应力场的耦合变化，虽然能模拟不同控压制度下单井的开发指标，但模型过于复杂且计算代价巨大，难以推广应用。

为理清页岩气压裂水平井采取控压生产制度延缓裂缝闭合与提高单井 EUR 的内在联系，明确控压生产的气藏工程意义，笔者以有限导流裂缝为基本流动单元，引入变应力敏感系数，建立地层—裂缝耦合的压裂水平井生产动态预测模型并求解，并与经典模型、Saphir 软件数值模型的计算结果进行对比，在此基础上，采用所建立的模型分析了应力敏感性特征参数、控压生产制度等因素对压裂水平井生产动态的影响，进而对某两口实例井在不同控压制度下的生产动态进行拟合和预测。

1 裂缝—地层流动模型

1.1 假设条件与参数说明

水平井进行水力压裂改造后，流体从地层到井筒的流动分为从地层到裂缝和裂缝内部这两个部分。裂缝与地层中的流量及压力在裂缝壁面处相等，由此将这两个部分流体的流动进行耦合，然后通过计算得到不同时刻地层中任一点的压力。具体假设如下：（1）矩形地层均质、等厚、四周及上下均封闭；（2）压裂裂缝纵向上穿透地层，裂缝垂直于水平井筒且关于井筒呈不对称分布，且由跟端到趾端依次排列第 1，…，n，…，n_f 条裂缝；（3）气井以定井底压力进行生产，整个流动过程均符合达西定律，不考虑井筒管流的影响；（4）气井生产引入拟压力，将气体流动问题等效转换为液体流动问题；（5）考虑人工裂缝具有强应力敏感性，未考虑储层基质的应力敏感性[4-6]。

拟压力表达式为：

$$p_\mathrm{p} = \frac{\mu_\mathrm{gi} Z_\mathrm{gi}}{p_\mathrm{i}} \int_{p_\mathrm{i}}^{p} \frac{p'}{\mu_\mathrm{g}(p') Z_\mathrm{g}(p')} \mathrm{d}p' \tag{1}$$

式中，p_p 为拟压力，Pa；μ_g 为气体黏度；Pa·s；Z_g 为气体偏差因子；p 为地层压力，Pa；p' 为压力积分符号，Pa；下标 i 为初始状态。

相对于渗透率的应力敏感性，孔隙度的应力敏感性通常可忽略[19]。随着裂缝承受的有效应力不断增加，裂缝发生闭合，渗透率/导流能力降低，这里设定裂缝渗透率与有效应力呈指数关系，如式（2）所示；同时，由于不同区域有效应力变化的速率不同，导致裂缝的闭合程度也不同，且随时间变化裂缝的导流能力将发生变化。

$$K_\mathrm{f} = K_\mathrm{fi} \exp\left[-d_\mathrm{f}\left(\sigma_\mathrm{eff} - \sigma_\mathrm{eff,i}\right)\right] \tag{2}$$

式中，K_f 为裂缝渗透率，10^{12}D；d_f 为渗透率模量，Pa^{-1}；σ_eff 为有效应力，Pa；下标 f 表示裂缝。

由于裂缝壁面粗糙度较大，即使裂缝已经不再随有效应力增加而闭合，裂缝仍具有一定渗流能力，此时所对应的最小裂缝渗透率（$K_{f,\min}$）仍远高于基质渗透率[20]。将式（2）改写为：

$$\frac{K_f - K_{f,\min}}{K_{fi} - K_{f,\min}} = \exp\left[-d_f\left(\sigma_{\text{eff}} - \sigma_{\text{eff},i}\right)\right] \tag{3}$$

式中，下标 min 为最小值。

为方便与渗流模型相结合，需将式（3）与孔隙压力建立起函数关系。由于有效应力、原地应力和孔隙压力满足的关系式为：

$$\sigma_{\text{eff}} = \sigma - \alpha p \tag{4}$$

式中，σ 为原地应力，Pa；α 为 Biot 系数。

应力差与孔隙压力差的关系式为[17]：

$$\Delta\sigma = \alpha \frac{1-2\upsilon}{1-\upsilon}\Delta p \tag{5}$$

式中，υ 为泊松比。

则式（3）转换为：

$$\frac{K_f - K_{f,\min}}{K_{fi} - K_{f,\min}} = \exp\left[-\gamma_f\left(p_i - p\right)\right] \tag{6}$$

其中

$$\gamma_f = \frac{d_f \upsilon \alpha}{1-\upsilon}$$

式中，γ_f 为应力敏感系数。

d_f 为常量或是与有效应力相关的变量，具体形式取决于介质类型。通常，对于硬地层，d_f 为常量，而对于软地层，d_f 则为变量。Alramahi 和 Sundberg[20] 提供了不同硬度地层中裂缝导流能力的实验数据，并进行最优拟合，结果显示对于硬地层，裂缝渗透率的自然对数与有效应力呈线性关系，而对于软地层，二者满足二项式关系。由长宁地区龙马溪组页岩人工裂缝的应力敏感性实验的测试结果，显示裂缝渗透率的自然对数与有效应力满足二项式关系（图1），渗透率模量是有效应力的函数，而不是常量。

基于渗透率模量与有效应力的函数关系式，再结合有效应力与孔隙压力的关系式，应力敏感系数的表达式可转换为：

$$\gamma_f = a(p_i - p_f) + b \tag{7}$$

式中，a、b 分别为应力敏感特征系数；p_f 为裂缝中压力，Pa。

为便于研究，定义无量纲时间（t_D）、裂缝长度（L_D）、导流能力（C_{fD}）和应力敏感系数（γ_{fD}），具体表达式为：

$$t_D = \frac{K_m t}{\phi_m \mu C_t L_{\text{ref}}^2} \tag{8}$$

$$L_D = \frac{L}{L_{\text{ref}}} \tag{9}$$

$$C_{fD} = \frac{K_f w_f}{K_m L_{ref}} \tag{10}$$

$$\gamma_{fD} = a_D p_{fD} + b_D \tag{11}$$

式中，K_m 为储层渗透率，10^{12}D；t 为时间，s；ϕ_m 为储层孔隙度；μ 为流体黏度，Pa·s；C_t 为综合压缩系数，Pa^{-1}；L_{ref} 为参考长度，m；L 为裂缝长度，m；w_f 为裂缝宽度，m；a_D、b_D 分别为无量纲应力敏感特征系数；下标 D 为无量纲。

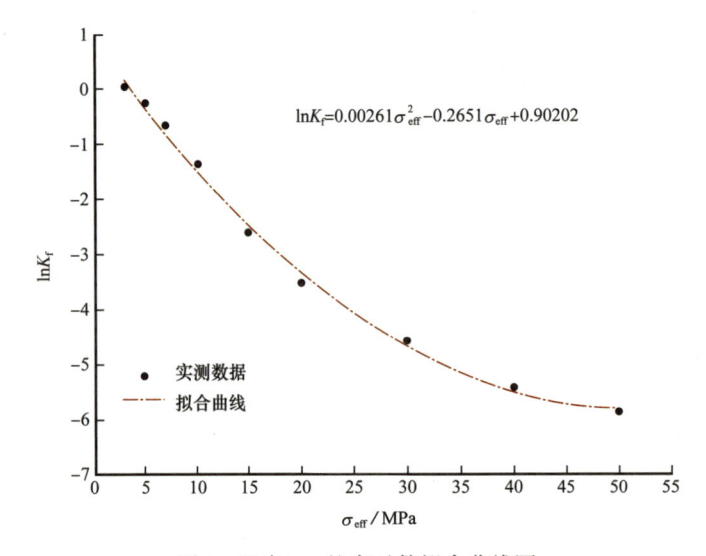

图 1 K_f 与 σ_{eff} 的半对数拟合曲线图

在定产量条件下，针对裂缝无量纲压力（p_{fD}）、应力敏感系数（γ_{fD}）、流量（q_{wfD}）及流量密度（ε_{fD}）的定义式为：

$$p_{fD} = \frac{2\pi K_m h(p_i - p_f)}{q_w \mu B} \tag{12}$$

$$\gamma_{fD} = \frac{q_w \mu B}{2\pi K_m h} \gamma_f \tag{13}$$

$$q_{wfD} = \frac{q_{wf}}{q_w} \tag{14}$$

$$\varepsilon_{fD} = \frac{2\varepsilon_f L_{ref}}{q_w} \tag{15}$$

式中，h 为储层厚度，m；q_w 为水平井流量，m^3/s；B 为流体体积系数；q_{wf} 为裂缝流量，m^3/s；ε_f 为裂缝流量密度，m^2/s。

在定井底压力条件下，针对裂缝无量纲压力、应力敏感系数、流量及流量密度的定义式为：

$$p_{fD} = \frac{p_i - p_f}{p_i - p_w} \tag{16}$$

$$\gamma_{fD} = (p_i - p_w)\gamma_f \tag{17}$$

$$q_{wfD} = \frac{q_{wf}\mu B}{2\pi K_m h(p_i - p_w)} \tag{18}$$

$$\varepsilon_{fD} = \frac{2\varepsilon_f L_{ref}\mu B}{2\pi K_m h(p_i - p_w)} \tag{19}$$

式中，p_w 为井底压力，Pa。

对于气井生产而言，需将式（12）、式（16）至式（19）中的压力替换为拟压力，相应式（8）、式（12）、式（18）至式（19）中的 μ、B 替换为 μ_{gi}、B_{gi}。

1.2 数学模型的建立

地层中流体在裂缝中的流动为变流量密度线性流，而后由裂缝汇入到井筒中的流动为近井筒径向流，因此，可将裂缝和井筒视为两个独立的流动空间（图2）。

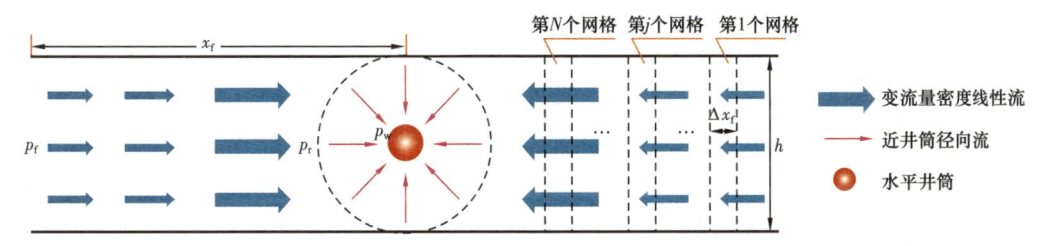

图 2 流体在裂缝和井筒中的流动模式示意图

p_r—井筒附近地层压力，Pa；x_f—裂缝半长，m；Δx_f—裂缝微元长度，m

将裂缝内流体的流动简化为一维稳态流动[21]，使用流量密度函数 $\varepsilon_f(x)$ 描述流体沿裂缝的流动、点源函数 δ 描述井筒的采出，则第 n 条裂缝内流体流动的无量纲方程为：

$$\begin{cases} \dfrac{\partial}{\partial x_{Dn}}\left[C_{fDn}(p_{fDn})\dfrac{\partial p_{fDn}}{\partial x_{Dn}}\right] - 2\pi\varepsilon_{fDn}(x_{Dn}) + \\ \qquad 2\pi q_{wfDn}(t_D)\delta(x_{Dn}, x_{wfDn}) = 0 \\ q_{cDn}(x_{Dn}) = \displaystyle\int_{x_{Dn}}^{L_{fDn}} \varepsilon_{fDn}(\varsigma)\mathrm{d}\varsigma \\ \left.\dfrac{\partial p_{fDn}}{\partial x_{Dn}}\right|_{x_{Dn}=0} = 0 \\ \left.\dfrac{\partial p_{fDn}}{\partial x_{Dn}}\right|_{x_{Dn}=L_{Dn}} = 0 \end{cases} \tag{20}$$

式中，x_D 为无量纲裂缝坐标；x_{wfD} 为无量纲井筒位置坐标；q_{cD} 为无量纲截面流量；L_{fD} 为无量纲裂缝长度；ς 为长度积分符号；下标 n 为第 n 条裂缝。

无量纲动态导流能力计算式为：

$$\frac{C_{fDn}(p_{fDn})}{C_{fDi}} = \left(1 - \frac{C_{fD,\min}}{C_{fDi}}\right)\exp[-\gamma_{fD}p_{fD}] + \frac{C_{fD,\min}}{C_{fDi}} \quad (21)$$

式中，$C_{fD,\min}$ 为最小无量纲导流能力。

式（20）是典型的非线性微分方程，无法直接求解。采用坐标转换法进行线性化处理，转换式为：

$$\xi_{Dn}(x_{Dn}) = \hat{C}_{fD}\int_0^{x_{Dn}}\frac{dx_D}{C_{fDn}(x_D)} \quad (22)$$

其中

$$\hat{C}_{fD} = \frac{L_{fDn}}{\int_0^{L_{fDn}}\frac{dx_D}{C_{fDn}(x_D)}}$$

式中，ξ_D 为变换后无量纲坐标；$\hat{C}_{fD}$ 为无量纲裂缝导流能力。

将式（22）代入式（20），再进行偏导数转换，可得新的压力控制方程，即

$$\frac{\partial}{\partial \xi_{Dn}}\left(\hat{C}_{fDn}\frac{\partial p_{fDn}}{\partial \xi_{Dn}}\right) + 2\pi\frac{\partial q_{cDn}(\xi_{Dn})}{\partial \xi_{Dn}} + 2\pi q_{wfDn}\frac{\partial H}{\partial \xi_{Dn}}\frac{\left(\frac{\partial \xi_{Dn}}{\partial x_{Dn}}\right)}{\left(\frac{\partial \xi_{Dn}}{\partial x_{Dn}}\right)} = 0 \quad (23)$$

式中，H 为 Heaviside 阶跃函数。

式（23）符合线性微分方程的形式，对该式进行双重积分，同时考虑近井筒附近的聚流效应，可以获得裂缝内的无量纲压力方程，即

$$\begin{aligned}p_{wD} - p_{fDn}(\xi_D) &= \frac{2\pi}{\hat{C}_{fD}}q_{wfDn}G(\xi_D, \xi_{wfDn}) \\ &\quad - \frac{2\pi}{\hat{C}_{fD}}\int_{\xi_{wfDn}}^{\xi_D}d\zeta\int_0^{\zeta}\varepsilon_{fDn}(\varsigma)d\varsigma + q_{wfDn}S_{cn}\end{aligned} \quad (24)$$

其中

$$S_{cn} = \frac{2h_D}{\hat{C}_{fD}L_{fD}}\left[\ln\left(\frac{h_D}{2r_{wD}}\right) - \frac{\pi}{2}\right]$$

$$\delta(\xi_{Dn}, \xi_{wfDn}) = \begin{cases}\infty, & \xi_{Dn} = \xi_{wfDn} \\ 0, & \xi_{Dn} \neq \xi_{wfDn}\end{cases}$$

$$H(\xi_{Dn}, \xi_{wfDn}) = \begin{cases}1, & \xi_{Dn} \geq \xi_{wfDn} \\ 0, & \xi_{Dn} < \xi_{wfDn}\end{cases}$$

式中，p_{wD} 为无量纲井底压力；G 为 Heaviside 阶跃函数的积分函数；ξ_{wfD} 为转换坐标下的无量纲井筒位置坐标；S_c 为裂缝内聚流表皮因子；h_D 为无量纲储层厚度；r_{wD} 为无量纲井筒半径。

对式（24）进行离散化处理，认为流体流量在裂缝网格上均匀分布，对应离散形式为：

$$p_{wD} - p_{fDn}\left(\xi_{Dn,j}\right) = \frac{2\pi}{\hat{C}_{fDn}} q_{wfDn} G\left(\xi_{Dn,j}, \xi_{wfDn}\right) \\ - \frac{2\pi}{\hat{C}_{fDn}} \sum_{j=1}^{N} q_{fDn,j} C_{n,j,n,i} + q_{wfDn} S_{cn} \tag{25}$$

式中，$q_{fDn,j} = \int_{\xi_{Dn,j-1}}^{\xi_{Dn,j-1}} \varepsilon_{fDn}(\varsigma)\mathrm{d}\varsigma$；$C$ 为裂缝导流影响函数；下标 i、j 为裂缝网格序号。

地层边界 Γ 满足的封闭条件为：

$$\left.\frac{\partial p_m}{\partial \boldsymbol{n}_B}\right|_{\Gamma} = 0 \tag{26}$$

式中，p_m 为储层压力，Pa；$\boldsymbol{n}_B$ 为沿边界的法向量，m。

以每个裂缝网格几何中心为计算点，根据位势叠加原理可以得到第 n 条裂缝第 j 个网格内的压力分布，即

$$\tilde{p}_{mD}(x_{wDn,j}, y_{wDn,j}) = \sum_{m=1}^{n_f}\sum_{i=1}^{N}\left[\left(\tilde{q}_{Dm,i} \frac{\Delta \xi_{Dm,i}}{\Delta x_{fDm,i}}\right) \cdot \tilde{p}_{uDn,j,m,i}(x_{wDn,j}, y_{wDn,j}, x_{wDm,i}, y_{wDm,i}, \Delta x_{fDm,i}, x_{eD}, y_{eD})\right] \tag{27}$$

式中，$\tilde{p}_{mD}$ 为 Laplace 空间储层的无量纲压力；x_{wD}、y_{wD} 分别为裂缝微元空间无量纲坐标；n_f 为裂缝条数；$\tilde{q}_{Dm,i}$ 为 Laplace 空间裂缝微元上的无量纲流量；x_{fD} 为无量纲裂缝微元长度；$\tilde{p}_{uD}$ 为 Laplace 空间单位流量强度下裂缝微元的扰动函数；x_{eD}、y_{eD} 分别为地层无量纲长度、宽度；下标 m 为第 m 条裂缝。

利用 Green 函数和 Newman 乘积法[22]计算裂缝微元对应的压力扰动解。在裂缝面上对储层与裂缝系统进行压力与流量的耦合，即

$$\tilde{p}_{fDm,i} = \tilde{p}_{mDm,i} \tag{28}$$

$$\tilde{q}_{fDm,i} = \tilde{q}_{Dm,i} \tag{29}$$

式中，$\tilde{p}_{fDm,i}$ 为裂缝微元对应裂缝内的无量纲压力；$\tilde{p}_{mDm,i}$ 为裂缝微元对应地层内的无量纲压力；$\tilde{q}_{fDm,i}$ 为裂缝微元对应裂缝内的无量纲流量，$\tilde{q}_{Dm,i}$ 为裂缝微元对应地层内的无量纲流量。

分别将式（25）、式（27）代入式（28）、式（29），得耦合流动方程组为：

$$\sum_{j=1}^{N}\left[\tilde{q}_{fDn,j}\left(\frac{-2\pi C_{n,j;n,i}}{\hat{C}_{fDn}}\right)\right] + \sum_{m=1}^{n_f}\sum_{i=1}^{N}\left[\tilde{q}_{fDm,i}\left(\frac{\Delta \xi_{Dm,i}}{\Delta x_{fDm,i}} \tilde{p}_{uDn,j;m,i}\right)\right] + \\ \tilde{q}_{wfDn} \frac{2\pi G\left(\xi_{Dn,j}, \xi_{wfDn}\right)}{\hat{C}_{fDn}} = \tilde{p}_{wD} \tag{30}$$

式（30）是关于裂缝网格流量密度、裂缝流量和井底压力的方程组，可写成如下复合矩阵形式，即

$$\begin{pmatrix} \bar{G}_{1,1} & \cdots & \bar{G}_{1,m} & \cdots & \bar{G}_{1,n_f} \\ \vdots & & \vdots & & \vdots \\ \bar{G}_{n,1} & \cdots & \bar{G}_{n,m} & \cdots & \bar{G}_{n,n_f} \\ \vdots & & \vdots & & \vdots \\ \bar{G}_{n_f,1} & \cdots & \bar{G}_{n_f,m} & \cdots & \bar{G}_{n_f,n_f} \end{pmatrix} \cdot \begin{pmatrix} \tilde{\varepsilon}_{fD1} \\ \vdots \\ \tilde{\varepsilon}_{fDm} \\ \vdots \\ \tilde{\varepsilon}_{fDn_f} \end{pmatrix} = \begin{pmatrix} \tilde{p}_{wD} \\ \vdots \\ \tilde{p}_{wD} \\ \vdots \\ \tilde{p}_{wD} \end{pmatrix} \quad (31)$$

其中

$$\tilde{\varepsilon}_{fDm} = \left(\tilde{q}_{fDm,1} \cdots \tilde{q}_{fDm,i} \cdots \tilde{q}_{fDm,N} \cdots \tilde{q}_{wfDm} \right)^T$$

第 n 条裂缝对第 m 条裂缝产生的子矩阵为：

$$\bar{G}_{n,m} = \begin{pmatrix} \dfrac{\Delta\xi_{Dm,1}\tilde{p}_{uD1,1,m,1}}{\Delta x_{fDm,1}} & \cdots & \dfrac{\Delta\xi_{Dm,1}\tilde{p}_{uD1,j,m,1}}{\Delta x_{fDm,1}} & \cdots & \dfrac{\Delta\xi_{Dm,1}\tilde{p}_{uD1,N,m,1}}{\Delta x_{fDm,1}} & 0 \\ \vdots & & \vdots & & \vdots & \vdots \\ \dfrac{\Delta\xi_{Dm,i}\tilde{p}_{uDn,1,m,i}}{\Delta x_{fDm,i}} & \cdots & \dfrac{\Delta\xi_{Dm,i}\tilde{p}_{uDn,j,m,i}}{\Delta x_{fDm,i}} & \cdots & \dfrac{\Delta\xi_{Dm,i}\tilde{p}_{uDn,N,m,i}}{\Delta x_{fDm,i}} & 0 \\ \vdots & & \vdots & & \vdots & \vdots \\ \dfrac{\Delta\xi_{Dm,N}\tilde{p}_{uDn,1,m,N}}{\Delta x_{fDm,N}} & \cdots & \dfrac{\Delta\xi_{Dm,N}\tilde{p}_{uDn,j,m,N}}{\Delta x_{fDm,N}} & \cdots & \dfrac{\Delta\xi_{Dm,N}\tilde{p}_{uDn,N,m,N}}{\Delta x_{fDm,N}} & 0 \\ 0 & \cdots & 0 & \cdots & 0 & 0 \end{pmatrix} - \\ \dfrac{2\pi}{\hat{C}_{fDm}} \begin{pmatrix} C_{m,1,m,1} & \cdots & C_{m,1,m,i} & \cdots & C_{m,1,m,N} & 0 \\ \vdots & & \vdots & & \vdots & \vdots \\ C_{m,j,m,1} & \cdots & C_{m,j,m,i} & \cdots & C_{m,j,m,N} & 0 \\ \vdots & & \vdots & & \vdots & \vdots \\ C_{m,N,m,1} & \cdots & C_{m,N,m,i} & \cdots & C_{m,N,m,N} & 0 \\ 0 & \cdots & 0 & \cdots & 0 & 0 \end{pmatrix} + \\ \dfrac{2\pi}{\hat{C}_{fDm}} \begin{pmatrix} 0 & \cdots & 0 & \cdots & 0 & G(\xi_{Dm,1},\xi_{wfDm}) \\ \vdots & & \vdots & & \vdots & \vdots \\ 0 & \cdots & 0 & \cdots & 0 & G(\xi_{Dm,i},\xi_{wfDm}) \\ \vdots & & \vdots & & \vdots & \vdots \\ 0 & \cdots & 0 & \cdots & 0 & G(\xi_{Dm,N},\xi_{wfDm}) \\ \Delta x_{fDm} & \cdots & \Delta x_{fDm} & \cdots & \Delta x_{fDm} & -1 \end{pmatrix} \quad (32)$$

式（32）等号右侧第 1 个矩阵为地层流动矩阵，只与裂缝网格空间位置有关，第 2 个矩阵为裂缝内部流动矩阵，第 3 个矩阵为井筒流出矩阵。

2 模型求解及验证

式（22）表明 ξ_D 是裂缝方向上与压力分布相关的函数，若给定第 k 个时间步的压力分布，则可获得该时间步下沿裂缝方向的导流能力分布。因此，可以由式（31）、式（25）分别计算得到第 k 个时间步沿裂缝的无量纲流量与压力，然后根据式（33），在新坐标系［由式（22）转换得到］下计算第 $k+1$ 个时间步沿裂缝的无量纲压力，结果由所编制的程序通过迭代计算得到。

$$C_{\text{fD}n}\left(p_{\text{fD}n}^{k},x_{\text{D}n}\right)\frac{\partial \tilde{p}_{\text{fD}n}^{k+1}}{\partial x_{\text{D}n}} + 2\pi \tilde{q}_{\text{cD}n}^{k+1}\left(x_{\text{D}n}\right) + \\ 2\pi\left[\tilde{q}_{\text{wfD}n}^{k+1} H\left(x_{\text{D}n},x_{\text{wfD}n}\right)\right] = 0 \tag{33}$$

本文所提出的半解析模型的优势在于能够快速模拟封闭地层中变应力敏感系数影响下的压裂井生产动态。将本文模型的计算结果与经典模型的计算结果进行对比，对于有限导流的对称裂缝，采用Cinco-Ley等[21]建立的模型进行计算，对于有限导流的非对称裂缝，采用Berumen等[22]建立的模型进行计算，均不考虑聚流表皮因子的影响，裂缝条数设为1，地层为无限大。如图3所示，本文模型与经典模型的计算结果基本吻合。

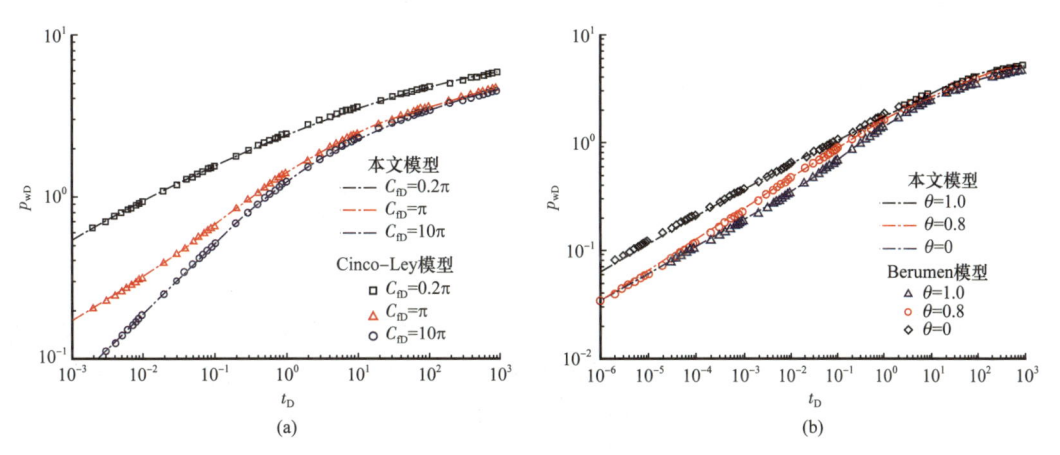

图3 本文模型与经典模型计算结果对比图

θ表示偏心程度，即井筒位置与裂缝中心的相对距离

对于存在多条裂缝的情况，将本文模型的计算结果与Saphir软件数值模型的计算结果进行对比，设置储层厚度为2.9m，基质渗透率为0.001mD，基质孔隙度为10%，裂缝宽度为0.0127m，裂缝半长为114.95m，裂缝孔隙度为35%，原始地层压力为50MPa，流体黏度为1mPa·s，综合压缩系数为$4.35\times10^{-4}\text{MPa}^{-1}$，裂缝条数为15条，初始无量纲导流能力为$10\pi$，最小无量纲导流能力为$0.1\pi$，模型长度为4598m，宽度为2299m，$a_\text{D}$、$b_\text{D}$选取4种不同的组合，其中组合1（$a_\text{D}=0$、$b_\text{D}=0$）对应于不存在应力敏感性的裂缝模型，组合3（$a_\text{D}=0$、$b_\text{D}=1$）对应于常应力敏感系数裂缝模型，组合2（$a_\text{D}=1$、$b_\text{D}=1$）、4（$a_\text{D}=1$、$b_\text{D}=0$）对应于变应力敏感系数裂缝模型。通过对比，发现本文模型与Saphir软件数值模型的计算结果基本吻合（图4）。

3 影响因素分析及实例应用

3.1 影响因素分析

影响压裂水平井生产动态的因素可分为裂缝参数（包括裂缝初始导流能力、长度、条数及空间位置等）和应力敏感性特征参数（包括a_D、b_D及最小导流能力）。本文建立一种理想模式，即在泄流区域的中心位置均匀部署等长、等导流能力、等应力敏感特征参数的多条裂缝进行影响因素分析。若 a_D、b_D 都等于 0，对应裂缝则为常导流能力裂缝（简称常导流缝），否则为动态导流能力裂缝（简称动态导流缝）。

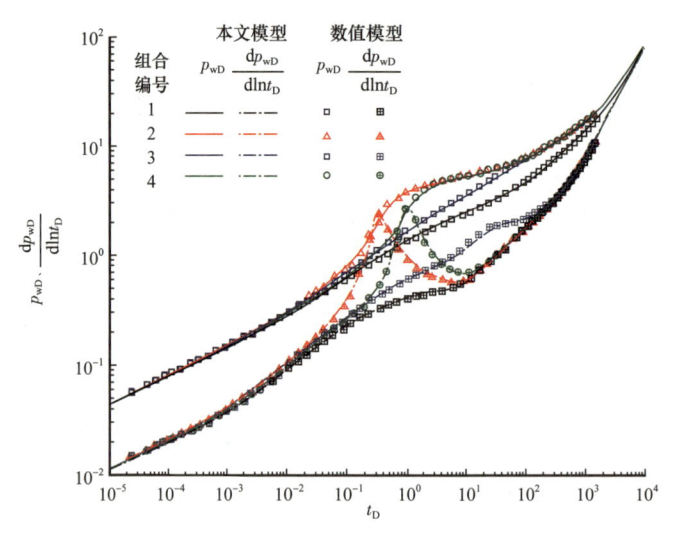

图 4 本文模型与 Saphir 软件数值模型计算结果对比图

对于动态导流缝，随着气井的生产，裂缝发生闭合，裂缝导流能力降低，压力波的传播受到抑制。相同时间内常导流缝压力波的传播速度要快于动态导流缝压力波的传播速度（图 5）；随时间的延续，裂缝导流能力逐渐降低，且趋近于井点处的裂缝导流能力（图 6）。

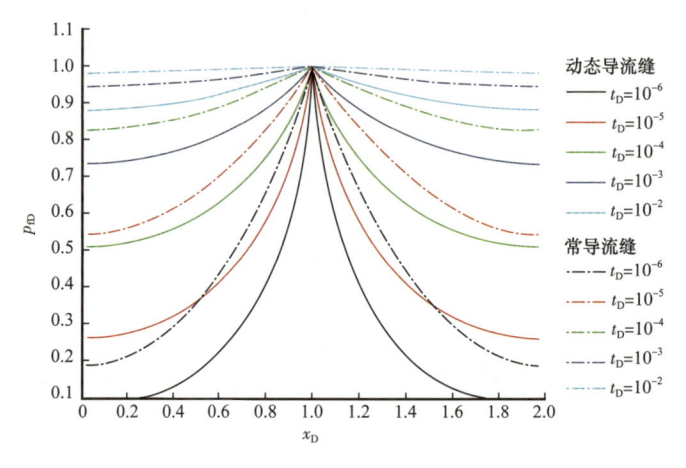

图 5 裂缝应力敏感性影响下沿裂缝 p_{fD} 分布图

如图 7 所示，随着 t_D 增加，气井的无量纲产气量 q_{wD}（由前述 q_{wfD} 求和得到）不断递减；对于动态导流缝，a_D、b_D 的不同组合决定了 q_{wD} 的下降幅度不同；应力敏感系数较小（$a_D=1$、$b_D=1$）时，q_{wD} 曲线趋近于常导流缝（$C_{fD}=100\pi$）对应的 q_{wD} 曲线，应力敏感系数较大（$a_D=1$、$b_D=10$）时，q_{wD} 曲线趋近于常导流缝（$C_{fD}=0.1\pi$）对应的 q_{wD} 曲线。

图 6 裂缝应力敏感性影响下沿裂缝 $\dfrac{C_{fD}}{C_{fDi}}$ 分布图（$a_D=b_D=1$，$C_{fDi}=100\pi$，$C_{fD,\ min}=0.1\pi$）

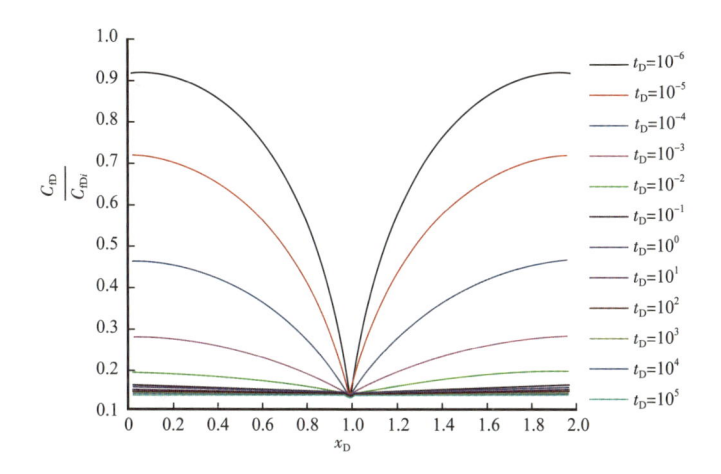

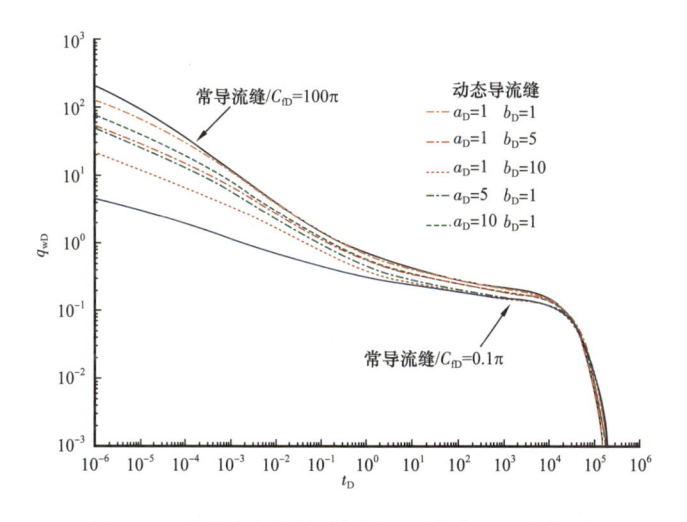

图 7 考虑裂缝应力敏感性影响的气井 q_{wD} 曲线图

如图 8 所示，在气井的整个生产周期内，$a_D=1$、$b_D=1$ 组合下，动态导流缝的 C_{fD} 都维持在较高水平（介于 42.78～314.00），与常导流裂缝（$C_{fD}=100\pi$）影响下气井的无量纲累计产气量（G_{pD}）曲线基本重合，这是由于高导流能力（$C_{fD}>300$）与中导流能力（$30<C_{fD}<300$）裂缝对气井累计产气量的贡献接近；$a_D=1$、$b_D=10$ 与 $a_D=10$、$b_D=1$ 组合下，动态导流缝的导流能力较低，且最后趋近于 0.315，相应 G_{pD} 曲线与常导流裂缝（$C_{fD}=0.1\pi$）影响下的 G_{pD} 曲线接近。

另外，裂缝参数（裂缝条数、缝长、缝宽等）也会对气井生产动态产生影响，尤其增加裂缝条数、缝长能够有效提高气井累计产气量的增长幅度，因此，合理设计裂缝参数以指导页岩气水平井的压裂实践意义重大。

3.2 控压增效潜力分析

气井生产采取控压方式时，对于存在动态导流缝的情况，气井的无量纲压力随着时间增加逐渐偏离存在常导流裂缝情况下气井的无量纲压力；而气井生产采取放压方式时，自开井伊始动态导流缝的导流能力即发生大幅度衰减，产量即偏离常导流缝存在时的气井产量，且定压时间越早，偏离程度越大。当无量纲井底压力较小时，累计产气量也较小，且无论裂缝导流能力为常量还是变量，对气井累计产气量的影响基本一致（图9）。

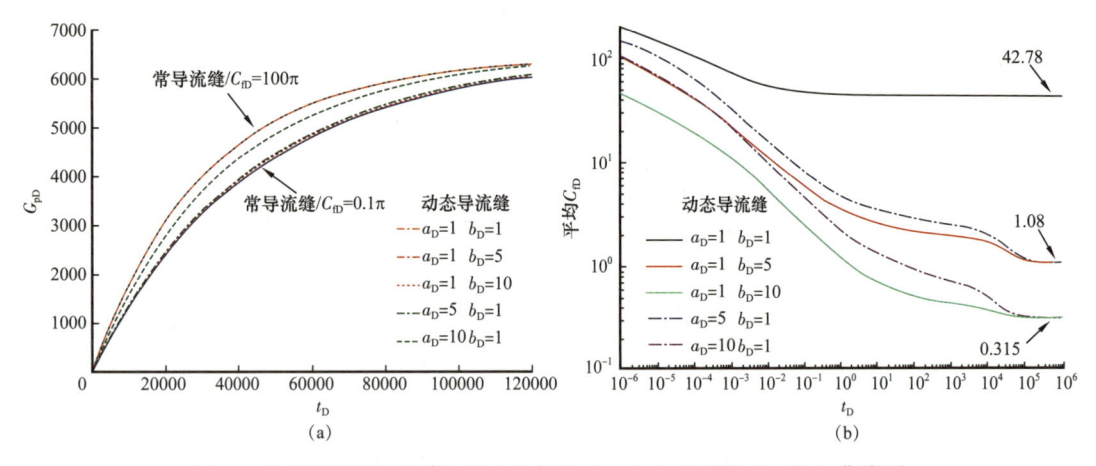

图8 考虑裂缝应力敏感性影响的气井 G_{pD}（a）及平均 C_{fD}（b）曲线图

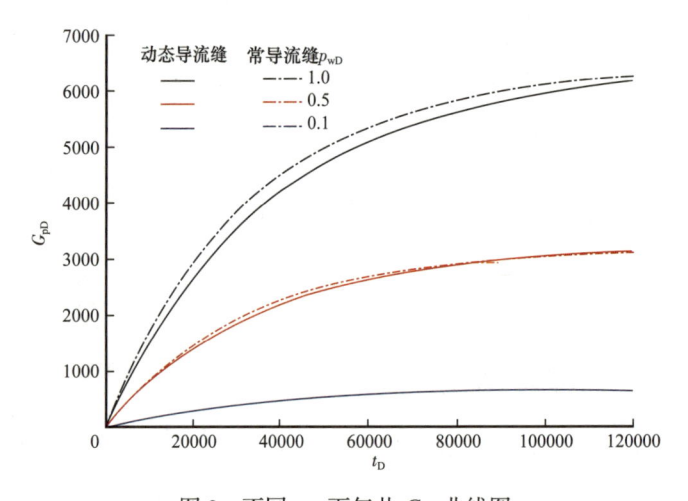

图9 不同 p_{wD} 下气井 G_{pD} 曲线图

因此，制订合理的井底压力，可以控制裂缝的闭合，缓解产量降低的幅度，进而提高气井的累计产气量。本文采用式（31）来表征控压生产中井底压力的变化，当平均衰减寿命取无穷小时，即相当于气井一开井则进入定压阶段。

井底压力与平均衰减寿命的关系式为：

$$p_w = (p_i - p_{wf})\exp\left(-\frac{t}{\tau}\right) + p_{wf} \qquad (34)$$

式中，p_{wf} 为定井底压力值，Pa；τ 为平均衰减寿命，s。

由式（7）定义应力敏感系数，模拟不同 τ_D 下气井的累计产气量，结果显示与放压方式相比，控压方式下的累计产气量虽在生产早期低于放压方式下的累计产气量，但最终的累计产气量更高，该结果也与采用流固耦合模型模拟的结果[4, 15, 17]趋势一致（图10）；同时，使用变应力敏感系数模型计算的累计产气量要低于常应力敏感系数模型计算的结果，这与Mirani等[16]使用流固耦合模型计算的结果也一致，进一步验证了所建立模型的正确性。

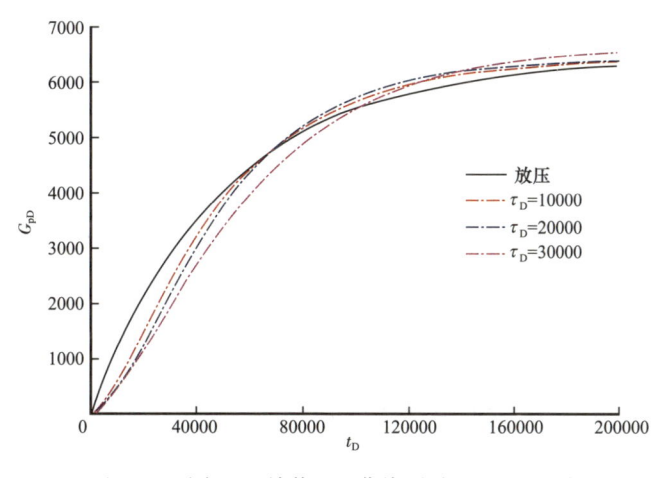

图 10　不同 τ_D 下单井 G_{pD} 曲线图（$a_D=1$、$b_D=1$）

如图 11 所示，在放压方式下，由于裂缝闭合严重，导致裂缝导流能力迅速递减至最小值，而气井一开井产气量即为最高，随后逐渐递减，无稳产期；在控压生产方式下，井底压差逐渐增加至最高值，延缓了裂缝导流能力衰减的速度，气井产气量呈现为逐渐增加至最高值后，再逐渐递减的变化趋势；在某个时间点后控压方式下的气井产气量将超过放压方式下气井的产气量，若控压制度制订合理，气井的生产将出现明显的稳产期。

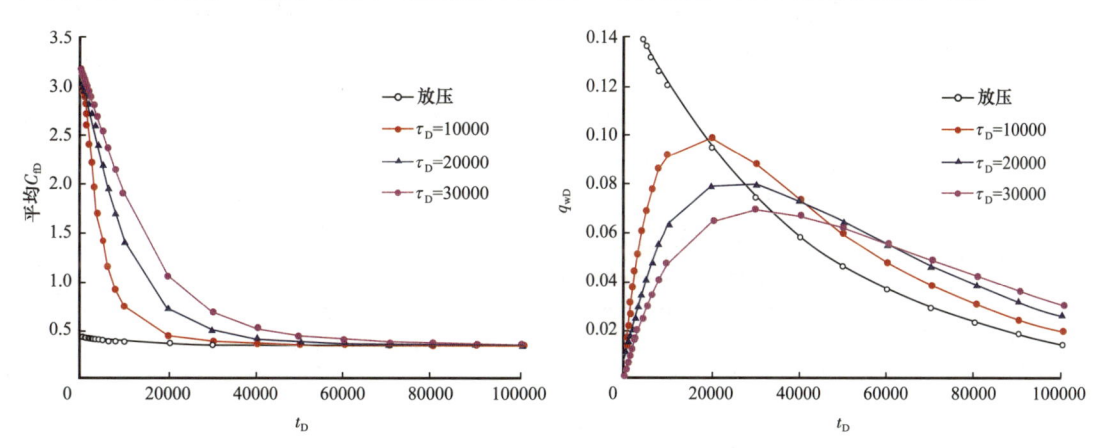

图 11　放压、控压方式下平均 C_{fD}、q_{wD} 曲线图

3.3　实例应用

压裂水平井的开发效果取决于地质条件、工程条件和生产制度等因素。此次选取某建产区地质参数和工程参数相近，但生产制度不同的两口井进行分析，其中 HD 井采用短期控压方式生产，MD 井采用长期控压方式生产。在地质方面，龙马溪组的优质页岩段为龙一$_1$亚段，其中又以龙一$_1^1$和龙一$_1^3$小层为最优，是水平井钻进的最优靶位，如表 1 所示，HD

井钻遇龙一$_1^1$和龙一$_1^3$小层的长度合计为717.7m，而MD井仅钻遇龙一$_1^3$小层513.6m；在工程方面，HD井的压裂段长度和压裂段数虽低于MD井，但其压裂加砂量是MD井加砂量的1.5倍（表2）。总体判断，若无其他因素的影响，HD井的开发效果应优于MD井。

表1　水平井页岩层位钻遇情况统计表

井名	层位	钻遇长度/m	井名	层位	钻遇长度/m
HD井	龙一$_1^4$	403	MD井	龙一$_1^4$	
	龙一$_1^3$	519.5		龙一$_1^3$	531.6
	龙一$_1^2$	389.3		龙一$_1^2$	1073.4
	龙一$_1^1$	198.2		龙一$_1^1$	

表2　水平井工程参数对比表

井名	生产方式	水平段长度/m	压裂段长度/m	压裂段数	加砂量/t
HD井	短期控压	1510	1318	18	1673
MD井	长期控压	1605	1568	22	1101

但从气井的实际生产动态来看，HD井前3个月的压降速率为0.278MPa/d，MD井仅为0.13MPa/d，但MD井前3个月单位压降采气量为64.7×10^{4}m^3/MPa，远高于HD井的37.8×10^{4}m^3/MPa，前者是后者的1.7倍；从目前生产情况来看，HD井的日产气量已低于MD井（图12）。造成此种差异的一个重要原因即在于HD井采用短期控压制度进行生产，而MD井采用长期控压制度进行生产。

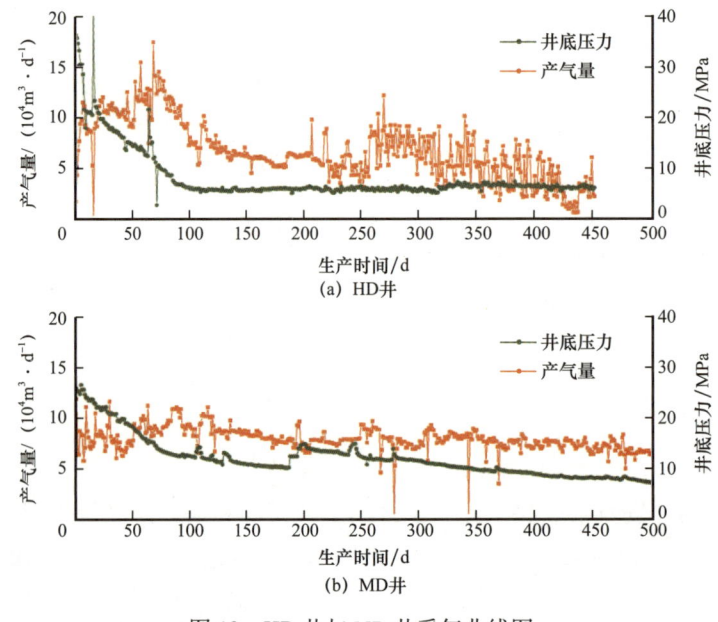

(a) HD井

(b) MD井

图12　HD井与MD井采气曲线图

通过比较单井 EUR 更能体现出生产制度差异带来的影响。使用 Laplace 正演算法将实时空域上井底压力转换为 Laplace 空间值[23]，代入本文模型，对气井产气量进行拟合，获取裂缝动态参数，在此基础上进行产量预测。如图 13 所示，生产时间在 4 年以内 HD 井累计产气量要高于 MD 井，而 5 年以后长期控压生产的优势逐渐显现，20 年末 MD 井的累计产气量为 $0.805 \times 10^8 m^3$，而 HD 井的累计产气量为 $0.636 \times 10^8 m^3$，前者是后者的 1.27 倍。可以看出，合理的控压生产制度可以使气井累计产气量增加 25% 以上。

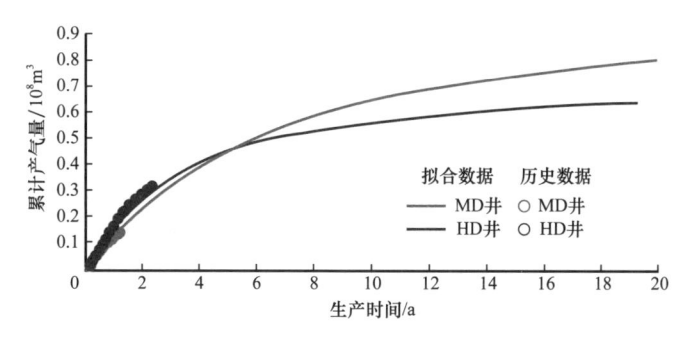

图 13 HD 井与 MD 井累计产气量拟合及预测曲线图

4 结论

（1）所建立的模型与经典模型、Saphir 软件数值模型计算的结果基本吻合；

（2）应力敏感性使得裂缝导流能力随着气井生产而递减，导致气井累计产气量降低，且应力敏感性越强，累计产气量降低的幅度越大；

（3）与放压方式相比，采用控压方式生产的页岩气压裂水平井的初期产气量及累计产气量虽偏低，但最终累计产气量却更高，可见页岩气压裂水平井长期控压生产更具合理性。

参 考 文 献

[1] 位云生, 齐亚东, 贾成业, 等. 四川盆地威远区块典型平台页岩气水平井动态特征及开发建议[J]. 天然气工业, 2019, 39（1）: 81-86.

[2] Eshkalak M O, Aybar U, Sepehrnoori K. Long term effect of natural fractures closure on gas production from unconventional reservoirs [C] // paper 171010-MS presented at the SPE Eastern Regional Meeting, 21-23 October 2014, Charleston, WV, USA. https://doi.org/10.2118/171010-MS.

[3] Wilson K. Analysis of drawdown sensitivity in shale reservoirs using coupled-geomechanics models [C] // paper 175029-MS presented at the SPE Annual Technical Conference and Exhibition, 28-30 September 2015, Houston, Texas, USA. https://doi.org/10.2118/175029-MS.

[4] Clarkson C R, Qanbari F, Nobakht M, et al. Incorporating geomechanical and dynamic hydraulic fracture-property changes into rate-transient analysis: Example from the Haynesville Shale [J]. SPE Reservoir Evaluation & Engineering, 2013, 16（3）: 86-95.

[5] Britt L K, Smith M B, Klein H H, et al. Production benefits form complexity-effects of rock fabric,

managed drawdown, and propped fracture conductivity［C］//Paper 179159-MS presented at the SPE Hydraulic Fracturing Technology Conference, 9-11 February 2016, The Woodlands, Texas, USA. https：//doi.org/10.2118/179159-MS.

［6］Fan L, Thompson J W, Robinson J R. Understanding gas production mechanism and effectiveness of well stimulation in the Haynesville shale through reservoir simulation［C］//Paper 136696-MS presented at the Canadian Unconventional Resources and International Petroleum Conference, 19-21 October 2010, Calgary, Alberta, Canada. https：//doi.org/10.2118/136696-MS.

［7］郭小哲，李景，张欣. 页岩气藏压裂水平井物质平衡模型的建立［J］. 西南石油大学学报（自然科学版），2017，39（2）：132-138.

［8］Mangha V O, Guillot F, San V, et al. Estimated ultimate recovery（EUR）as a function of production practices in the Haynesville shale［C］//Paper 147623-MS presented at the SPE Annual Technical Conference and Exhibition, 30 October-2 November 2011, Denver, Colorado, USA. https：//doi.org/10.2118/147623-MS.

［9］贾爱林，位云生，金亦秋. 中国海相页岩气开发评价关键技术进展［J］. 石油勘探与开发，2016，43（6）：1-13.

［10］王志刚. 涪陵页岩气勘探开发重大突破与启示［J］. 石油与天然气地质，2015，36（1）：1-6.

［11］Akande J A, Spivey J. Considerations for pore volume stress effects in over-pressured shale gas under controlled drawdown well management strategy［C］//paper 162666-MS presented at the SPE Canadian Unconventional Resources Conference, 30 October-1 November, Calgary, Alberta, Canada. https：//doi.org/10.2118/162666-MS.

［12］Chen Z M, Liao X W, Zhao X L, et al. Performance of horizontal wells with fracture networks in shale gas formation［J］. Journal of Petroleum Science and Engineering, 2015, 133: 646-664.

［13］欧阳伟平，孙贺东，张冕. 考虑应力敏感的致密气多段压裂水平井试井分析［J］. 石油学报，2018，39（5）：570-577.

［14］刘洪，廖如刚，李小斌，等. 页岩气"井工厂"不同压裂模式下裂缝复杂程度研究［J］. 天然气工业，2018，38（12）：70-76.

［15］El-Din A S, Elbanbi A H, Abdelwaly A A. A fully coupled geomechanics and fluid flow model for well performance modeling in stress-dependent gas reservoirs［C］//paper 143439-MS presented at the SPE EUROPEC/EAGE Annual Conference and Exhibition, 23-26 May 2011, Vienna, Austria. https：//doi.org/10.2118/143439-MS.

［16］Mirani A, Marongiu-Porcu M, Wang H Y, et al. Production pressure drawdown management for fractured horizontal wells in shale gas formations［C］//paper 181365-MS presented at the SPE Annual Technical Conference and Exhibition, 26-28 September, Dubai, UAE. https：//doi.org/10.2118/181365-MS.

［17］Wilson K, Hanna Alla RR. Efficient stress characterization for real-time drawdown managment［C］//paper 2721192-MS presented at the SPE/AAPG/SEG Unconventional Resources Technology Conference, 24-26 July 2017, Austin, Texas, USA. https：//doi.org/10.15530/URTEC-2017-2721192.

［18］Aybar U, Eshkalak M O, Sepehrnoori K, et al. Long term effect of natural fractures closure on gas production from unconventional reservoirs [C] //paper 171010-MS presented at the SPE Eastern Regional Meeting, 21-23 October, Charleston, WV, USA. https : //doi.org/10.2118/171010-MS.

[19] Berumen S. Evaluation of fractured wells in pressure-sensitive formations [D]. Tulsa : University of Oklahoma, 1995.

[20] Alramahi B, Sundberg M I. Proppant embedment and conductivity of hydraulic fractures in shales [C] // paper ARMA-2012-291 presented at the 46th U.S. Rock Mechanics/Geomechanics Symposium, 24-27 June 2012, Chicago, Illinois.

[21] Cinco-L H, Samaniego V F, Dominguez A N. Transient pressure behavior for a well with a finite-conductivity vertical fracture [J]. Society of Petroleum Engineers Journal, 1978, 18 (4): 253-264.

[22] Berumen S, Tiab D, Rodriguez F. Constant rate solutions for a fractured well with an asymmetric fracture [J]. Journal of Petroleum Science and Engineering, 2000, 25: 49-58.

[23] Roumboutsos A, Stewart G. A direct deconvolution or convolution algorithm for well test analysis [C] // paper 18157-MS presented at the SPE Annual Technical Conference and Exhibition, 2-5 October 1988, Houston, Texas, USA. https : //doi.org/10.2118/18157-MS.

Transient pressure analysis for multi-fractured horizontal well with the use of multi-linear flow model in shale gas reservoir

Wang Guangdong[1], Jia Ailin[1], Wei Yunsheng[1], Xiao Cong[2]

(1 Research Institute of Petroleum Exploration and Development;
2 Delft Institute of Applied Mathematics, Delft University of Technology)

Abstract: Shale gas reservoirs (SGR) has been a central supply of carbon hydrogen energy consumption and hence widely produced with the assistance of advanced hydraulic fracturing technologies. On the one hand, due to the inherent ultra-low permeability and porosity, there is stress sensitivity in the reservoirs generally. On the other hand, hydraulic fractures and the stimulated reservoir volume (SRV) generated by the massive hydraulic fracturing operation have contrast properties with the original reservoirs. These two phenomena pose huge challenges in SGR transient pressure analysis.

Limited works have been done to take the stress sensitivity and spatially varying permeability of SRV zone into consideration simultaneously. This paper first idealizes the SGR to be four linear composite regions. What's more, SRV zone is further divided into sub-sections on the basis of non-uniform distribution of proppant within SRV zone which easily yields spatially varying permeability away from the main hydraulic fracture. By means of perturbation transformation and Laplace transformation, an analytical multi-linear flow model (MLFM) is obtained and validated as a comparison with the previous models. The flow regimes are identified and the sensitivity analysis of critical parameters are conducted to further understand the transient pressure behaviors. The research results provided by this work are of significance for an effective recovery of SGR resources..

Key words: shale gas reservoirs; stress sensitivity; multi-fractured horizontal well; spatially varying permeability; pressure transient analysis

Due to extremely low permeability and porosity of shale gas reservoir (SGR), multistage hydraulic fracturing has become an integral tool to improve the gas recovery. The economic feasibility of shale gas reservoirs has a strong relationship with the fracture system permeability near the wellbore. Considered to be the most effective way to produce gas resources, multistage fractured horizontal well can create several high-conductivity hydraulic fractures as flow paths, at the same time, activate and connect existing natural fractures so as to develop large fracture network system[1].

The zone containing the main high-conductivity hydraulic fractures and large spatial network system which can effectively improve well performance is defined as SRV (stimulated reservoir volume), and the remaining zone which hardly influenced by the treatment of hydraulic fracturing is similarly defined as USRV (un-stimulated reservoir volume)[2].

The presence of a complex fracture network in the SRV has a significant impact on the pressure transient analysis of unconventional reservoirs. Analytical and semi-analytical approaches have been used to model the transient flow behavior in such systems. Zhao et al[3-4] and Wang[5] have established semi-analytical solutions with the use of Laplace transformation. The point source function or line source function, coupled with superposition principle, was utilized to mathematically incorporate the interference among hydraulic fractures. Alternatively, multi-Linear flow modes have been extensively developed to simulate the gas flow in SGR. The SGR is generally divided into some coupled zones and linear flow patterns are assumed. These models also assumed that continuous pressure drops along the hydraulic fractures exist to push the hydrogen gas to the wellbore. El-Banbi[6], Hasan and Al-Ahmadi[7], Xu et al[8], Ekaterina Stalgorova[9] simplified SGR as linear composition reservoirs and the governing equations of each zone can be derived.

As we all know, after the hydraulic fracturing stimulation of shale gas reservoirs, the induced fracture and frac-formation can be sub-divided into several categories based on the fracturing pattern and proppant distribution. Therefore, it is too idealistic to simply assume that the induced hydraulic fracture is bi-wing. In order to concisely describe the fracture network (natural or induced) around the hydraulic fractures, some aforementioned methods (including analytical methods, semi-analytical methods and even time-consumption numerical methods) are proposed. With assuming uniform distribution of identical hydraulic fractures along the length of the horizontal well, Ozkan et al.[2, 10] utilizes the concept of tri-linear model with inner reservoir of naturally fractured to represent the MFH well performance in unconventional reservoirs. Brown et al.[11] presented an analytical tri-linear flow model that incorporates transient fluid transfer from matrix to fracture to simulate the pressure transient and production behavior of fractured horizontal wells in unconventional reservoirs. However, those proposed analytical models are lack of the ability to explicityly consider the medium conductivity secondary fracture, which absolutely induces certan errous to some degree. And then, Zeng et al[12], Ekaterina et al[9], Wang et al[13] used a unstructural-grid simulator to analyze the type curves of an MFHW with fracture complexity, the generation of complex unstructured grid is time consuming, moreover, the adjacency of the high conductivity fractures is refined with fine grids, it is necessary to increase the complication and economical consumption of computation while this latest technology can make the simulation of fracture complexity more accurate.

Aforementioned analytical models significantly simplify the geometry of complex fracture system, e.g., regular orthogonal hydraulic fractures, and stress sensitivity phenomenon. To resolve these shortages, a plenty of semi-analytical models without above simplifications have been established. Zhao et al.[4] proposed a radial compound model which treated the SRV as a circular area with high permeability. This model was used to simulate the performance of multi-fractured horizontal wells in SGR. According to source functions, Jiang et al.[13] analyzed pressure and gas rate transient laws of multistage fractured horizontal well for tight oil reservoirs with considering SRV, although the fractures were still confined to be vertical to the wellbore and the stress sensitivity was also ignored in their study. Ren et al.[14-15] amalgamated the perturbation technique with linear source function method to consider the effect of stress sensitivity when they researched transient pressure behaviors of horizontal wells. However, the complexity of fracture networks was still neglected. Jia et al.[16] proposed a new semi-analytical model through combining finite-difference and line source function to study the flow behaviors of horizontal well after hydraulic fracturing, unfortunately, the stress sensitivity effect was still ignored. Given the stress sensitivity effect of fractures and reservoirs, Wang et al.[17] established a semi-analytical model suitable to study transient flow behaviors of multistage fractured horizontal well with complex fracture networks without considering the property contrast between the stimulated zone and un-stimulated zone.

Recently, some works tend to consider the well interference using semi-analytical models, such as Xiao et al.[18] and Jia et al.[19].

Although semi-analytical radial flow model could be employed to characterize hydraulic fractures with complex topology, the multi-linear flow models are easy to be used in the real-field application. By means of perturbation transformation and Laplace transformation, we proposed a new analytical multi-linear flow model to systematically investigate the effects of stress sensitivity in SGR. In addition, the SRV zone is further divided into sub-sections on the basis of non-uniform distribution of proppant within SRV zone which easily yields spatially varying permeability away from the main hydraulic fracture. This paper also discusses the influence of relevant parameters on the of fractured horizontal wells in stress sensitive SGR, including stress sensitivity, mobility ratio of the SRV and the outer region, SRV size, coefficient of permeability variation. Corresponding solutions can be useful for fracturing design and well test interpretation in field practice.

1 Mathematical model and analytical solution

1.1 Model descriptions

The micro-seismic data maps show [Fig.1(a)] that hydraulic fracturing treatments create irregular fracture geometry. A simplified physical model of multi-fractured horizontal well in SGR is illustrated in Fig.1b. The entire reservoir after hydraulic fracturing can be conceptually divided into four coupled linear flow zones with different properties, including main hydraulic fracture zone Ⅰ, SRV zone Ⅱ, SRV zone Ⅲ and the other outer zone without stimulation zone Ⅳ. The fracture half-length is x_f, the width and length of entire zone are x_e and y_e, respectively. The length of SRV zone Ⅱ is l. All these variables have been depicted in Fig.1a. The initial permeability of these four zones are separately denoted as k_1, k_2, k_3 and k_4.

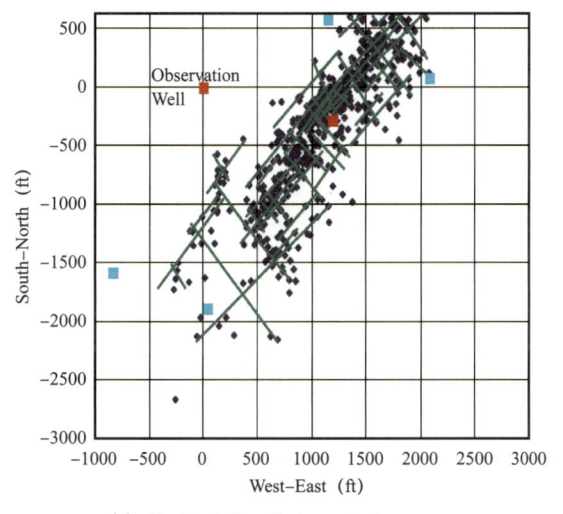

(a) The illustration of micro-seismic response

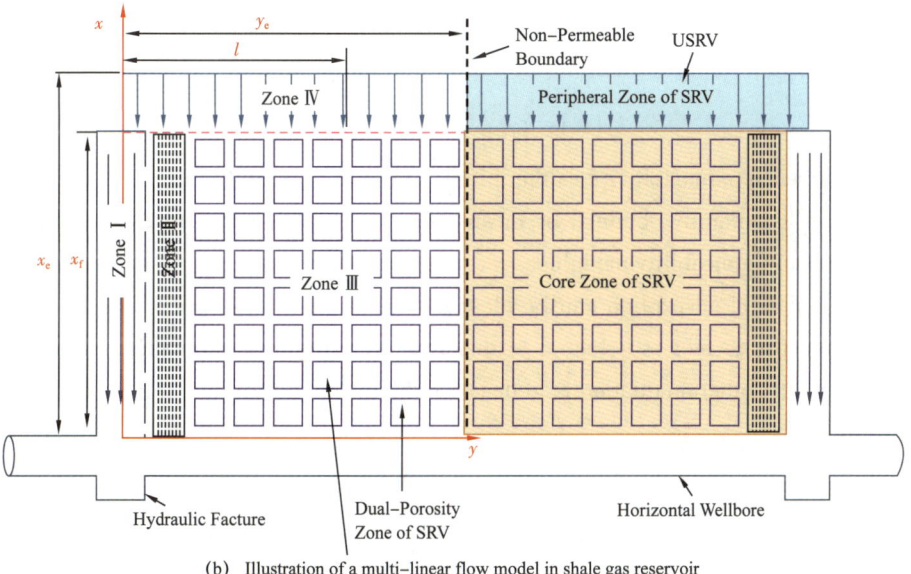

(b) Illustration of a multi-linear flow model in shale gas reservoir

(c) Illustration of a further subdivision of Zone II

Fig.1 Physical model of multi-fractured horizontal well in shale gas reservoir

As far as we known, the non-uniform distribution of proppant within SRV zone easily yields spatially varying permeability away from the main hydraulic fracture. In this paper, we further

− 127 −

divide Zone Ⅲ into N small zones as illustrated in Fig.1（b）. All zones are idealized as dual-porosity media with different permeability. The stress sensitivity of permeability is taken into consideration. Single-phase and micro-compressible gas is assumed to flow in SGR, which follows the Darcy's Law.

1.2 Mathematical Formula of Multi-Linear Model

To begin with, we define the pseudo-pressure as follows：

$$m(p) = \frac{(\mu_g Z_g)_i}{p_i} \int_{p_{ref}}^{p} \frac{p}{\mu_g Z_g} dp \qquad (1)$$

And then

$$m_{nD} = \frac{2\pi k_{ref} h [m(p_i) - m(p_n)]}{q_{gsc} \mu_{gi} B_{gi}} \qquad (2)$$

According to the theory proposed by Pedrosa. O.A., stress-sensitive permeability could be described as follows

$$k_i(m) = k_0 e^{-\gamma_i(m_0 - m_i)} \quad i = 1,2,3,4 \qquad (3)$$

Where, the reference permeability k_0 is the initial permeability in SGR, m^2; γ_i is the permeability modulus, which is generally obtained by means of indoor experiments. We also assume that four zones have different permeability modulus.

The mechanism of adsorption can be classified into instant adsorption and time-dependent adsorption, and the former is selected to describe adsorption phenomenon in this paper. At present, Langmuir isotherm equation is to describe the instant adsorption process of shale gas[20], and its expression is：

$$V_E = V_L \frac{m}{m_L + m} \qquad (4)$$

Where, V_E is the adsorption equilibrium concentration, sm^3/m^3; V_L is the Langmuir adsorption concentration, sm^3/m^3; P_L is the Langmuir pressure, MPa; m_L is Langmuir pressure, MPa^2/（mPa·s）.

For the convenience and simplicity of formulas deducing, some dimensionless parameters are introduced first.

$$m_i = \frac{k_{ref} h [m_0 - m_i]}{1.842 q_{gsc} \mu_{gi} B_{gi}}, i = 1,2,3,4 \qquad t_D = \frac{3.6 \times 10^{-3} k_{ref} t}{\mu_{gi} (\phi C_g)_{ref} x_f^2}$$

$$x_D = \frac{x}{x_f}, y_D = \frac{y}{x_f}, l_{j,D} = \frac{l_j}{x_f}, w_{fD} = \frac{w_f}{x_f}, \gamma_{iD} = \frac{1.842 q_{sc} \mu_g B_g}{k_{ref} h} \gamma_i$$

$$C_{\mathrm{fD}} = \frac{k_1 w_{\mathrm{f}}}{k_{\mathrm{ref}} x_{\mathrm{f}}}, k_{i\mathrm{D}} = \frac{k_i}{k_{\mathrm{ref}}}, \omega_{i\mathrm{D}} = \frac{\phi_i C_{gi}}{(\phi C_g)_{\mathrm{ref}}}$$

$$\omega_{3m\mathrm{D}} = \frac{\phi_{3m} C_{gi}}{(\phi C_g)_{\mathrm{ref}}}, \lambda = \sigma \frac{k_{3m}}{k_{\mathrm{ref}}} L_{\mathrm{ref}}^2$$

$$m_{\mathrm{DL}} = \frac{k_{\mathrm{ref}} h(m_{\mathrm{L}} - m_i)}{1.842 q_{\mathrm{gsc}} \mu_{gi} B_{gi}}, V_{\mathrm{DL}} = 2 \frac{p_{\mathrm{sc}} T}{\mu_i T_{\mathrm{SC}}} \frac{k_{\mathrm{ref}} h[m_{\mathrm{L}} - m_i] \phi_{3m} V_{\mathrm{L}}}{1.842 q_{\mathrm{gsc}} \mu_{gi} B_{gi}}$$

A multi-linear flow model (MLFM) will be used to derive the mathematical equations. In the following parts, the governing equations are separately established for each zone.

1.2.1 Zone I

In this zone, the gas supplies from the adjacent reservoir Zone II to the main hydraulic fractures, and then flow to wellbore with a linear flow pattern. Gas pseudo-pressure is used to consider the effects of gas compressibility. When the stress sensitivity is considered, the governing function can be presented as follows,

$$\frac{\partial^2 m_1}{\partial x^2} + \gamma_1 \left(\frac{\partial m_1}{\partial x} \right)^2 + \frac{2}{w_{\mathrm{f}}} \frac{k_{2,1}}{k_1} \frac{\partial m_{2,1}}{\partial y} \bigg|_{y=\frac{w_{\mathrm{f}}}{2}} = e^{\lambda_1 (m_0 - m_1)} \frac{\phi_1 \mu C_g}{k_1} \frac{\partial m_1}{\partial t} \tag{5}$$

With the inner and outer boundary conditions:

$$e^{-\gamma_1 (m_0 - m_1)} \frac{\partial m_1}{\partial x} \bigg|_{x=0} = \frac{q_{\mathrm{gsc}} \mu_{gi} B_{gi}}{2 w_{\mathrm{f}} h k_1}, \frac{\partial m_1}{\partial x} \bigg|_{x=x_{\mathrm{f}}} = 0 \tag{6}$$

Substituting predefined dimensionless variables to Eq. (5) and Eq. (6), their dimensionless formula is as follows.

$$\frac{\partial^2 m_{1\mathrm{D}}}{\partial x_{\mathrm{D}}^2} + \gamma_{1\mathrm{D}} \left(\frac{\partial m_{1\mathrm{D}}}{\partial x_{\mathrm{D}}} \right)^2 + \frac{2 k_{2,1\mathrm{D}}}{C_{1\mathrm{D}}} \frac{\partial m_{2,1\mathrm{D}}}{\partial y_{\mathrm{D}}} \bigg|_{y_{\mathrm{D}} = \frac{w_{f\mathrm{D}}}{2}} = e^{\gamma_{1\mathrm{D}} m_{1\mathrm{D}}} \frac{\omega_{1\mathrm{D}}}{k_{1\mathrm{D}}} \frac{\partial m_{1\mathrm{D}}}{\partial t_{\mathrm{D}}} \tag{7}$$

With $e^{-\gamma_{1\mathrm{D}} m_{1\mathrm{D}}} \dfrac{\partial m_{1\mathrm{D}}}{\partial x_{\mathrm{D}}} \bigg|_{x_{\mathrm{D}} = 0} = -\dfrac{\pi}{C_{1\mathrm{D}}}$, $\dfrac{\partial m_{1\mathrm{D}}}{\partial x_{\mathrm{D}}} \bigg|_{x_{\mathrm{D}} = 1} = 0$

Eq. (7) is strongly nonlinear partial differential equation, which are not convenient to be solve analytically. A perturbation transformation proposed by Pedrosa O.A. can be used to eliminate the non-linearity. The new dimensionless variables $\eta_{j\mathrm{D}}$ related to the dimensionless pressure are introduced as follows:

$$m_{i\mathrm{D}} = -\frac{1}{\gamma_{i\mathrm{D}}} \ln(1 - \gamma_{i\mathrm{D}} \eta_{i\mathrm{D}}), i = 1, 2, 3, 4 \tag{8}$$

Due to the dimensionless permeability modulus γ_{D} is usually a small value, $\eta_{j\mathrm{D}}$ can be expanded as a power series in the parameter γ_{D}.

$$\eta_{j\mathrm{D}} = \eta_{j\mathrm{D}0} + \gamma_{\mathrm{D}} \eta_{j\mathrm{D}1} + (\gamma_{\mathrm{D}})^2 \eta_{j\mathrm{D}2} + (\gamma_{\mathrm{D}})^3 \eta_{j\mathrm{D}3} + \ldots \tag{9}$$

$$\frac{1}{1 - \gamma_{\mathrm{D}} \eta_{j\mathrm{D}}} = 1 + \gamma_{\mathrm{D}} \eta_{j\mathrm{D}} + (\gamma_{\mathrm{D}})^2 (\eta_{j\mathrm{D}})^2 + (\gamma_{\mathrm{D}})^3 (\eta_{j\mathrm{D}})^3 + \ldots \tag{10}$$

Substituting Eq. (9) and (10) into Eq. (7), we can get a sequence of linear problem that can be solved for η_{jD0}, η_{jD1}, and so on. According to Liu et al [21] the zero-order approximation η_{jD0} was accurate enough for pressure analysis.

$$\frac{\partial^2 \eta_{1D}}{\partial x_D^2} + \frac{2k_{2,1D}}{C_{1D}} \frac{\partial \eta_{2,1D}}{\partial y_D}\bigg|_{y_D=\frac{w_{fD}}{2}} = \frac{\omega_{1D}}{k_{1D}} \frac{\partial \eta_{1D}}{\partial t_D} \quad (11)$$

With $\dfrac{\partial \eta_{1D}}{\partial x_D}\bigg|_{x_D=0} = -\dfrac{\pi}{C_{1D}}$, $\dfrac{\partial \eta_{1D}}{\partial x_D}\bigg|_{x_D=1} = 0$

Finally, Laplace transformation can be used to transform these equations from time domain to Laplace domain so as to eliminate the effects of time domain.

$$\frac{d^2 \overline{\eta_{1D}}}{dx_D^2} + \frac{2k_{2,1D}}{C_{1D}} \frac{d\overline{\eta_{2,1D}}}{dy_D}\bigg|_{y_D=\frac{w_{fD}}{2}} = s\frac{\omega_{1D}}{k_{1D}} \overline{\eta_{1D}} \quad (12)$$

With $\dfrac{d\overline{\eta_{1D}}}{dx_D}\bigg|_{x_D=0} = -\dfrac{1}{s}\dfrac{\pi}{C_{1D}}$, and $\dfrac{d\overline{\eta_{1D}}}{dx_D}\bigg|_{x_D=1} = 0$

1.2.2 Zone II

Zone II is a highly permeable SVR zone which is adjacent to the main hydraulic fracture as illustrated in Fig.2 (a). This area is significantly stimulated by the massive hydraulic fracturing operation, as a result, it can be assumed to single porous media with high permeability due to the support by the proppant. When the stress sensitivity is considered, the governing function can be presented as follows,

$$\frac{\partial^2 m_2}{\partial y^2} + \gamma_2 \left(\frac{\partial m_{m_2}}{\partial y}\right)^2 + \frac{k_4}{k_2}\frac{\partial m_4}{\partial y}\bigg|_{x=\frac{x_f}{2}} = e^{\gamma_2(m_0 - m_2)}\frac{\phi_2 \mu C_g}{k_2}\frac{\partial m_2}{\partial t}, \frac{w_f}{2} < x < l_1 \quad (13)$$

The left condition is adjacent to Zone I, while the right condition is connected to Zone III. Specifically speaking, these two conditions can be presented.

$$m_2\big|_{y=w_f/2} = m_1\big|_{y=w_f/2}, \quad m_2\big|_{y=l_1} = m_3\big|_{y=l_1} \quad (14)$$

Their dimensionless formulas with the consideration of stress sensitivity effects can be as follows

$$\frac{\partial^2 \eta_{2D}}{\partial y_D^2} + \frac{k_{4D}}{k_{2D}}\frac{\partial \eta_{4D}}{\partial x_D}\bigg|_{x_D=1} = \frac{\omega_{2D}}{k_{2D}}\frac{\partial \eta_{2D}}{\partial t_D}, w_{fD}/2 < y_D < l_{1,D} \quad (15)$$

With the boundary conditions:

$$\gamma_{1D}\eta_{1D}\big|_{y=w_{fD}/2} = \gamma_{2D}\eta_2\big|_{y=w_{fD}/2} \quad \gamma_{2D}\eta_{2D}\big|_{y=l_{1,D}} = \gamma_{3,1D}\eta_{3,1D}\big|_{y=l_{1,D}} \quad (16)$$

Similarly, Laplace transformation can be used to transform these equations from time domain to Laplace domain.

$$\frac{d^2\overline{\eta_{2D}}}{dy_D^2} + \frac{k_{4D}}{k_{2D}}\frac{d\overline{\eta_{4D}}}{dx_D}\bigg|_{x_D=1} = s\frac{\omega_{2D}}{k_{2D}}\overline{\eta_{2D}}, w_{fD}/2 < y_D < l_{1,D} \quad (17)$$

With the boundary conditions:

$$\gamma_{1D}\overline{\eta_{1D}}\Big|_{y=w_{fD}/2}=\gamma_{2D}\overline{\eta_{2D}}\Big|_{y=w_{fD}/2} \quad \gamma_{2D}\overline{\eta_{2D}}\Big|_{y=l_{1,D}}=\gamma_{3,1D}\overline{\eta_{3,1D}}\Big|_{y=l_{1,D}} \quad (18)$$

1.2.3 Zone III

In this work, Zone III is also assumed to be a stimulated area induced by massive hydraulic fracturing operation. Unlike Zone II, this zone is partially supported by the proppant. In addition, the pre-existing natural fractures are stimulated as well. On condition to these facts, Zone III is idealized as a dual-porosity media where the fracture and matrix are coupled. As we have mentioned above, to systematically investigate the influences of spatially varying permeability away from the main hydraulic fracture, we have divided Zone III into N small zones as illustrated in Fig.2(b). We need to establish flow equations for each small zones and coupled them together on the basis of continuity condition at the interface of neighboring zones. Here, we take the j-th small zone as an example to derive a set of generic governing equations. We also should separately derive the governing equations for fracture system and matrix system. The effects of adsorption and pseudo-steady state inter-porosity flow from matrix to micro fracture system are considered simultaneously. When the stress sensitivity is considered, the governing function can be presented as follows

Fracture system:

$$\frac{\partial^2 m_{3,j}}{\partial y^2}+\gamma_{3,j}\left(\frac{\partial m_{m_{3,j}}}{\partial y}\right)^2+\frac{k_4}{k_{3,j}}\frac{\partial m_4}{\partial y}\bigg|_{x=x_f}+\sigma\frac{k_{3m}}{k_{3,j}}\left(m_{3m}-m_{3,j}\right)=e^{\gamma_{3,j}(m_0-m_{3,j})}\frac{\phi_{3,j}\mu C_g}{k_{3,j}}\frac{\partial m_{3,j}}{\partial t}, l_j<x<l_{j+1} \quad (19)$$

Matrix system:

$$-\sigma k_{3m}\left(m_{3m}-m_{3,j}\right)=\phi_{3m}\mu C_g\frac{\partial m_{3m}}{\partial t}+\frac{p_{sc}T}{T_{sc}}\frac{\partial V}{\partial t} \quad (20)$$

The boundary conditions are derived from the interface of neighboring two sections, where the pressure and gas flux should be strictly equal between those two sections.

$$m_{3,j}\Big|_{y=l_j}=m_{3,j+1}\Big|_{y=l_j}, k_{3,j}e^{-\gamma_{3,j}(m_0-m_{3,j})}\frac{\partial m_{3,j}}{\partial y}\bigg|_{y=l_j}=k_{3,j+1}e^{-\gamma_{3,j+1}(m_0-m_{3,j+1})}\frac{\partial m_{3,j+1}}{\partial y}\bigg|_{y=l_j} \quad j=2,\dots,N-1 \quad (21)$$

Especially, the first small zone will directly connect to the Zone II, and the right side of N-th small zone is considered to be impermeable. The boundary conditions could be explicitly presented as follows,

$$m_2\Big|_{y=l1}=m_{3,1}\Big|_{y=l1}, k_2 e^{-\gamma_2(m_0-m_2)}\frac{\partial m_2}{\partial y}\bigg|_{y=l_1}=k_{3,1}e^{-\gamma_{3,1}(m_0-m_{3,1})}\frac{\partial m_{3,1}}{\partial y}\bigg|_{y=l_1}, \frac{\partial m_{3,N}}{\partial y}\bigg|_{y=y_e}=0 \quad (22)$$

Obtained from Fick's first law of diffusion, the diffusion rate of shale gas can be expressed as:

$$\frac{\partial V}{\partial t}=-\phi_{3m}\frac{\partial V_E}{\partial t} \quad (23)$$

Substituting Langmuir isotherm equation Eqs. (4) into Eqs. (23), we can obtain:

$$\frac{\partial V}{\partial t} = -\phi_{3m} V_L \frac{m_L}{(m_L + m_{3m})^2} \frac{\partial m_{3m}}{\partial t} \quad (24)$$

Substituting predefined dimensionless variables to Eq. (19) ~ (24) could obtain their dimensionless formulas, and then aforementioned perturbation transformation proposed by Pedrosa O.A is used to eliminate the non-linearity. Finally, their dimensionless formulas with the consideration of stress sensitivity effects can be as follows

Fracture system:

$$\frac{\partial^2 \eta_{3,jD}}{\partial y_D^2} + \frac{k_{4D}}{k_{3,jD}} \frac{\partial m_{4D}}{\partial y_D}\bigg|_{x_D=1} + \frac{\lambda}{k_{3,jD}} (\eta_{3mD} - \eta_{3,jD}) = \frac{\omega_{3,jD}}{k_{3,jD}} \frac{\partial \eta_{3,jD}}{\partial t_D}, l_{j,D} < y_D < l_{j+1,D} \quad (25)$$

Matrix system:

$$-\lambda (\eta_{3mD} - \eta_{3,jD}) = \omega_{3mD} \frac{\partial \eta_{3mD}}{\partial t_D} + V_{DL} \frac{m_{DL}}{(m_{DL} + m_{Di} - m_{3mD})^2} \frac{\partial \eta_{3mD}}{\partial t_D} \quad (26)$$

With the boundary conditions:

$$\gamma_{3,jD} \eta_{3,jD}\bigg|_{x_D=l_{j,D}} = \gamma_{3,j+1D} \eta_{3,j+1D}\bigg|_{x_D=l_{j,D}}, \frac{\partial \eta_{3,jD}}{\partial y_D}\bigg|_{y_D=l_{j,D}} = \frac{k_{3,j+1D}}{k_{3,jD}} \frac{\partial \eta_{3,j+1D}}{\partial y_D}\bigg|_{y_D=l_{j,D}}, j = 2, ..., N-1$$

$$\gamma_{3,1D} \eta_{3,1D} = \gamma_{2D} \eta_{2D}, \frac{\partial \eta_{3,1D}}{\partial y_D}\bigg|_{y_D=l_{1,D}} = \frac{k_{2D}}{k_{3,1D}} \frac{\partial \eta_{2D}}{\partial y_D}\bigg|_{y_D=l_{2,D}}, \frac{\partial \eta_{3,ND}}{\partial y_D}\bigg|_{y_D=l_{N,D}} = 0$$

$$(27)$$

Similarly, Laplace transformation can be used to transform these equations as follows:

Fracture system:

$$\frac{d^2 \overline{\eta_{3,jD}}}{dy_D^2} + \frac{k_{4D}}{k_{3,jD}} \frac{d \overline{m_{4D}}}{dy_D}\bigg|_{x_D=1} + \frac{\lambda}{k_{3,jD}} (\overline{\eta_{3mD}} - \overline{\eta_{3,jD}}) = \frac{\omega_{3,jD}}{k_{3,jD}} s \overline{\eta_{3,jD}}, l_{j,D} < y_D < l_{j+1,D} \quad (28)$$

Matrix system:

$$-\lambda (\overline{\eta_{3mD}} - \overline{\eta_{3,jD}}) = \omega_{3mD} s \overline{\eta_{3mD}} \quad (29)$$

With the boundary conditions:

$$\gamma_{3,jD} \overline{\eta_{3,jD}}\bigg|_{x_D=l_{j,D}} = \gamma_{3,j+1D} \overline{\eta_{3,j+1D}}\bigg|_{x_D=l_{j,D}}, \frac{d \overline{\eta_{3,jD}}}{dy_D}\bigg|_{y_D=l_{j,D}} = \frac{k_{3,j+1D}}{k_{3,jD}} \frac{d \overline{\eta_{3,j+1D}}}{dy_D}\bigg|_{y_D=l_{j,D}} \quad j = 2,...,N-1$$

$$\gamma_{3,1D} \overline{\eta_{3,1D}} = \gamma_{2D} \overline{\eta_{2D}}, \frac{d \overline{\eta_{3,1D}}}{dy_D}\bigg|_{y_D=l_{1,D}} = \frac{k_{2D}}{k_{3,1D}} \frac{d \overline{\eta_{2D}}}{dy_D}\bigg|_{y_D=l_{2,D}}, \frac{d \overline{\eta_{3,ND}}}{dy_D}\bigg|_{y_D=l_{N,D}} = 0$$

$$(30)$$

1.2.4 Zone IV

Zone IV is assumed to be an USVR area without any fracturing stimulation as illustrated in Fig.2 (a). As we have shown in the previous process of model derivations, Zone IV simultaneously supplies gas to Zone II and Zone III. When the stress sensitivity is considered, the governing function can be presented as follows,

$$\frac{\partial^2 m_4}{\partial x^2} + \gamma_4 \left(\frac{\partial m_{m_4}}{\partial x}\right)^2 = e^{\gamma_4(m_0-m_4)} \frac{\phi_4 \mu C_g}{k_4} \frac{\partial m_4}{\partial t}, x_f < x < x_e \quad (31)$$

The upper condition is assumed to be impermeable

$$\left.\frac{\partial m_4}{\partial x}\right|_{x=x_e} = 0 \quad (32)$$

The lower boundary is adjacent to Zone II and Zone III along with the y direction. Specifically, this lower boundary condition can be presented.

$$m_4\big|_{x=x_f} = \begin{cases} m_2\big|_{x=x_f}, & w_f/2 < y < l_1 \\ m_{3,j}\big|_{x=x_f}, & l_{j-1} < y < l_j \end{cases} \quad (33)$$

Their dimensionless formulas with the consideration of stress sensitivity effects can be as follows

$$\frac{\partial^2 \eta_{4D}}{\partial x_D^2} = \frac{\omega_{4d}}{k_{4D}} \frac{\partial \eta_{4D}}{\partial t_D} \quad 1 < x_D < x_{eD} \quad (34)$$

With the upper and lower boundary conditions

$$\left.\frac{\partial \eta_{4D}}{\partial x_D}\right|_{x_D=x_{eD}} = 0, \quad \gamma_{4D}\eta_{4D}\big|_{x_D=1} = \begin{cases} \gamma_{2D}\eta_{2D}\big|_{x_D=1}, & w_{fD}/2 < y_D < l_{1,D} \\ \gamma_{3,jD}\eta_{3,jD}\big|_{x_D=1}, & l_{j-1,D} < y_D < l_{j,D} \end{cases} \quad (35)$$

Laplace transformation is used to transform these equations Laplace domain

$$\frac{d^2 \overline{\eta_{4D}}}{dx_D^2} = s\frac{\omega_{4D}}{k_{4D}} \overline{\eta_{4D}} \quad 1 < x_D < x_{eD} \quad (36)$$

$$\left.\frac{d\overline{\eta_{4D}}}{dx_D}\right|_{x_D=x_{eD}} = 0, \quad \gamma_{4D}\overline{\eta_{4D}}\big|_{x_D=1} = \begin{cases} \gamma_{2D}\overline{\eta_{2D}}\big|_{x_D=1}, & w_{fD}/2 < y_D < l_{1,D} \\ \gamma_{3,jD}\overline{\eta_{3,jD}}\big|_{x_D=1}, & l_{j-1,D} < y_D < l_{j,D} \end{cases} \quad (37)$$

1.3 Solution of Multi-Linear Model

On the basis of solution methods proposed by Ozkan et al [2], and Stalgorova et al [22], this coupled multi-linear models from Eq. (5) to Eq. (37) can be solved starting the Zone IV. The general solutions of partial differential Eq. (36) and Eq. (37) can be presented,

$$\overline{\eta_{4D}} = A_4 e^{\sqrt{s\frac{\omega_{4D}}{k_{4D}}}x_D} + B_4 e^{-\sqrt{s\frac{\omega_{4D}}{k_{4D}}}x_D} \quad (38)$$

The upper boundary condition can derive the relationship between A_4 and B_4,

$$B_4 = A_4 e^{2\sqrt{s\frac{\omega_{4D}}{k_{4D}}}x_{eD}} \quad (39)$$

Two lower boundary conditions as shown in Eq. (32) could get different solutions

$$\overline{\eta_{4D}} = \begin{cases} \dfrac{\gamma_{2D}\cosh\left[\sqrt{s\frac{\omega_{4D}}{k_{4D}}}(x_D-x_{eD})\right]}{\gamma_{4D}\cosh\left[\sqrt{s\frac{\omega_{4D}}{k_{4D}}}(1-x_{eD})\right]} \overline{\eta_{2D}} & w_{fD}/2 < y_D < l_{1,D} \\[2ex] \dfrac{\gamma_{3,jD}\cosh\left[\sqrt{s\frac{\omega_{4D}}{k_{4D}}}(x_D-x_{eD})\right]}{\gamma_{4D}\cosh\left[\sqrt{s\frac{\omega_{4D}}{k_{4D}}}(1-x_{eD})\right]} \overline{\eta_{3,jD}} & l_{j-1,D} < y_D < l_{j,D} \end{cases} \quad (40)$$

After obtaining the solutions in Zone Ⅳ, the equations for Zone Ⅱ and Zone Ⅲ should be coupled together to simultaneously obtain the analytical solutions. After substituting Eq. (40) to Eq. (17) and Eq. (28), the effects of Zone Ⅳ can be incorporated into Zone Ⅱ and Zone Ⅲ.

Zone Ⅱ:

$$\frac{d^2 \overline{\eta_{2D}}}{dy_D^2} = \left\{ \frac{\gamma_{2D} k_{4D}}{\gamma_{4D} k_{2D}} \sqrt{s \frac{\omega_{4D}}{k_{4D}}} \tanh\left[\sqrt{s \frac{\omega_{4D}}{k_{4D}}}(1-x_{eD})\right] + s \frac{\omega_{2D}}{k_{2D}} \right\} \overline{\eta_{2D}}, w_{fD}/2 < y_D < l_{1,D} \quad (41)$$

Zone Ⅲ:

$$\frac{d^2 \overline{\eta_{3,jD}}}{dy_D^2} = \left\{ \frac{\gamma_{3jD} k_{4D}}{\gamma_{4D} k_{3,jD}} \sqrt{s \frac{\omega_{4D}}{k_{4D}}} \tanh\left[\sqrt{s \frac{\omega_{4D}}{k_{4D}}}(1-x_{eD})\right] + \frac{\lambda \omega_{3mD} s}{(\lambda + \omega_{3mD} s) k_{3,jD}} + s \frac{\omega_{3,jD}}{k_{3,jD}} \right\} \overline{\eta_{3,jD}}, l_{j,D} < y_D < l_{j+1,D} \quad (42)$$

To simplify the notation,

$$\beta = \sqrt{\frac{\gamma_{2D} k_{4D}}{\gamma_{2D} k_{3D}} \sqrt{s \frac{\omega_{4D}}{k_{4D}}} \tanh\left[\sqrt{s \frac{\omega_{4D}}{k_{4D}}}(1-x_{eD})\right] + s \frac{\omega_{2D}}{k_{2D}}}$$

$$\alpha_j = \sqrt{\frac{\gamma_{3jD} k_{4D}}{\gamma_{4D} k_{2,jD}} \sqrt{s \frac{\omega_{4D}}{k_{4D}}} \tanh\left[\sqrt{s \frac{\omega_{4D}}{k_{4D}}}(1-x_{eD})\right] + \frac{\lambda \omega_{3mD} s}{(\lambda + \omega_{3mD} s) k_{3,jD}} + s \frac{\omega_{3,jD}}{k_{3,jD}}}, j = 1, 2, \ldots, N$$

Similarly, the general solutions for Eq. (39) and Eq. (40) can be described as follow,

Zone Ⅱ:

$$\overline{\eta_{2D}} = A_2 e^{\beta x_D} + B_2 e^{-\beta x_D}, w_{fD}/2 < y_D < l_{1,D} \quad (43)$$

Zone Ⅲ:

$$\overline{\eta_{3,jD}} = A_{3,j} e^{\alpha_j x_D} + B_{3,j} e^{-\alpha_j x_D}, l_{j,D} < y_D < l_{j+1,D} \quad (44)$$

There are in total 2(N+1) coefficients needed to be determined. That is to say, we should write 2(N+1) equations to obtain an unique solutions. Specifically, there are N interfaces in the coupled area of Zone Ⅱ and Zone Ⅲ which allows us to write $2N$ equations.

At the interface of neighboring small sub-sections in Zone Ⅲ:

$$\gamma_{3,jD}\left(A_{3,j} e^{\alpha_j l_{j,D}} + B_{3,j} e^{-\alpha_j l_{j,D}}\right) = \gamma_{3,j+1D}\left(A_{3,j+1} e^{\alpha_{j+1} l_{j,D}} + B_{3,j+1} e^{-\alpha_{j+1} l_{j,D}}\right) \quad (45)$$

and

$$k_{3,jD}\left(\alpha_j A_{3,j} e^{\alpha_j l_{j,D}} - \alpha_j B_{3,j} e^{-\alpha_j l_{j,D}}\right) = k_{3,j+1D}\left(\alpha_{j+1} A_{3,j+1} e^{\alpha_{j+1} l_{j,D}} - \alpha_{j+1} B_{3,j+1} e^{-\alpha_{j+1} l_{j,D}}\right) \quad (46)$$

At the interface of Zone Ⅱ and Zone Ⅲ:

$$\gamma_{3,1D}\left(A_{3,1} e^{\alpha_1 l_{1,D}} + B_{3,1} e^{-\alpha_1 l_{1,D}}\right) = \gamma_{2D}\left(A_2 e^{\beta l_{1,D}} + B_2 e^{-\beta l_{1,D}}\right) \quad (47)$$

and

$$k_{3,1D}\left(\alpha_1 A_{3,1} e^{\alpha_1 l_{1,D}} - \alpha_1 B_{3,1} e^{-\alpha_1 l_{1,D}}\right) = k_{2D}\left(\beta A_2 e^{\beta l_{1,D}} - \beta B_2 e^{-\beta l_{1,D}}\right) \quad (48)$$

Other two equations can be further obtained at the inner boundary of Zone II and outer boundary of Zone III.

$$\gamma_{1D}\overline{\eta_{1D}} = \gamma_{2D}\left(A_2 e^{\alpha_1 \frac{w_{fD}}{2}} + B_2 e^{-\alpha_1 \frac{w_{fD}}{2}}\right) \quad (49)$$

and

$$A_{3,N} e^{\beta y_{eD}} - B_{3,N} e^{-\beta y_{eD}} = 0 \quad (50)$$

All these equations can be used to determine the required coefficients. After obtaining the solution at Zone II, we substitute this equation into Eqs. (13) as follows

$$\frac{d^2 \overline{\eta_{1D}}}{dx_D^2} = \left[s\frac{\omega_{1D}}{k_{1D}} - \frac{2k_{2D}}{C_{1D}}\beta\left(A_2 e^{\beta \frac{w_{fD}}{2}} - B_2 e^{-\beta \frac{w_{fD}}{2}}\right)\right]\overline{\eta_{1D}} \quad (51)$$

The general solutions should be

$$\overline{\eta_{1D}} = A_1 e^{\chi x_D} + B_1 e^{-\chi x_D}, 0 < x_D < 1 \quad (52)$$

Where,

$$\chi = \sqrt{s\frac{\omega_{1D}}{k_{1D}} - \frac{2k_{2D}}{C_{1D}}\beta\left(A_2 e^{\beta \frac{w_{fD}}{2}} - B_2 e^{-\beta \frac{w_{fD}}{2}}\right)}$$

Conditioning to the inner and outer boundary conditions, these two coefficients are as follows

$$A_1 = \frac{\pi}{sC_{1D}\chi\left(e^{2\chi} - 1\right)}, \quad B_1 = \frac{\pi e^{2\chi}}{sC_{1D}\chi\left(e^{2\chi} - 1\right)} \quad (53)$$

Finally, we can obtain the bottom-hole pressure

$$\overline{\eta_{wD}} = \overline{\eta_{1D}}\big|_{x_D=0} = \frac{\pi\left(e^{2\chi} + 1\right)}{sC_{1D}\chi\left(e^{2\chi} - 1\right)} \quad (54)$$

Following the methods used in Xu et al.[8] the zero-order perturbation solution of the bottom-hole pressure in the Laplace space considering the wellbore storage C_D and the skin factor S is obtained:

$$\overline{\eta}_{wD}(S, C_D) = \frac{s\overline{\eta}_{wD} + S}{s + C_D s^2\left[s\overline{\eta}_{wD} + S\right]} \quad (55)$$

Finally, by using Stefest numerical inversion methods proposed by Stefest[23], the zero-order perturbation solution of the bottom hole pressure in real space is obtained, and the real bottom hole pressure m_{wD} can be obtained by Eq. (8).

2 Model verification

In this section, we will simplify our proposed model to make a comparison with other models. Wu et al.[24] have established a similar multi-linear flow model of multi-fractured horizontal wells in stress-sensitive SGR. Further analysis of the proposed model in this work reveals that if we do not subdivide the inner SRV subsections, our model can be simplified to be a four-linear flow model as proposed by Wu et al.[24]. To verify our model, comparison is made with the model proposed by Wu et al.[24]. Some basic parameter settings are listed in Tab.1. As shown in Fig.2, there is a perfect agreement between the results of the two models.

Tab.1 Basic data used for model

Parameters	Value	Units
Reference length, L_{ref}	0.1	m
Hydraulic fracture permeability in zone I, k_1	1000	mD
Permeability in SRV zone II, k_2	0.1	mD
Permeability in SRV zone III, k_3	0.01	mD
Permeability in zone IV, k_4	0.001	mD
Initial formation pressure, p_i	30	MPa
Fluid viscosity	2.7×10^{-3}	mPa·s
Compressibility in zone I, Ct_1	5×10^{-4}	MPa^{-1}
Compressibility in SRV zone II, Ct_2	5×10^{-4}	MPa^{-1}
Compressibility in SRV zone III, Ct_3	5×10^{-4}	MPa^{-1}
Compressibility in zone IV, Ct_4	5×10^{-4}	MPa^{-1}
Porosity in zone I	0.15	fraction
Porosity in SRV zone II	0.15	fraction
Porosity in SRV zone III	0.15	fraction
Porosity in zone IV	0.15	fraction
Half length of the fractures, x_f	100	m
Width of the SRV zone, y_e	100	m
Length of the SRV zone, x_e	150	m

In addition, our new model has the ability to subdivide the inner SRV zones into small subsections with different permeability. Our model can be equivalent to standard four-linear flow models if all subsections have the same permeability. In numerical experiment, we subdivide the SRV zones III into 1 (without division), 5, 20 subsections. Fig.3 shows the the evolution of dimensionless pressure and its derivative curves. The number of subsections has no any influences on the pressure curves, Which indicates that our proposed models can be generalized to be any multi-linear flow models (MLFM).

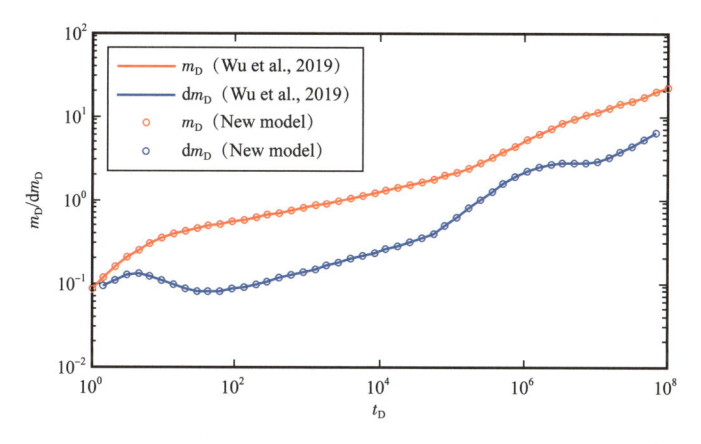

Fig.2 Comparison between the model in this paper and that in Wu et al. (2019)

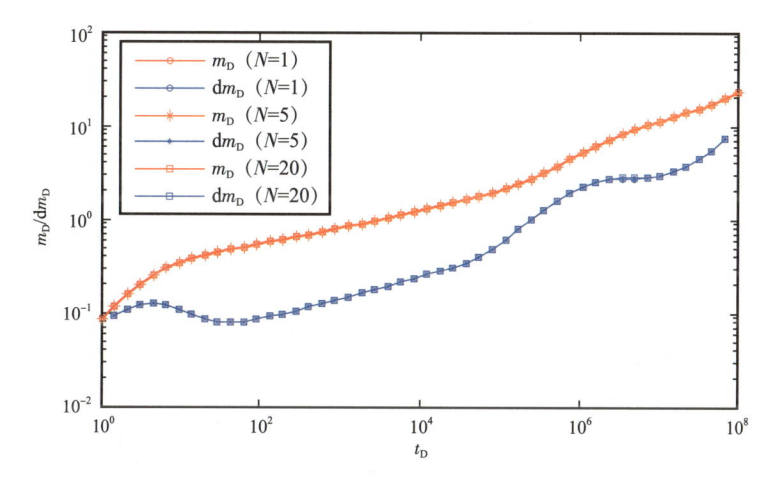

Fig.3 Investigation of number of subdivision of SRV zones Ⅲ on the type curves

3 Results and discussions

3.1 Transient pressure behaviors analysis

Fig.4 illustrates the typical pressure and pressure derivative curves of multi-fractured horizontal well in stress sensitive SGR on the basis of proposed MLFM. The physical models of SGR consisting of SRV and USRV can be seen in Fig.2. We take the reference length L_{ref} as 0.1m. The values of the relevant dimensionless variables used in this numerical experiments are as follows: $C_{fD}=10$, $k_{2D}=100$, $k_{3D}=10$, $k_{4D}=1$, $\omega_{3D}=0.01$, $\lambda=0.15$, $m_{DL}=0.12$, $V_{DL}=0.1$, $\gamma_D=0.05$, $C_D=0.001$, $S=0.1$. We can approximately observe five flow stages from this log-log of dimensionless pseudo-pressure and its derivative curves.

Stage Ⅰ: Wellbore storage flow regime. This is a very common flow regime at the early stage of hydrogen extraction through wells. The wellbore storage coefficient and the skin can be obtained by fitting the plotted pressure curves to the real data. The detailed explanation of this stage can be referred to the literature [25-27].

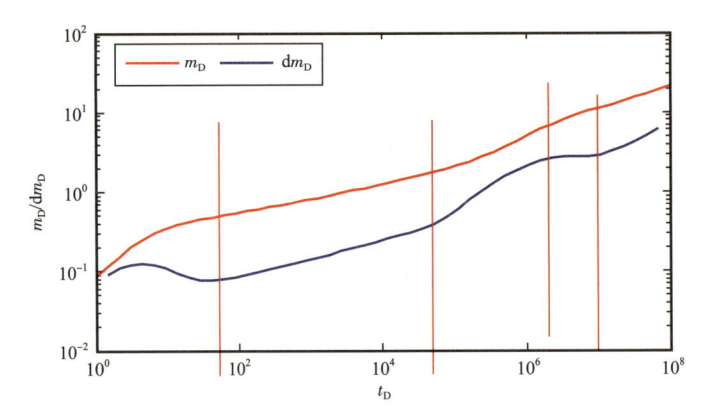

Fig.4　Typical curves of multi-fractured horizontal well in stress-sensitive SGR formation

Stage II: The formation bi-linear flow in SRV zone I. This flow regime is a transition between fracture linear flow and formation linear flow. The existence of high permeable hydraulic fractures speeds up the liquid supply from formation to the hydraulic fractures at the direction of being perpendicular to the fracture faces. In generally a 1/4-slope straight line (Fig.4) will be observed on the dimensionless derivative pressure curve. We also should note that due to the high-permeability of hydraulic fractures, the fracture linear flow is also blurred in this stage. This phenomenon is also consistent with the observations from the realistic applications.

Stage III: The formation linear flow in SRV zone II and the transition flow between SRV zone II and zone IV. In this stage, a 1/2-slope straight line (Fig.4) can be clearly identified from the dimensionless pseudo-pressure derivative curve. In addition, the fluid in the zone IV also starts to supply SRV zone II. The fluid supply of the zone IV delay the pressure depletion evidently. We can noticeably observe that a shape of "recess" exists in the pressure derivative curve.

Stage IV: the transition flow between SRV zone II and SRV zone III. The stress sensitivity begins to take effects in this regime, and the front of pressure wave reaches the boundary of the SRV. However, due to the lower permeability of the un-stimulated zone IV, there is no enough fluid supply from the outer region. This stage can be used to identify the boundary-dominated flow.

Stage V: The formation linear flow in outer SRV zone III. We have idealized the outer SRV zone as dual-porosity media in SGR. The fluid will feed from the matrix to the natural fracture in this region. In this stage, the typical feature of this flow behavior is that a 1/2-slope straight line (Fig.4) occurs on the dimensionless derivative pressure curve. The effects of stress sensitivity on the transient pressure response become evident in this stage, which will be systematically analyzed in the following part.

3.2　Sensitivity analysis

In this section, on the basis of our proposed multi-linear flow model, comprehensive sensitivity analysis of key parameters to well performance is implemented to systematically investigate the transient pressure behaviors of the multi-fractured horizontal well in stress sensitive SGR.

3.2.1 Effect of stress sensitivity

Fig.5 shows the effect of stress sensitivity on the pressure transient curves. The values of relevant parameters are listed as follows: $C_{fD}=10$, $k_{2D}=100$, $k_{3D}=10$, $k_{4D}=1$, $\omega_{3D}=0.01$, $\lambda=0.15$, $m_{DL}=0.12$, $V_{DL}=0.1$, $C_D=0.001$, $S=0.1$. This newly proposed multi-linear flow model can be used to analyze two situations: (a) there is a constant permeability module for all zones; (b) each zone has different permeability module. We will separately investigate these two situations.

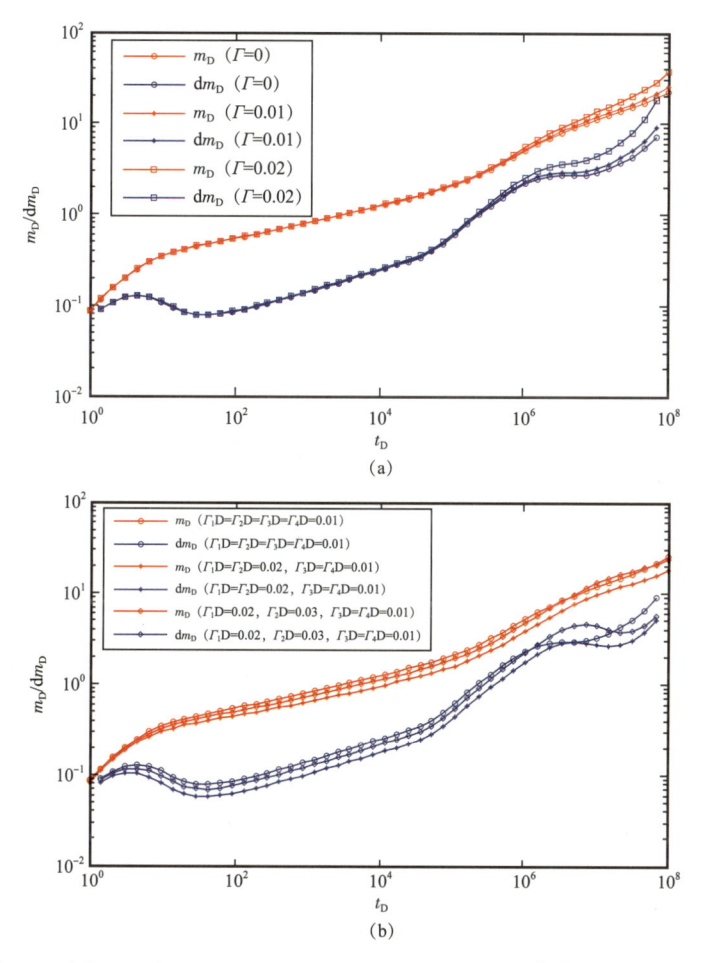

Fig.5 The effect of permeability modulus on pressure and derivative curves, (a) constant permeability module for all zones; (b) each zone has different permeability module

When we make an assumption that there is a constant permeability module for all zones, three cases were studied in which the dimensionless permeability modulus γ_D is equal to 0, 0.01 and 0.02, respectively as illustrated in Fig.5(a). It can be seen from Fig.5(a) that as the dimensionless permeability modulus increases, the dimensionless pressure and its derivative curves rise gradually, the stress sensitivity mainly affect the flow behaviors in intermediate and later flow stages. In stress sensitive shale gas reservoirs, as the fluid is produced, the gradual reduction of formation pressure will result in a decrease of the permeability of the system and a growing of pressure depletion. When the dimensionless permeability modulus increases to a certain value, the pressure derivative curve rises up significantly in later periods, showing the characteristic of closed boundary.

When we assume that each zone has different permeability module, three cases also were studied as illustrated in Fig.5 (a). As we can see from Fig.5 (b), the stress sensitivity will have significant influences on the entire flow stages, which is significantly different from the previous situation. This finding is very important for us to enhance the gas recovery from shale gas reservoir. we need to design reasonable production scheme to maintain the formation pressure. Both zone I and SRV zone II are closed to the wellbore, as a result, the pressure drops are significant and hence causes severe stress sensitivity. Thus, larger stress sensitivity coefficients in SRV zone generally induce larger pressure drops.

3.2.2 Effect of the mobility capacity of the zone I and SRV zone II

Transient pressure and pressure derivative curves for different mobility capacity of the zone I and SRV zone II are illustrated in Fig.6. The values of relevant parameters are listed as follows: $k_{3D}=10$, $k_{4D}=1$, $\omega_{3D}=0.01$, $\lambda=0.15$, $m_{DL}=0.12$, $V_{DL}=0.1$, $\gamma_D=0.05$, $C_D=0.001$, $S=0.1$. By analyzing, the mobility capacity of the zone I and SRV zone II are represented by dimensionless fracture conductivity C_{fD} and dimensionless formation permeability k_{2D}. In this numerical experiment, the SRV zone III is not divided into subsections, e.g., $N=1$. It can be seen from Fig.6a that dimensionless fracture conductivity C_{fD} has significant influences on the flow behaviors of the earl time, especially the wellbore storage and skin effect flow. Other flow stages are almost not affected. This result is consistent with our previous observations that fracture linear flow is easily blurred by the wellbore storage and skin effect flow stage due to its high mobility capacity. Specifically, as C_{fD} decreases, the duration of wellbore storage and skin effect flow period will be shortened and the starting time of formation bi-linear flow period in the SRV zone II will be advanced. It is mainly because the mobility capacity determines flow capacity contrast of the SRV and hydraulic fracture. Therefore, the pressure and derivatives responses associated SRV zone will surely enlarge; the curves of the transient response will rise in wellbore storage and skin effect flow.

The mobility capacity k_{2D} determines gas flow capacity in the SRV zone II. We set mobility capacity k_{2D} separately to be 100, 200 and 300 to investigate its effects on the pressure curves. It can be seen from Fig.6 (b) that the dimensionless formation permeability k_{2D} almost influences the entire flow stages. Specifically, the flow regimes of the shape of type curves are not distorted, the pressure only raise up as the the dimensionless formation permeability k_{2D} decreases. This means that a large value of k_{2D} increases the gas flow capacity in SRV zone and therefore leads to a small pressure drop. An effective maintenance of formation pressure is very significant to yield a long-term gas production from shale gas reservoir. In addition, we also find a very interesting phenomenon that a "concave" will occur at the transition flow stage between SRV zone II and zone III as the k_{2D} increases. Our proposed multi-linear flow model will show a special flow stage for the dual-porosity model. The occurrence of a "concave" also can be an indicator that the permeability of induced SRV zone is much larger than that of outer-zone.

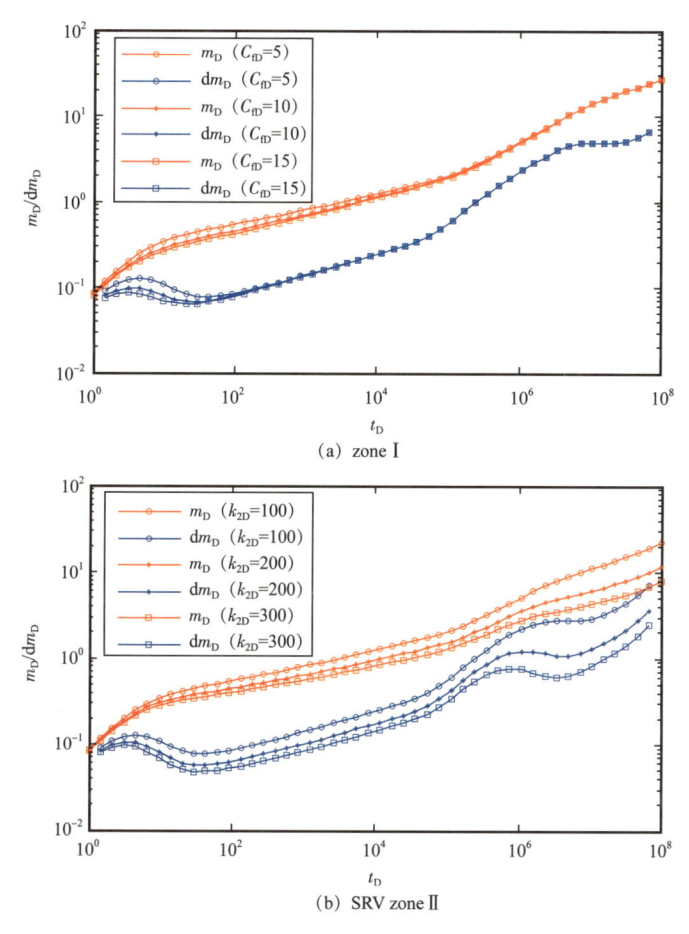

Fig.6 The effect of mobility capacity of zone Ⅰ and the SRV zone Ⅱ on pressure and derivative curves

3.2.3 Effect of the mobility capacity of the zone Ⅲ and zone Ⅳ

Transient pressure and pressure derivative curves for different mobility capacity of the SRV zone Ⅲ and the outer region zone Ⅳ are illustrated in Fig.7. The values of relevant parameters are listed as follows: $C_{fD}=10$, $k_{2D}=100$, $\omega_{3D}=0.01$, $\lambda=0.15$, $m_{DL}=0.12$, $V_{DL}=0.1$, $\gamma_D=0.05$, $C_D=0.001$, $S=0.1$. In this numerical experiment, the SRV zone Ⅲ is not divided into subsections as wells, e.g., $N=1$. We can observe from Fig.7 (a) that the mobility capacity of the zone Ⅲ almost has no any influences on the flow behaviors of the early and intermediate flow stages, on the contrary, the transition flow and formation linear flow regimes are severely dominated. This results are also in agreement with our common understanding. Specifically, as the mobility capacity of zone Ⅲ k_{3D} decreases, the duration of transition flow period will be enlarged and the starting time of formation linear flow period and boundary-dominated flow stage in the SRV zone Ⅲ will be delayed.

The mobility capacity k_{4D} determines gas flow capacity in the outer zone Ⅳ. We set mobility capacity k_{4D} separately to be 10, 20 and 30 to investigate its effects on the pressure curves. Fig.7 (b) apparently demonstrates that the dimensionless formation permeability k_{4D} has great influence on the formation bi-linear flow stage in SRV zone Ⅱ. This is very different from the effects of mobility capacity k_{3D} in SRV zone Ⅲ. In the process of model development, we assume that the outer zone Ⅳ simultaneously supplies gas to SRV zone Ⅱ and SRV zone Ⅲ. Because SRV zone Ⅱ is adjacent to the wellbore and thus yields larger pressure drop, the outer zone Ⅳ is more prone to supply gas to SRV zone Ⅱ, as illustrated in Fig.7 (b). As the mobility capacity k_{4D} decreases, the gas supply will be reduced as well, as a result, the dimensionless pressure curves will be move upward.

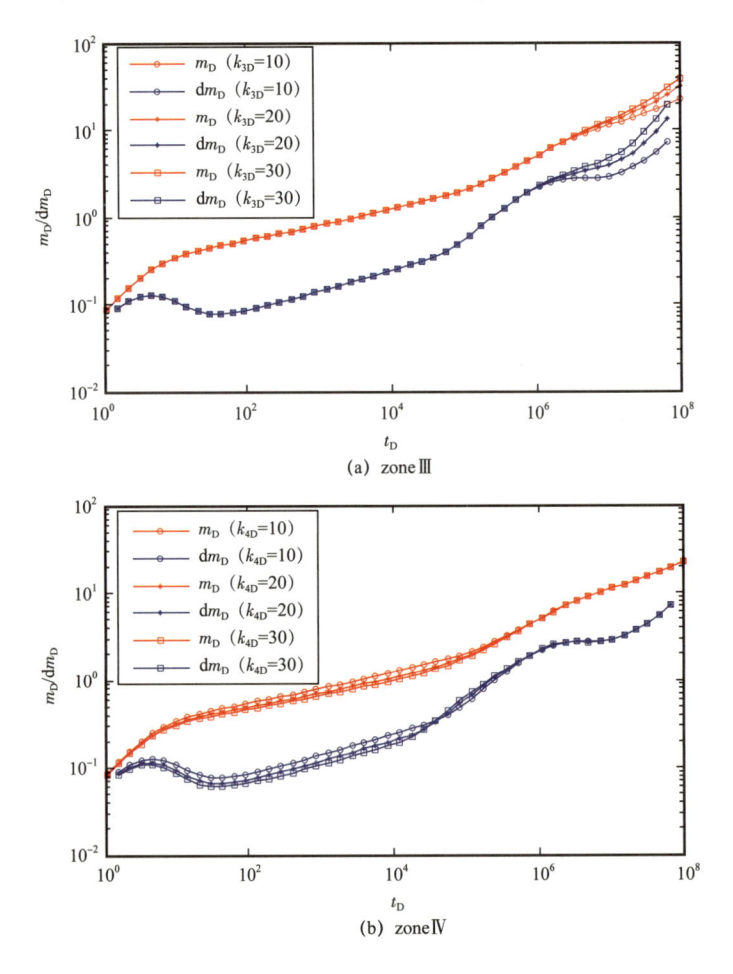

Fig.7 The effect of mobility capacity of the zone Ⅲ and zone Ⅳ on pressure and derivative curves

3.2.4 Effect of the Size of Outer Region

Three cases with different size of outer regions shown in Fig.8 are studied in this section to identify the effect of size of outer regions on the transient behaviors. The values of relevant parameters are listed as follows $C_{fD}=10$, $k_{2D}=100$, $k_{3D}=10$, $k_{4D}=1$, $\omega_{3D}=0.01$, $\lambda=0.15$, $m_{DL}=0.12$, $V_{DL}=0.1$, $\gamma_D=0.05$, $C_D=0.001$, $S=0.1$. As shown in Fig.8, the width and length of outer-region have mainly affected different flow regimes in SRV zone Ⅱ and SRV zone Ⅲ. Specifically speaking, the width of outer-region mainly has influences on the formation linear flow regime in SRV zone Ⅲ, while the width of outer-region mainly impacts the formation bi-linear and transition flow regimes. As the size of outer-region decreases, the boundary conditions will be interfere the flow regimes in SRV zone in advance, as a result, the dimensionless pressure and its derivatives will be upper. This phenomenon is very similar to the effects of the mobility capacity of the SRV zone III and zone IV. Enlarging the size of outer-region can decrease the negative effects of boundary condition and increasing the mobility capacity of the zone III and zone IV can decrease the flow resistance in SRV zone, which is beneficial to obtain high production rate.

(a) Length of outer region

(b) Width of outer region

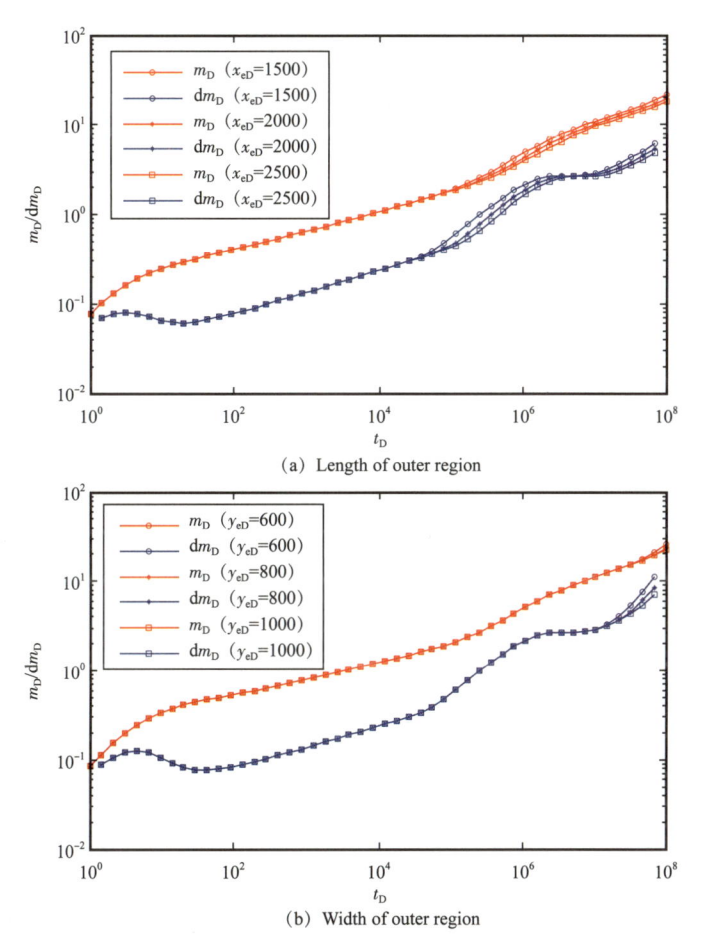

Fig.8 The effect of size of the outer region on pressure and derivative curves

3.2.5 Effect of the size of SRV zone II

Fig.9 illustrates the effect of the size of SRV zone II on the transient behaviors. The values of relevant parameters are listed as follows: $C_{fD}=10$, $k_{2D}=100$, $k_{3D}=10$, $k_{4D}=1$, $\omega_{3D}=0.01$, $\lambda=0.15$, $m_{DL}=0.12$, $V_{DL}=0.1$, $\gamma_D=0.05$, $C_D=0.001$, $S=0.1$, and the dimensionless size of SRV zone II L_D is 300, 500, 700, respectively (Fig.9). It can be seen from Fig.9 that the dimensionless size of SRV zone II can affect all the flow stages after the formation bi-linear flow regimes. The larger size of SRV zone II has (1) a later end of formation bi-linear in the SRV zone II, (2) a postponed beginning of formation linear flow in SRV zone III, (3) the lower values of dimensionless pressure and its derivatives in transition flow regime between SRV zone II and SRV zone III. All these phenomena might provide useful information to identify the sizes of SRV after massive hydraulic fracturing. Fig.9 also shows that the larger the SRV size, the smaller the dimensionless pressure and its derivatives. Smaller dimensionless pressure indicates lower pressure depletion in the formation, which is beneficial to obtain a high production rate. In this paper, Zone Ⅱ is a highly permeable SVR zone which is adjacent to the main hydraulic fracture as illustrated in Fig.2（a）. Zone Ⅱ is the gas flow path from SGR to wellbore. Increasing the size of this zone will delay the negative effects of outer boundary, which is beneficial to obtain high production rate as well.

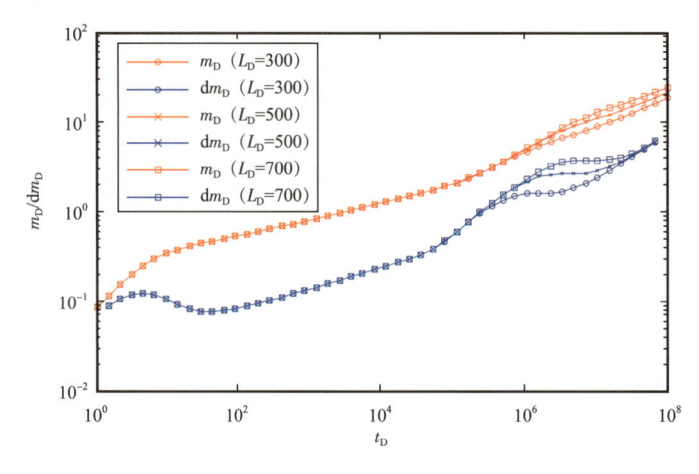

Fig.9　The Effect of the size of SRV zone Ⅱ on pressure and derivative curves

3.2.6　Effect of coefficient of permeability variation

In this paper, to characterize spatially varying permeability due to the non-uniform distribution of proppant within SRV Zone Ⅲ, we further divide Zone Ⅲ into N small zones as illustrated in Fig.1b. Fig.10 show the effect of coefficient of permeability variation in SRV zone Ⅲ on the transient behavior, the values of the parameters related to the Fig.10 are listed as follows: $C_{fD}=10$, $k_{2D}=100$, $k_{3D}=10$, $k_{4D}=1$, $\omega_{3D}=0.01$, $\lambda=0.15$, $m_{DL}=0.12$, $V_{DL}=0.1$, $\gamma_D=0.05$, $C_D=0.001$, $S=0.1$. To systematically represent the spatial variation of permeability in Zone Ⅲ, two mathematical formulas, including linear and logarithmic functions, are used to describe the reduction of permeability away from the main hydraulic fracture. Fig.10（a）and Fig.10（b）show the dimensionless permeability for each sub-sections. Fig.10（c）and Fig.10（d）depict the dimensionless permeability as a function of distance to the main hydraulic fracture. In contrast to the linear function, logarithmic function yields a more rapid reduction of dimensionless permeability. What's more, increasing the number of subsections can generate a continuous reduction of permeability.

Fig.10（e）and Fig.10（f）show the effect of coefficient of permeability variation in SRV zone Ⅲ on the transient behavior. Because the linear function generates a continuous reduction of permeability, large parts of the SRV zone Ⅲ still preserves relatively high permeability, and only a small part of regions which is adjacent to the outer-boundary will get small permeability. Under this condition, the formation linear flow regime will be mainly impacted as Fig.10（e）, as the number of subsections increases, the reduction rate of permeability will be decreases, as a result, the negative of boundary condition will be delayed. On the contrary, the logarithmic function easily leads to a rapid reduction of permeability as illustrated in Fig.10 (d). Almost the entire flow regimes are severely influenced. This phenomenon can be considered to be reliable indicator to judge whether the variation of permeability follows a logarithmic function or not. In the real-field applications, a linear variation of permeability induced by massive hydraulic fracturing operation has the potential to obtain high production rate.

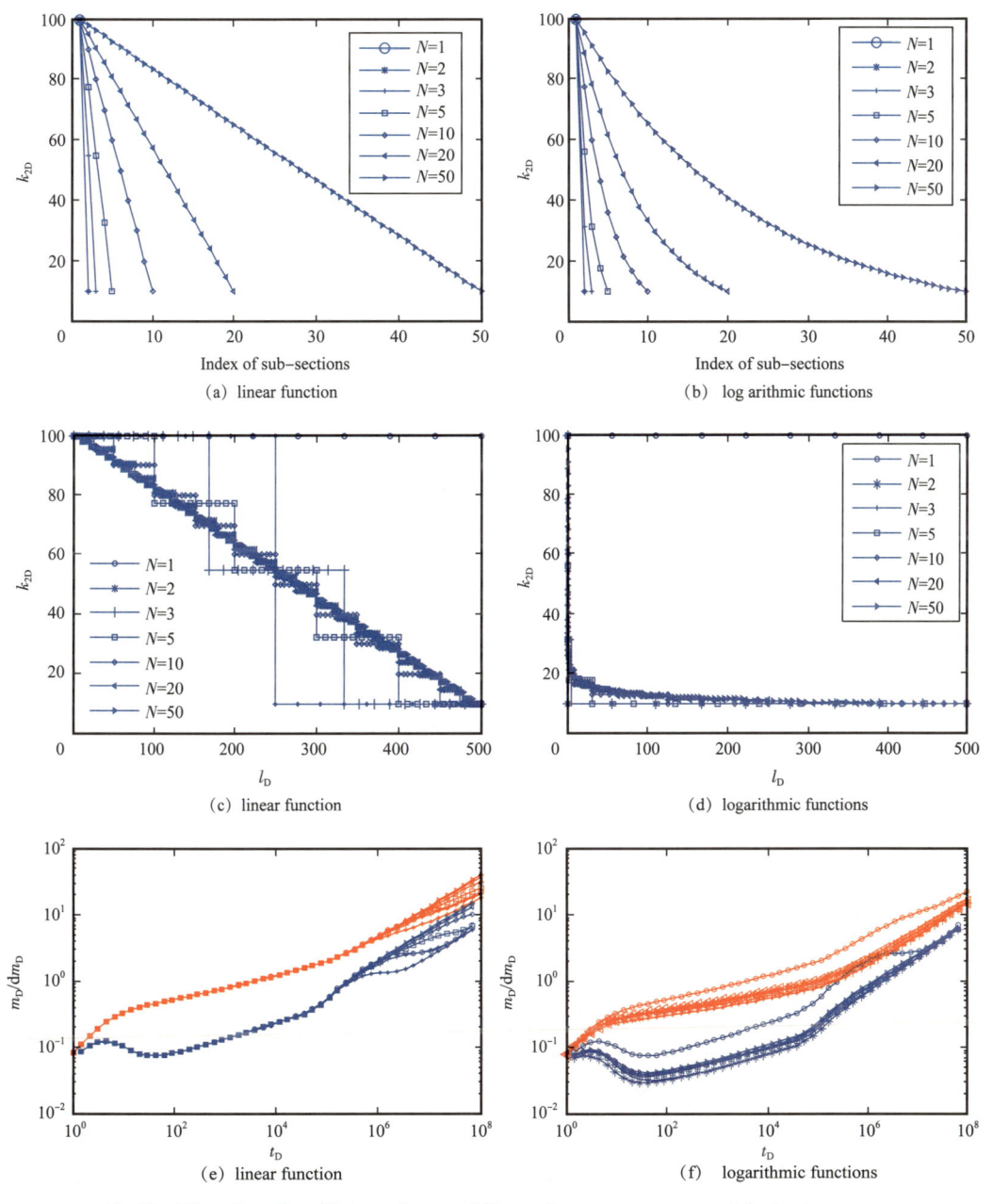

Fig.10 The effect of coefficient of permeability variation on pressure and derivative curves

4 Conclusion

An analytical multi-linear flow model is proposed for multi-fractured horizontal wells with SRV in shale gas reservoir. The transient pressure curves are also established to analyze the performance of this new model. Some key results could be summarized as follows:

On the basis of perturbation technique and Laplace transformation, an easy-to-simulation model with the consideration of stress sensitivity and SRV is obtained in Laplace space, and then the solution of transient pressure behaviors for multi-fractured horizontal well with SRV is finally obtained by Stefest numerical inversion algorithm.

This newly proposed mulit-linear flow model (MLFM) is validated by comparing with an existing model and a perfect agreement has been obtained.

Approximately five flow stages can be identified: Wellbore storage and skin effect flow; The formation bi-linear flow in the inner SRV; The formation linear flow in the SRV zone and the transition flow between SRV zone Ⅱ and zone Ⅳ; the transition flow between SRV zone Ⅱ and zone Ⅲ; The formation linear flow in zone Ⅲ.

SRV zone Ⅱ provides the main flow path to supply gas from shale gas reservoir to wellbore. Increasing the size of outer-region could delay the negative influences of boundary conditions. The permeability of SRV zone Ⅲ mainly impacts the formation linear flow regimes, while the permeability of zone Ⅳ has significant influences on the formation linear flow regime in SRV zone Ⅱ.

To systematically represent the spatial variation of permeability in Zone Ⅲ, both linear and logarithmic functions are used to describe the reduction of permeability. In contrast to the linear function, logarithmic function yields a more rapid reduction of dimensionless permeability. As the number of subsections increases, the reduction rate of permeability will be decreases, as a result, the negative of boundary condition will be delayed. The logarithmic function easily leads to a rapid reduction of permeability and hence almost the entire flow regimes are severely influenced.

Conflicts of Interest

The author (s) declared no potential conflicts of interest with respect to the research, authorship, and/or publication of this article.

Funding Statement

The authors acknowledge the financial support provided by the National Science and Technology Major Project "Characterization of Shale Gas Production Law and Optimization of Development Technology Strategy" (No.2017ZX05037-002); National Science and Technology Major Project "Shale Gas Geological Evaluation and Development Technology Optimization" (No.2017ZX05062-002); CNPC Major Science and Technology Project "Evaluation and Optimization of Shale Gas Production and Favorable Areas in Sichuan Basin and Optimization of Development Technology Policy Research and Application" (No.2016E-0611).

References

[1] Clarkson C R. Production data analysis of unconventional gas wells: Review of the theory and best practices [J]. Int. J, Coal Geoal, 2013, 109-110: 101-149.

[2] Ozkan E, Brown M, Raghavan R, et al. Comparison of fractured horizontal well performance in tight sand and shale gas reservoirs [J]. SPE Reservoir Evaluation & Engineering, 2011, 14 (2): 248-259.

[3] Zhao Y L, Zhang L H, Luo J X et al. Performance of fractured horizontal well with stimulated reservoir volume in unconventional gas reservoir [J]. Journal of Hydrology, 2014, 512: 447-456.

[4] Zhao Y L, Zhang L H, Xu B Q, et al. Analytical solution and flow behaviors of horizontal well in stress-sensitive naturally fractured reservoirs [J]. International Journal of Oil, Gas and Coal Technology, 2016, 11 (4): 350-369.

[5] Wang X D, Luo W J, Hou X C, et al. Pressure transient analysis of multi-stage fractured horizontal wells in boxed reservoirs [J]. Petrol. Explor. Develop., 2014, 41 (1): 82-87.

[6] El-Banbi A H. Analysis of Tight Gas Wells [D]. Texas, 1998.

[7] Al-Ahmadi H A, Wattenbarger R A. Triple porosity Models: One Future Step Towards Capturing Fractured Reservoirs Heterogeneity [C] // SPE/DGS Saudi Arabia Section Technical Symposium and Exhibition, Al-Khobar, Saudi Arabia, 2011, 15-18 May.

[8] Xu B X, Haghighi M, Li X F, et al. Development of new type curves for production analysis in naturally fractured shale gas/tight gas reservoirs [J]. Journal of Petroleum Science and Engineering, 2013, 105: 107-115.

[9] Ekaterina S, Louis M. Analytical Model for Unconventional Multifractured Composite Systems [J]. SPE Reservoir and Evaluation & Engineering, 2013, 16 (3): 246-256.

[10] Ozkan S, Koseler R. Multi-dimensional students' evaluation of e-learning systems in the higher education context: An empirical investigation. 2009, 53 (4): 1285-1296.

[11] Brown M, Ozkan E, Raghavan R. Practical solutions for pressure transient responses of fractured horizontal wells in unconventional reservoirs [J]. 2009, 14 (6): 4271-4288.

[12] Zeng F, Zhao G, Liu H. A new model for reservoirs with a discrete-fracture system [J]. J. Can. Pet. Technol., 2012, 51 (2): 127-136.

[13] Jiang R, Xu J, Sun Z, et al. Rate transient analysis for multistage fractured horizontal well in tight oil reservoirs considering stimulated reservoir volume [J]. Mathematical Problems in Engineering, 2014.

[13] Wang J, Jia A, Wei Y, et al. Semi-analytical simulation of transient flow behavior for complex fracture network with stress-sensitive conductivity [J]. Journal of Petroleum Science and Engineering, 2018, 171: 1191-1210.

[14] Ren Z X, Wu X D, Liu D D, et al. Semi-analytical model of the transient pressure behavior of complex fracture networks in tight oil reservoirs [J]. Journal of Natural Gas Science and Engineering, 2016, 35: 497-508.

[15] Ren Z X, Wu X D, Han G Q, et al. Transient pressure behavior of multi-stage fractured horizontal wells in stress-sensitive tight oil reservoirs [J]. Journal of Petroleum Science and Engineering, 2017, 157: 1197-1208.

[16] Jia P, Cheng L, Huang S, et al. A comprehensive model combining Laplace-transform finite difference and boundary-element method for the flow behavior of a two-zone system with discrete fracture network [J]. Journal of Hydrology, 2017, 551: 453-469.

[17] Wang J H, Wang X D, Dong W X. A semianalytical model for multiple-fractured horizontal wells with SRV in tight oil reservoirs [J]. Geofluids, 2017 (1): 1-15.

[18] Xiao C, Dai Y, Tian L, et al. A semianalytical methodology for pressure-transient analysis of multiwell-pad-production scheme in shale gas reservoirs, Part 1: new insights into flow regimes and multiwell interference [J]. SPE Journal, 2018, 23 (3): 885-905.

[19] Jia P Cheng L, Clarkson C R, et al. A Laplace-domain hybrid model for representing flow behavior of multifractured horizontal wells communicating through secondary fractures in unconventional reservoirs [J]. SPE Journal, 2017, 22 (6): 1-856.

[20] Chen Z, Liao X, Yu W. Pressure-Transient Behaviors of Wells in Fractured Reservoirs With Natural-and Hydraulic-Fracture Networks [J]. SPE Journal, 2018, 24 (1).

[21] Liu M, Xiao C, Wang Y, et al. Sensitivity analysis of geometry for multi-stage fractured horizontal wells with consideration of finite-conductivity fractures in shale gas reservoirs [J]. Journal of Natural Gas Science and Engineering, 2015, 22: 182-195.

[22] Stalgorova K, Mattar L. Analytical model for unconventional multi-fractured composite systems [J]. SPE Reservoir Eval. Eng., 2013, 16 (3): 246-256.

[23] Stehfest H. Numerical inversion of Laplace transforms [J]. Commun. ACM, 1970, 13 (1): 47-49.

[24] Wu Z W, Cui C, Lv G, et al. A multi-linear transient pressure model for multistage fractured horizontal well in tight oil reservoirs with considering threshold pressure gradient and stress sensitivity [J]. Journal of Petroleum Science and Engineering, 2019, 172: 839-854.

[25] Agarwal R G, Al-Hussainy R, Ramey H J. An investigation of wellbore storage and skin effect in unsteady liquid flow: I. Analytical treatment [J]. Society of Petroleum Engineers Journal, 1970, 10 (03): 279-290.

[26] Kuchuk F J, Onur M, Hollaender F. Pressure transient formation and well testing: convolution, deconvolution and nonlinear estimation (Vol. 57). Elsevier, 2010.

[27] Song B, Economides M J, Ehlig-Economides C A. Design of multiple transverse fracture horizontal wells in shale gas reservoirs [C] // SPE Hydraulic Fracturing Technology Conference, The Woodlands, Texas, USA, 2011, 24-26 January.

[28] Wang H T. Performance of multiple fractured horizontal wells in shale gas reservoirs with consideration of multiple mechanisms [J]. Journal of Hydrology, 2014, 510: 299-312.

Appendix:

Here, we take the zone I as an example to derive the governing equations as shown in the main content. Based on the mass conservation principle, Darcy law, the shale gas flow along the fracture direction can be presented as follows:

$$-\frac{\partial}{\partial x}\left[\rho_g\left(-\frac{k_{1f}}{\mu_g}\frac{\partial p_{1f}}{\partial x}\right)\right]-\rho_g\left(-\frac{k_{2f}}{\mu_g}\frac{\partial p_{2f}}{\partial y}\right)\frac{2}{w_f}=\frac{\partial(\rho_g\phi_{1f})}{\partial t} \quad (1)$$

After introducing the pseudo-pressure

$$\frac{k_{1f}}{\mu_g}\frac{\partial^2 m(p_{1f})}{\partial x^2}+\frac{2}{w_f}\frac{k_{2f}}{\mu_g}\frac{\partial m(p_{2f})}{\partial y}\bigg|_{y=\frac{w_f}{2}}=\phi_{1f}C_{gi}\frac{\partial m(p_{1f})}{\partial t} \quad (2)$$

The inner boundary condition:

$$\frac{1}{2}q_{gsc}B_g=w_f h\frac{k_{1f}}{\mu_g}\frac{\partial p_{1f}}{\partial x}\bigg|_{x=0} \quad (3)$$

$$\frac{\partial p_{1f}}{\partial x}\bigg|_{x=0}=\frac{q_{gsc}B_g\mu_g}{2w_f hk_{1f}}=\frac{q_{gsc}\mu_g}{2w_f hk_{1f}}\frac{Z_g T p_{sc}}{p_{1f} T_{sc}} \quad (4)$$

$$\frac{\partial m(p_{1f})}{\partial x}\bigg|_{x=0}=\frac{q_{gsc}\mu_{gi}B_{gi}}{2w_f hk_{1f}} \quad (5)$$

Introducing the dimensionless definition, the term $-\dfrac{2\pi k_{ref}hx_f}{q_{gsc}\mu_{gi}B_{gi}}$ is multiplied to the two sides:

$$-\frac{2\pi k_{ref}h}{q_{gsc}\mu_{gi}B_{gi}}\frac{\partial m(p_{1f})}{\frac{1}{x_f}\partial x}\bigg|_{x=0}=-\frac{q_{gsc}\mu_{gi}B_{gi}}{2w_f hk_{1f}}\cdot\frac{2\pi k_{ref}hx_f}{q_{gsc}\mu_{gi}B_{gi}} \quad (6)$$

$$\frac{\partial m_{1fD}}{\partial x_D}\bigg|_{x_D=0}=-\frac{\pi k_{ref}x_f}{k_{1f}w_f}=-\frac{\pi k_{2f}x_f}{k_{1f}w_f}\frac{k_{ref}}{k_{2f}}=-\frac{\pi}{C_{fD}k_{2fD}} \quad (7)$$

The outer boundary condition:

$$w_f h\frac{k_{1f}}{\mu_g}\frac{\partial p_{1f}}{\partial x}\bigg|_{x=x_f}=0 \quad (8)$$

$$\frac{\partial p_{1f}}{\partial x}\bigg|_{x=x_f}=0 \quad (9)$$

$$\frac{\partial m_{1fD}}{\partial x_D}\bigg|_{x_D=1}=0 \quad (10)$$

Finally, the dimensionless governing functions for zone Ⅰ can be summarized as follows

$$\frac{\partial^2 m_{1fD}}{\partial x_D^2}+\frac{2}{C_{fD}}\frac{\partial m_{2fD}}{\partial y_D}\bigg|_{y_D=\frac{w_{fD}}{2}}=\frac{\omega_{1f}}{k_{1fD}}\frac{\partial m_{1fD}}{\partial t_D} \quad (11)$$

$$\frac{\partial m_{1fD}}{\partial x_D}\bigg|_{x_D=0}=-\frac{\pi}{C_{fD}k_{2fD}} \quad (12)$$

$$\frac{\partial m_{1fD}}{\partial x_D}\bigg|_{x_D=1}=0 \quad (13)$$

页岩气井 EUR 快速评价新方法

赵玉龙[1]，梁洪彬[1]，井 翠[2]，商绍芬[3]，李成勇[4]

（1. 西南石油大学油气藏地质及开发工程国家重点实验室；
2. 四川长宁天然气开发有限责任公司；
3. 中国石油西南油气田分公司蜀南气矿；
4. 成都理工大学能源学院）

摘　要：页岩气井通常采用控压进行生产，这既可降低储层应力敏感效应，又可改善气井生产效果，提高气井最终可采储量（EUR）。同时，频繁变更工作制度又会导致气井生产数据波动剧烈，给准确预测气井 EUR 带来了新的挑战。因此，通过对现有气井生产数据处理方法的综合分析，首先开展了变工作制度下波动数据降噪方法的研究，结果表明双对数坐标系下的压力规整化产量—物质平衡时间图版能使波动数据得到有效的收敛；然后基于降噪处理后的数据形态特征，建立了页岩气井 EUR 评价新模型，提出了一套页岩气井 EUR 评价新方法。实例分析表明，新方法与经典方法的评价结果相近，确保了新方法的可靠性；同时，新方法操作更简便，计算更高效，避免了经典方法中参数多、拟合难的问题，可更加快速地完成页岩气井 EUR 评价。

关键词：页岩气；最终可采储量；控压生产；产量递减模型；数据处理

页岩气藏在水平井压裂改造后常进行控压生产，这既有助于抑制支撑剂回流，降低储层应力敏感效应，又可改善气井生产效果，提高气井最终可采储量（EUR）[1]。但与此同时，由于频繁变更气井工作制度，导致气井生产数据波动剧烈[2-3]，增大了数据拟合难度，降低了 EUR 计算效率。因此，如何快速、准确预测控压生产下的页岩气井 EUR 显得尤为重要，而现有手段和方法仍存在一定的不足。

在页岩气井数据处理时通常会去除关井阶段数据，有时为降低数据波动频率还会剔除明显偏离总体递减规律的数据点[4-5]，该过程人为因素影响较大，特别是当气井存在明显的多段递减时，数据点不合理的剔除可能会改变气井递减规律。尽管规整化方法[6-8]能在一定程度上降低数据的波动，但对波动剧烈的页岩气井生产数据仍难以取得较好的效果，需进一步开展波动数据降噪方法研究。

页岩气井 EUR 评价方法主要有解析模型法[9-11]、数值模拟法[12-14]、物质平衡法[15-16]、现代产量递减法[17]、经验产量递减法（经验法）[18-20]和概率法[21-22]。由于各种方法的理论不同，对数据要求也不同，如表 1 所示。其中，前 4 种方法涉及的物理参数多，所需数据体大，模型复杂，且计算量较大；经验法虽然形式简便，计算量小，但计算结果受数据波动性影响较大，这些都降低了页岩气井 EUR 评价精度。因此，本文首先开展了波动数据降噪处理方法的研究，提出了能够有效地降低数据波动的处理方法，然后基于降噪处理后的数据特征建立了一个新的页岩气井 EUR 评价模型，最终形成了一套快速评价页岩气井 EUR 新方法，并结合实际典型页岩气井生产数据开展了新方法的可行性分析与应用。

1 页岩气井经验产量递减模型分析

经验法由于仅考虑产量与时间的关系，使得该方法应用简便，现已被广泛应用于现场。基于经验法建立的页岩气井 EUR 评价模型可归纳为两类：一类是对经典递减模型的修正，如广义 Arps 模型[20]、SEPD 模型[24]；一类是基于页岩气流动特征建立的新模型，但模型中仍未涉及产量与时间以外的物理参数，如 Duong 模型[25]、EEDCA 模型[26]、组合模型[27]等。其中，广义 Arps 模型、SEPD 模型和 Duong 模型是目前在页岩气井评价中应用最为广泛的 3 个经典模型。

表 1 页岩气井 EUR 评价方法

方法	特征	数据要求
解析模型法	基于页岩气复杂流动机理建立的，能够分析各作用机理对产量的影响，指导气藏合理高效的开发，但现阶段页岩气流动机理尚未完全弄清[23]，且模型建立与求解难度较大，普适性较差	流体物性参数＋地质参数＋工艺参数＋生产动态资料
数值模拟法		
物质平衡法	基于质量守恒定律建立的，即：原始地质储量＝采出量＋地下剩余储量。通常只研究采出气量与地层压力之间的关系	流体物性参数＋地质参数＋生产动态资料（至少两个测压点）
经验产量递减法	不考虑复杂渗流理论，仅根据大量生产井实测数据进行规律性分析	生产动态资料
现代产量递减法	将不稳定渗流理论与经验法相结合来对气井产量进行规律性分析	流体物性参数＋地质参数＋生产动态资料
概率法	考虑到部分影响气井产量的物理参数值具有不确定性，因此，根据不确定性大小来开展气井 EUR 评价，其计算方法与其他方法无差异	根据评价方法确定

1.1 广义 Arps 模型

经典 Arps 模型递减指数 b 通常介于 0 到 1 之间，并根据 b 值范围可将递减划分为指数递减（$b=0$）、双曲递减（$0<b<1$）和调和递减（$b=1$）3 种类型，如式（1）～式（3）所示。然而，由于页岩气在储层内流动机理极为复杂，使得 b 值通常大于 1。因此，通过扩展 b 值范围提出了广义 Arps 模型[20]。该模型中的指数递减模型和调和递减模型均为双参数模型，双曲递减模型则为三参数模型。

指数递减（$b=0$）

$$q_g = q_{gi} e^{-D_i t} \tag{1}$$

双曲递减（$0<b<1$）

$$q_g = \frac{q_{gi}}{\left(1+bD_i t\right)^{1/b}} \tag{2}$$

调和递减（$b=1$）

$$q_g = \frac{q_{gi}}{1+D_i t} \tag{3}$$

式中　q_g——气井日产量，m^3；
　　　q_{gi}——气井初始日产量，m^3；
　　　b——递减指数，无因次；
　　　D_i——初始递减率，1/d；
　　　t——生产时间，d。

1.2　SEPD 模型

Valko[24]在 2010 年提出了 SEPD 模型，该模型在经过大量井的检验后已成为页岩气井递减分析的常用方法之一。与 Arps 模型中的指数模型对比可知，SEPD 模型是指数递减的修正模型，但属于三参数模型。

$$q_g = q_{gi}\exp\left[-(t/\tau)^n\right] \qquad (4)$$

式中　τ——特征松弛时间，d；
　　　n——时间指数，无因次。

1.3　Duong 模型

Duong 模型是基于页岩气井生产过程中以裂缝主导流为主，长期处于线性流阶段，较难达到晚期稳定流阶段的假设而提出来的[25]。该模型形式较复杂，为三参数模型，且评价结果多偏高。

$$q_g = q_{gi}t^{-m}\exp\left[a\left(t^{1-m}-1\right)/(1-m)\right] \qquad (5)$$

式中　m——幂函数指数，无因次；
　　　a——递减系数，1/d。

2　页岩气井 EUR 评价模型的建立

2.1　页岩气井生产动态数值模拟

由于实际页岩气井生产数据点多且波动大，难以对其进行有效分析，因此，基于长宁区块基础资料（表 2）建立了页岩气藏压裂水平井数值模拟理论模型，模拟了气井控压生产中 4 次变压 5 个阶段的生产情况，如图 1 所示，从而获得了气井生产动态数据。

表 2　页岩气藏压裂水平井数值模拟理论模型物理参数表

油藏面积/（m×m）	储层厚度/m	储层顶深/m	储层温度/℃	原始地层压力/MPa	裂缝半长/m	压裂段数/段
2500×600	60	3000	90	40	60	18

最小井底流压/MPa	基质孔隙度/无因次	裂缝孔隙度/无因次	基质渗透率/mD	裂缝渗透率/mD	Langmuir 体积/（$m^3 \cdot t^{-1}$）	Langmuir 压力/MPa
4	0.3	0.003	5×10^{-5}	5×10^{-3}	2	6.5

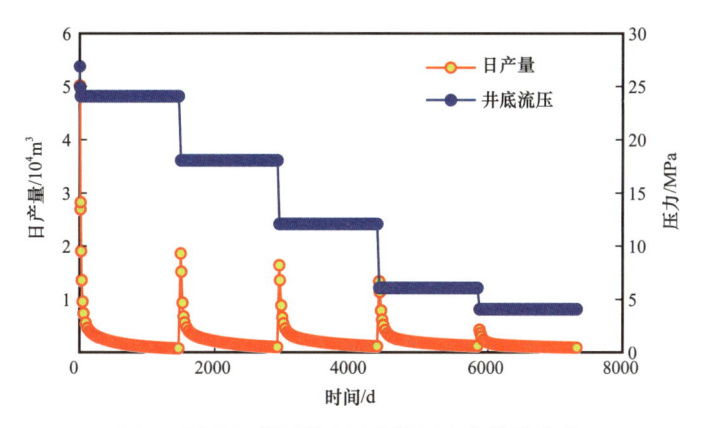

图1 页岩气藏压裂水平井控压生产模拟曲线

2.2 数据处理与模型建立

2.2.1 数据处理

由于规整化产量可纵向放缩产量，时间函数转换可延长横向真实时间[28]。因此，结合压力规整化产量和物质平衡时间，分别对模拟产量和时间进行处理，发现处理后的数据在半对数坐标系下和双对数坐标系下均得到了有效的收敛，如图2和图3所示。特别是双对数坐标系下数据可收敛于一条直线，更有利于数据分析。因此，采用双对数坐标系下压力规整化产量—物质平衡时间图版来进行气井生产数据分析。

$$\bar{q}_g = q_g / (P_i - P_{wf}) \tag{6}$$

$$t_m = Q / q_g \tag{7}$$

式中　$\bar{q}_g$——规整化产量，m³/(d·MPa)；

　　　P_i——初始地层压力，MPa；

　　　P_{wf}——井底流压，MPa；

　　　Q——气井累计产量，m³；

　　　t_m——物质平衡时间，d。

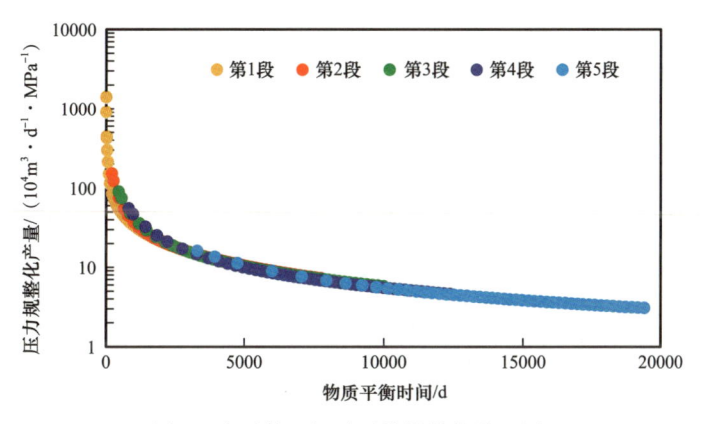

图2 半对数坐标系下模拟数据处理图

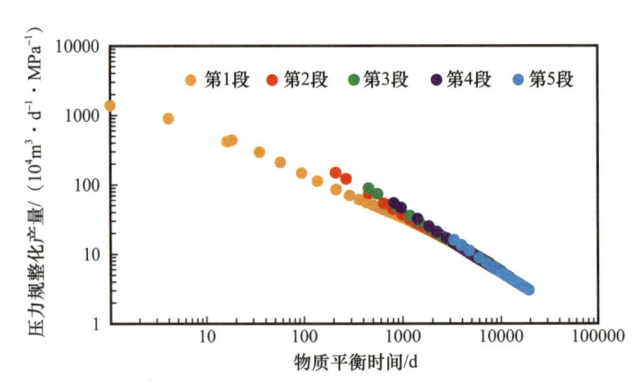

图 3　双对数坐标系下模拟数据处理图

2.2.2 模型建立

根据对双对数坐标下的处理后数据点晚期线性特征，可以建立如下的特征线方程：

$$\lg\frac{q_g}{p_i - p_{wf}} = A\lg t_m + B \tag{8}$$

式中　A——直线斜率，无因次；
　　　B——直线截距，无因次。

由式（6）~式（8）可得页岩气井累计产量与日产量间的关系为：

$$Q = q_g \cdot 10^{\left(\lg\frac{q_g}{P_i - P_{wf}} - B\right)/A} \tag{9}$$

由于页岩气井是定井底流压下的控压生产，因此气井最小井底流压 P_{wf_min} 是已知的。当气井产量达到废弃产量 q_{g_ab} 时，此时气井累计产量 Q 即为气井的 EUR，由此可建立页岩气井 EUR 评价模型为：

$$Q_{EUR} = q_{g\_ab} \cdot 10^{\left(\lg\frac{q_{g\_ab}}{P_i - P_{wf\_min}} - B\right)/A} \tag{10}$$

式中　P_{wf_min}——最小井底流压，MPa；
　　　q_{g_ab}——气井废弃产量，m³/d；
　　　Q_{EUR}——气井最终累计产量，m³。

3　页岩气井 EUR 评价模型应用

3.1　页岩气井 EUR 快速评价流程的建立

根据建立的数据处理方法与 EUR 评价模型，提出了一套页岩气井 EUR 评价流程，即在原始数据中去除关井数据点后，仅需在双对数坐标系下获取压力规整化产量–物质平衡时间图版中线性部分数据点的直线斜率与截距便可完成页岩气井 EUR 评价，如图 4 所示。该流程较基于多参数非线性经验模型（如 Duong 模型）所建立的评价流程[5]不仅降低了数据波动造成的影响，同时也降低了参数拟合难度。因此，相较于以往的 EUR 评价流程，本文提出的评价流程操作更简便，计算更高效。

原始生产动态数据 → 去除生产数据中关井点 → 建立双对数坐标系下规整化产量-物质平衡-时间图版 → 拟合图版中线性数据点，获取直线斜率与截距 → 基于页岩气井EUR评价新模型完成气井的EUR评价

图 4　页岩气井 EUR 快速评价流程

3.2　页岩气井 EUR 评价算例

以长宁区块 3 口典型页岩气井生产数据为例进行气井 EUR 评价分析，如图 5 所示。其中，H1 井累计投产 1468 天，生产数据波动剧烈；H2 井累计投产 512 天，当生产第 256 天以后，气井日产量不到 $5 \times 10^4 m^3$，生产至 272 天时因工作制度调整，气井日产量最大增至 $11.62 \times 10^4 m^3$，使得生产数据呈现为两段式形态；类似地，H3 井在投产的 472 天中，因生产工作制度更加频繁的变动导致气井生产数据呈现更为复杂的多段式形态。

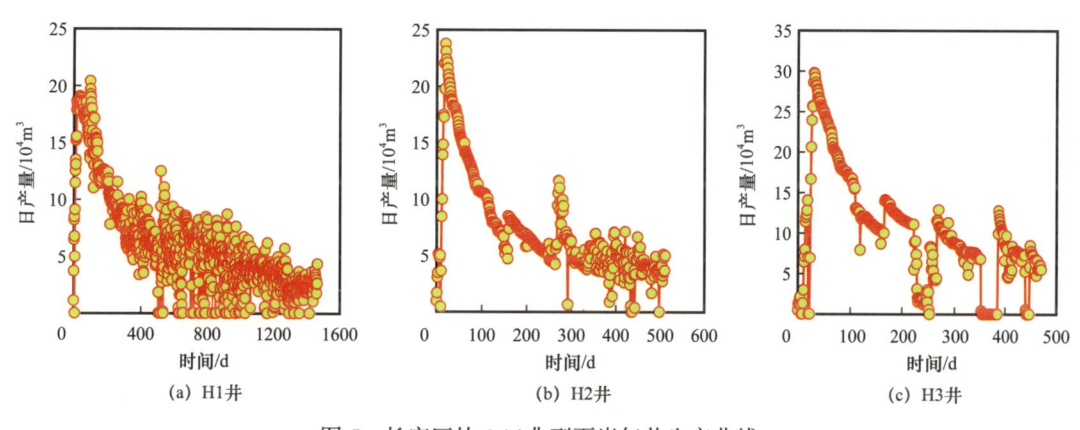

(a) H1井　　(b) H2井　　(c) H3井

图 5　长宁区块 3 口典型页岩气井生产曲线

利用本文提出的页岩气井 EUR 快速评价流程，首先通过对 3 口页岩气井的实测产量及时间分别进行规整化处理和物质平衡时间转换，然后再进行对数处理，由此可得图 6，不难发现 3 口典型气井的生产数据在处理后均取得了较好的收敛。由图 7 可知，直线对后半部分数据点的拟合系数 R^2 均大于 90%，表明此时 3 口气井均已达到了基质向裂缝供给的线性流阶段。

将 3 口井拟合得到的直线斜率与截距分别代入页岩气井 EUR 评价新模型（式 10）中进行气井的 EUR 评价，其评价结果与广义 Arps 模型、SEPD 模型和 Duong 模型 3 种经典模型评价结果相近（表 3），由此确保了评价新方法的可靠性。

表 3　不同方法对 3 口典型页岩气井 EUR 评价结果表

井名	EUR/$10^8 m^3$			
	新方法	广义 Arps	SEPD	Duong
H1	1.44	1.76	1.23	1.78
H2	0.77	1.08	0.49	1.15
H3	0.92	0.77	0.50	1.40

(a) H1井　　　　　　　　　(b) H2井　　　　　　　　　(c) H3井

图6　处理后的3口典型页岩气井生产曲线

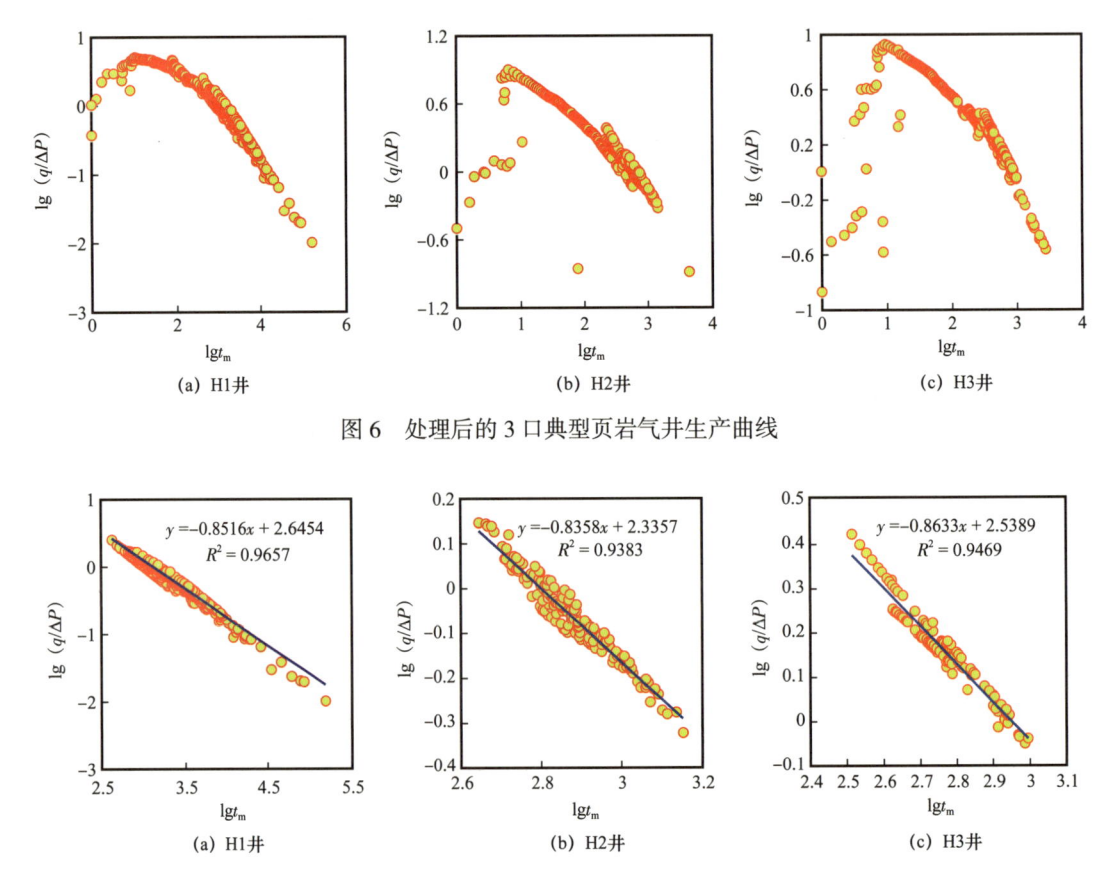

(a) H1井　　　　　　　　　(b) H2井　　　　　　　　　(c) H3井

图7　数据点线性部分拟合结果图

4　结论

（1）针对页岩气井生产数据波动剧烈的问题，提出了双对数坐标系下建立压力规整化产量—物质平衡时间图版的数据处理方法，实例应用表明该方法能使波动数据获得较好的收敛，具有可行性。

（2）建立了一个新的页岩气井EUR评价模型。通过对实际页岩气井进行EUR评价表明，新模型与经典模型计算结果相近，确保了新模型的可靠性。

（3）提出了一套页岩气井EUR评价新方法。该方法操作简便，易计算，避免了经典方法中参数多、拟合难的问题，可用于页岩气井EUR快速评价。

参 考 文 献

[1] 张德良，吴建发，张鉴，等.北美页岩气规整化产量递减分析方法应用——以长宁—威远示范区为例[J].科学技术与工程，2018，18（34）：56-61.

[2] Pang W, Du J, Zhang T Y. Production data analysis of shale gas wells with abrupt gas rate or pressure changes[C]// SPE 195134-MS, 2019. doi: https://doi.org/10.2118/195134-MS.

[3] Wang K, Li H T, Wang J C, et al. Predicting production and estimated ultimate recoveries for shale gas

wells: A new methodology approach [J]. Applied Energy, 2017, 206: 1416-1431.

［4］宋海敬，苏云河，熊小林，等. 页岩气井EUR评价流程及影响因素分析 [J]. 天然气地球科学，2019, 30（7）: 1-7.

［5］白玉湖，陈桂华，徐兵祥，等. 页岩油气产量递减典型曲线预测推荐做法 [J]. 非常规油气, 2017, 21（1）: 49-53.

［6］Christopher R C, Morteza N, Danial K, et al. Production analysis of tight gas and shale gas reservoirs using the dynamic-slippage concept [C] // SPE 144317-MS, 2011. doi: https://doi.org/10.2118/144317-MS.

［7］Liang P, Rahmanian M, Mattar L. Superposition-rate as an alternative to superposition-time [C] // SPE 167124-MS, 2013. doi: https://doi.org/10.2118/167124-MS.

［8］Chigozie A, Fabian V. RTA Assisted production forecasting in shale reservoir development [C] // URTEC-2870785-MS, 2018. doi: https://doi.org/10.15530/URTEC-2018-2870785.

［9］Zhang L H, Gao J, Hu S Y, et al. Five-region flow model for MFHWs in dual porous shale gas reservoirs [J]. Journal of Natural Gas Science and Engineering, 2016, 33: 1316-1323.

［10］吴永辉，程林松，黄世军，等. 考虑页岩气赋存及非线性流动机理的产能预测半解析方法 [J]. 中国科学：技术科学, 2018, 48（6）: 691-700.

［11］Wu M L, Ding M C, Yao J, et al. Production-Performance Analysis of composite shale-gas reservoirs by the boundary-element method [J]. SPE Reservoir Evaluation & Engineering, 2019, 22: 238-252.

［12］Zhang L H, Shan B C, Zhao Y L, et al. Gas transport model in organic shale nanopores considering langmuir slip conditions and diffusion: pore confinement, real gas, and geomechanical effects [J]. Energies, 2018, 11: 1-23.

［13］梁洪彬. CN-WY页岩气藏产能预测模型研究及应用 [D]. 重庆：重庆科技学院，2018.

［14］Chai D, Fan Z Q, Li X L. A new unified gas-transport model for gas flow in nanoscale porous media [J]. SPE Journal, 2019, 24（2）: 1-22.

［15］梅海燕，何浪，张茂林，等. 考虑多因素的页岩气藏物质平衡方程 [J]. 新疆石油地质, 2018, 39（4）: 456-461.

［16］郭小哲，李景，张欣. 页岩气藏压裂水平井物质平衡模型的建立 [J]. 西南石油大学学报（自然科学版）, 2017, 39（2）: 132-138.

［17］魏明强，段永刚，方全堂，等. 基于物质平衡修正的页岩气藏压裂水平井产量递减分析方法 [J]. 石油学报, 2016, 37（4）: 508-515.

［18］Tan L, Zuo L H, Wang B B. Methods of decline curve analysis for shale gas reservoirs [J]. Energies, 2018, 11（3）: 1-18.

［19］Omar M, Mazher I, Chester P, et al. EUR Prediction for Unconventional Reservoirs: State of the Art and Field Case [C] // SPE 191160-MS, 2018. doi: https://doi.org/10.2118/191160-MS.

［20］于荣泽，姜巍，张晓伟，等. 页岩气藏经验产量递减分析方法研究现状 [J]. 中国石油勘探, 2018, 23（1）: 109-116.

［21］Fanchi J R, Cooksey M J, Lehman K M, et al. Probabilistic decline curve analysis of Barnett, Fayetteville, Haynesville, and Woodford gas shales [J]. Journal of Petroleum Science and

Engineering, 2013, 109: 308-311.

[22] 徐兵祥, 白玉湖, 陈桂华, 等. 页岩油气不确定性产量预测方法及应用[J]. 兰州大学学报(自然科学版), 2017, 53(6): 757-763.

[23] Zhang L H, Shan B C, Zhao Y L, et al. Review of micro seepage mechanisms in shale gas reservoirs[J]. International Journal of Heat and Mass Transfer, 2019, 13: 144-179.

[24] Valko P P, Lee W J. A better way to forecast production from unconventional gas wells[C]// SPE 134231-MS, 2010. doi: https://doi.org/10.2118/134231-MS.

[25] Duong A N. Rate-Decline Analysis for fracture-dominated shale reservoirs[J]. SPE Reservoir Evaluation & Engineering, 2011, 14(3): 377-387.

[26] Zhang He, Nelson Eric, Olds Dan, et al. Effective Applications of Extended Exponential Decline Curve Analysis to both Conventional and Unconventional Reservoirs[C]// SPE 181536-MS, 2016. doi: https://doi.org/10.2118/181536-MS.

[27] 王怒涛, 陈仲良, 祝明谦, 等. 页岩气压裂水平井产量递减组合模型分析[J]. 大庆石油地质与开发, 2018, 37(5): 135-140.

[28] 徐兵祥. 变产—变压情况下的页岩油气生产数据分析方法[J]. 天然气工业, 2017, 37(11): 70-76.

Comprehensive optimization of managed drawdown for a well with pressure-sensitive conductivity fractures : workflow and case study

Wei Yunsheng, Jia Ailin, Wang Junlei, Qi Yadong

(PetroChina Research Institute of Petroleum Exploration and Development)

Abstract: The effect of pressure drawdown schedule on well performance is generally related to stress sensitivity in the fractures. The purpose of this paper is to develop a detailed workflow of optimizing drawdown schedule to improve long-term production by delaying fracture closure and conductivity degradation in hydraulically fractured well. The workflow consists of the following three steps : based on experimental data of propped fracture in shale samples, an alternative relationship between conductivity and pressure drawdown is developed to mimic approximately the change of conductivity with effective stress during reservoir depletion. Based on an existing semi-analytical modeling approach, transient inflow performance relationship is introduced to build a direct link between drawdown and well productivity. The value of BHP drawdown on the reversal behavior of productivity is defined as the optimum drawdown in the specified IPR, and the optimum profile of BHP-drawdown verse time could be achieved. Finally, we demonstrate the effectiveness of this workflow with a field case from Zhaotong shale in China. The proposed workflow could improve the ability to improve the long-term performance of the well, thereby not only mitigating the productivity loss but maintaining enough driving force.

Key words: pressure sensitivity ; conductivity fracture ; transient IPR

In the last decades, the wide application of horizontal drilling and fracturing-stimulation treatment has gained great success in economic development of unconventional resources. A large volume of fluids is pumped to create a huge contact area between fractures and tight formation [1]. Although large amount of proppants are pumped to prevent fracture closure to some extent, the increases in effective stress caused by the extraction of fluids inevitably results in fracture closure. Fracture closure is also referred to as "geomechanics effect" [2-3]. Given that fractures are more deformable than the matrix and the conductivity of fractures dominates the overall flow behavior, productivity loss is often attributed to geomechanics-related factors in conductivity of propped/ unpropped fractures due to proppant embedment, crushing, and fracture-face creep. It was found that a proper managed drawdown schedule can improve EUR by reducing the effective stress on the fractures. Field practices in Haynesville shale demonstrate that production wells using restricted drawdown have an average performance with a first-year decline of only 38%, lower than 83% for these wells on unrestricted drawdown[4]. Inversely, an unrestricted drawdown strategy may cause a reduction in EUR of up to 20% in Vaca Muerta shale[5].

Most of experimental studies investigated the fracture conductivity under changing stress conditions, not pore pressure. The results found that the relationship between the effective stress and fracture conductivity is non-linear. According to the poroelastic theory that the effective normal stress is the difference between the total normal stress and the pore pressure when Biot coefficient is assumed to be 1, the change in effective stress is typically about 0.4–0.7 times the change in formation pore pressure[6]. For the fracture, the change in closure stress can be also related to the drop in pore pressure by the relationship:

$$\Delta\sigma_{closure} = \alpha \frac{1-2\upsilon}{1-\upsilon} \Delta p \quad (1)$$

where α is the Biot coefficient, υ is the Poisson's ratio. Noted that the relationship is not a straightforward concept in physics. Traditional reservoir simulation approaches widely incorporate compaction tables (pore pressure and fracture conductivity), relating pore pressure to effective stress, to mimic the fundamental physics of fracture closure. The key step is to couple reservoir flow with pressure-dependent permeability by using pseudo-pressure[7-8] and the permeability modulus techniques[2, 5, 7-10]. These approaches assume that the change in pore pressure has a predefined relationship with effective stress, which ignores the path-dependency of the stress changing caused by different depletion schedules. To capture the critical process of stress path dependency that governs the effective stress on the fractures, various coupled-geomechanics reservoir simulation models such as fully-coupled, iteratively coupled and explicit schemes were constructed[11-16]. Although the dynamic stress changes can be calculated during reservoir depletion, the coupled-geomechnics simulations are very specialized and time-consuming.

In this work, based on the assumption of dynamic permeability-decay modulus, we develop an efficient and accurate method using semi-analytical models to predict fracture closure and the resulting productivity. The model is able to approximately reproduce the coupled-geomechanical results across a wide range of drawdown management strategies with an acceptable error. Then, by use of transient inflow performance relation (IPR), a workflow is calibrated to find the optimal path of drawdown strategy. A field case study from a shale gas reservoir in China is used to demonstrate the effectiveness of the workflow.

1 Integrated Approach

1.1 Permeability-decay coefficient

The permeability in the fracture is related with effective stress, which is given by

$$K_f = K_{fi} \exp\left[-d_f(\sigma_{\text{eff}} - \sigma_{\text{eff},i})\right] \quad (2)$$

where k_{fi} is the initial permeability, σ_{eff} is effective stress and d_f is the permeability modulus. The changing in the effective stress controls fracture aperture and resulting conductivity. Noted that due to the mismatch of asperities as shown in Fig.1, fractures can still retain part of their conductivity even if fracture walls have come into contact. The residual fracture conductivity is denoted as $k_{f\min}$, which is still considerably larger than matrix permeability. After further using the relation given in Eq.1, Eq.2 is rewritten by

$$\frac{K_f - K_{f\min}}{K_{fi} - K_{f\min}} = \exp\left[-\gamma_f(p_i - p)\right] \quad (3)$$

where permeability-decay coefficient γ_f satisfies the relationship: $\gamma_f = d_f \upsilon \alpha / (1-\upsilon)$.

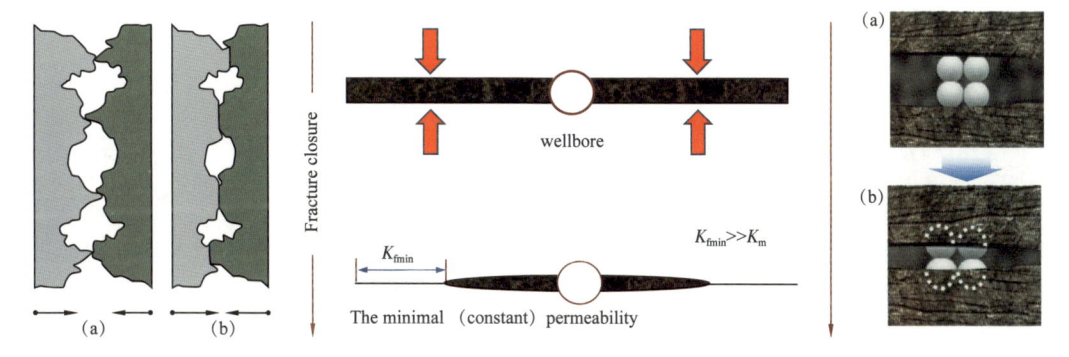

Fig.1 Schematic of fracture closure: initial condition (a) and final condition (b)

According to the lab measurement data for shale samples in Sichuan Basin China, Fig.2 (a) presents the relationship between normalized fracture conductivity and closure pressure due to proppant embedment for different shale samples. As may be analyzed from Fig.2 (b), the relationship is nonlinear. Therefore, it is acceptable to regard the permeability-decay coefficient to be a function of closure pressure. The permeability-decay coefficient is a line function of closure pressure (pressure drawdown) with non-zero intercept.

$$\gamma_f = a\Delta p_f + b \quad (4)$$

where a and b are the characteristic parameters determined by experimental data.

1.2 Transient IPR based on semi-analytical modeling

For multiple-fractured horizontal well with stress-sensitivity conductivity, a semi-analytical model is developed to efficiently simulate the production performance across a wide range of drawdown strategies. For flow in the fracture, the pressure/stress dependency is taken into account in the revisited diffusivity equation as following dimensionless form (dimensionless variables are defined in Appendix A):

$$\begin{cases} \dfrac{\partial}{\partial x_{Dn}}\left(C_{fDn}(p_{fDn})\dfrac{\partial p_{fDn}}{\partial x_{Dn}}\right)-2\pi q_{fDn}(x_{Dn})+2\pi q_{wfDn}(t_D)\delta(x_{Dn},x_{wfDn})=0 \\ q_{cDn}(x_{Dn})=\displaystyle\int_{x_{Dn}}^{L_{fDn}} q_{fDn}(\varsigma)d\varsigma, \left.\dfrac{\partial p_{fDn}}{\partial x_{Dn}}\right|_{x_{Dn}=0}=\left.\dfrac{\partial p_{fDn}}{\partial x_{Dn}}\right|_{x_{Dn}=L_{fDn}}=0 \end{cases} \quad (5)$$

where dynamic conductivity is given by

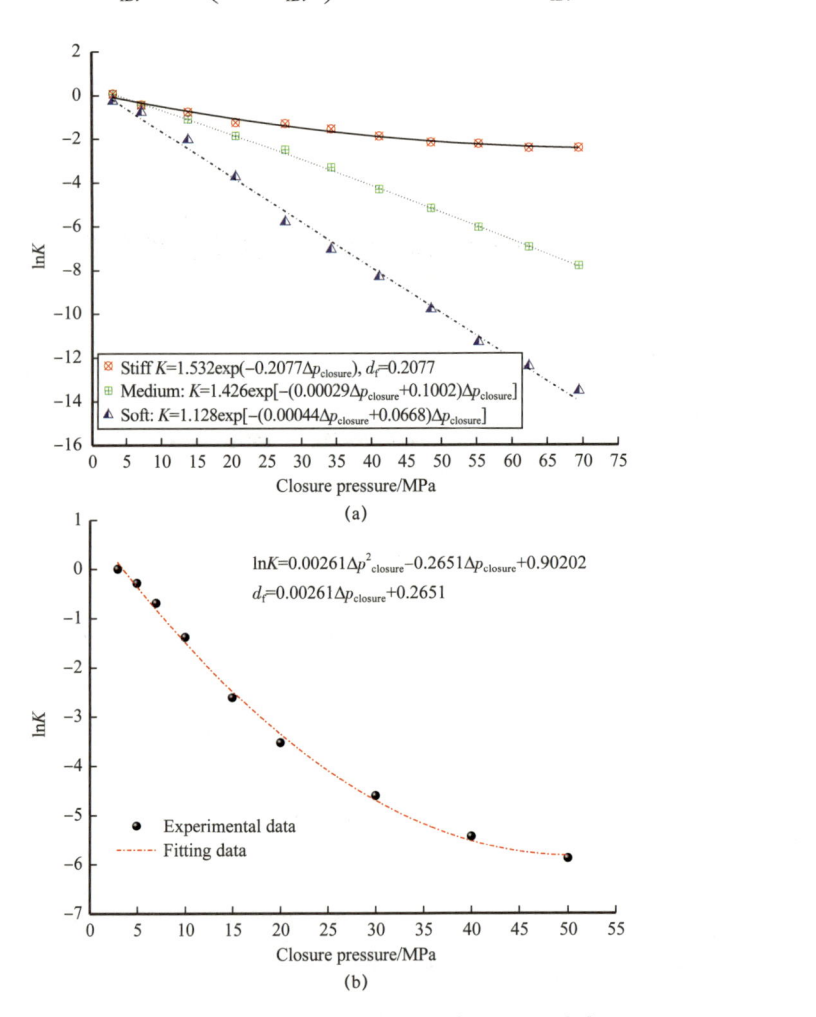

$$\frac{C_{fDn}(p_{fDn})}{C_{fDi}}=\left(1-\frac{C_{fD\min}}{C_{fDi}}\right)\cdot\exp[-\gamma_{fD}p_{fD}]+\frac{C_{fD\min}}{C_{fDi}} \quad (6)$$

Fig.2 The relationship between normalized fracture conductivity and closure pressure

The semi-analytical model utilizes the Fredholm integral equation to solving the nonuniform-influx-density linear flow in the fracture [17] and an analytical instantaneous point solution [18] for describing the flow in the matrix. The dependence of dynamic conductivity on pressure makes the diffusion equation nonlinear. The dimension transformation method renders the nonlinear equation amenable to linear analytical treatment [19]. Besides, the operation condition of BHP-drawdown or production-rate management is incorporated by modifying inner boundary condition of the model. More details about the model development can be found in our previous work [20-21].

On the basis of semi-analytical modeling, the production performance can be represented using inflow performance relationship (IRP) (Fig.3). The IPR curve for a well is the relationship between the production rate and the BHP. Conventional IPR methods are based on a knowledge of the average reservoir pressure and an assumption of stabilized-flow condition. Slope of the IPR is equal to the inverse of the productivity index

$$\text{PI} = \frac{q_D}{p_{wD} - p_{avgD}} = \left(\frac{1}{2}\ln\frac{4A}{e^\gamma C_A r_{\text{eff}}^2}\right)^{-1} \tag{7}$$

where A is the drainage area, γ is the Eular constant (=0.5772), r_{eff} is the effective wellbore radius, and C_A is the shape factor depending on the geometry of drainage area and location of well.

For a slightly-compressible fluid under single phase flow condition, the fluid obey simple Darcy's law, and the IPR curve behaves a straight-line relationship between the production rate and BHP over the whole BHP range. For pressure-dependent fractured well, as shown in Fig.4, when BHP approximates to initial pressure IPR is a straight line and when BHP further decreases IPR exhibits a downward curvature.

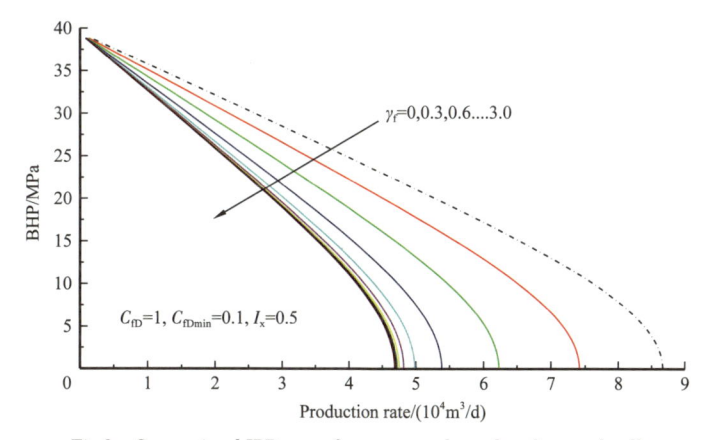

Fig.3 Conventional IPR curve for pressure-dependent fractured well

For low-permeability reservoir, transient flow response lasts long period, and the average pressure is time dependent. Assuming that average pressure equals to initial pressure during transient flow, we can achieve a transient IPR. Semi-analytical model presented previously allows transient IPRs to be readily generated. First, a representative fracture-reservoir model is selected. Next, the model is used to forecast the production rate (or BHP) for a specified time under different BHP (or production rate) conditions. Fig.4 illustrates the IPRs during transient flow. With the increase of BHP drawdown, the production rate is also increasing, but the production gain is lower than predicted straight-line behavior. In other words, the productivity index (defined as dq/dp_w) still increases with drawdown increasing. Noted that the IPR passes through initial pressure at zero rate. At the time of interest, the rate and BHP pairs for a given time are used to generate transient IPR.

The permeability-decay coefficient in Eq.4 is incorporated into semi-analytical model. different from the phenomenon in Fig.4, an illustration in Fig.5 is presented that for a specified time decreasing BHP causes the production rate to increase up to a certain point; once beyond that point, the reverse behavior would appear and production rate also decreases with the drawdown increasing. Put another way, there is an optimum operational condition at which the maximum production rate can be obtained.

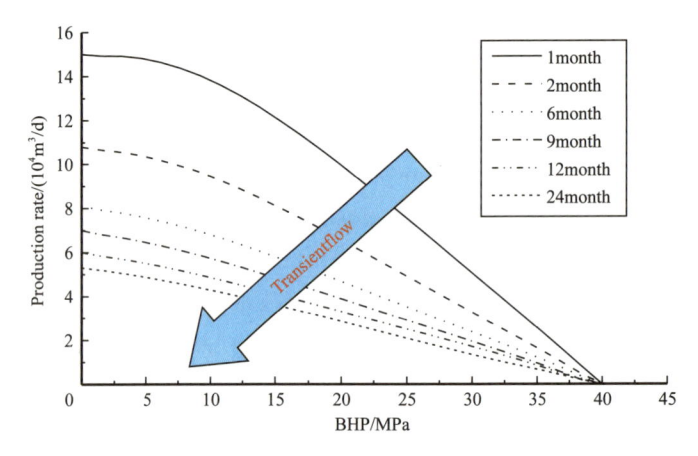

Fig.4 Transient IPR curves for pressure-dependent fractured well

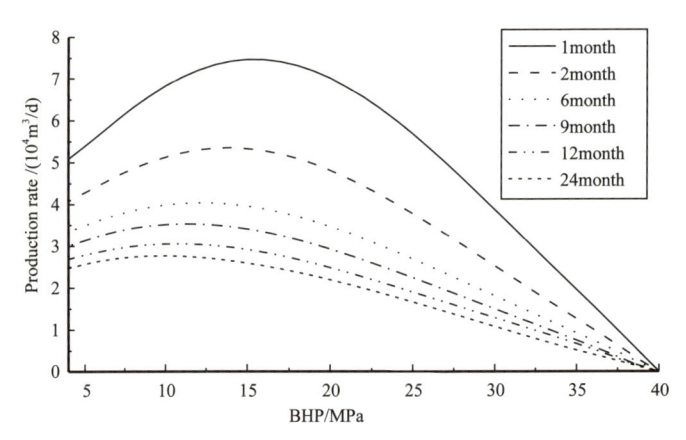

Fig.5 Reversal of productivity index on transient IPR

1.3 Optimization Workflow

A workflow of optimizing BHP drawdown strategy is presented as shown in Fig.6. First, based on experimental data, constructing custom compaction table and getting the pressure-dependent permeability decay coefficient by fitting experimental data. Second, the key inputs including reservoir & fracture & fluid properties are collected. Then, for each specified time, these variables are put into semi-analytical model to simulate production rate under different BHP-drawdown conditions.

As shown in Fig.7 (a), the operating point is defined as the optimum BHP-drawdown at that given time. The corresponding reversal behavior is attributed to the serious degradation of productivity, which depends on the BHP drawdown. Finally, the optimum profile of BHP-drawdown can be achieved by integrating operating points on transient IPR curves corresponding to different times. Noted that the optimum BHP drawdown is increasing with time.

Fig.7 (b) illustrates the effect of the order of pressure-dependency magnitude on optimum drawdown profile. The larger the permeability-decay coefficient, the smaller the drawdown at the same time. It indicates that the fractured well with more intense stress effect requires a more conservative drawdown schedule to maximize the production rate. In other words, the optimum schedule for the case of intense stress sensitivity has a BHP profile that declined very gradually until line BHP was reached, while the optimum schedule for the case of weak stress sensitivity has a rapid BHP-decline profile.

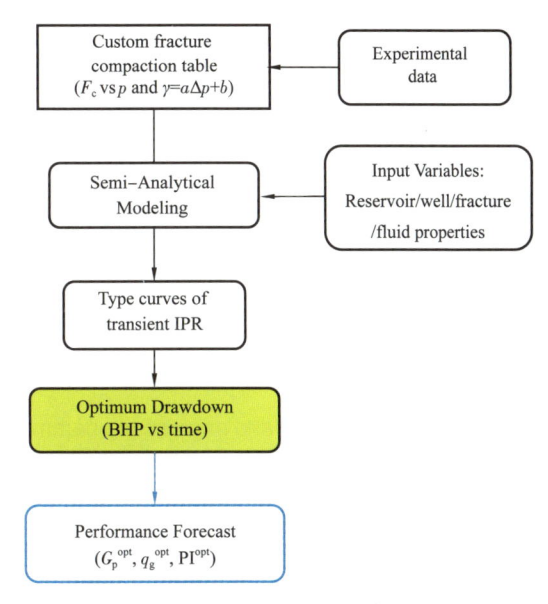

Fig.6 An integrated approach to optimize BHP drawdown schedule modified from the work of Wang et al.[21]

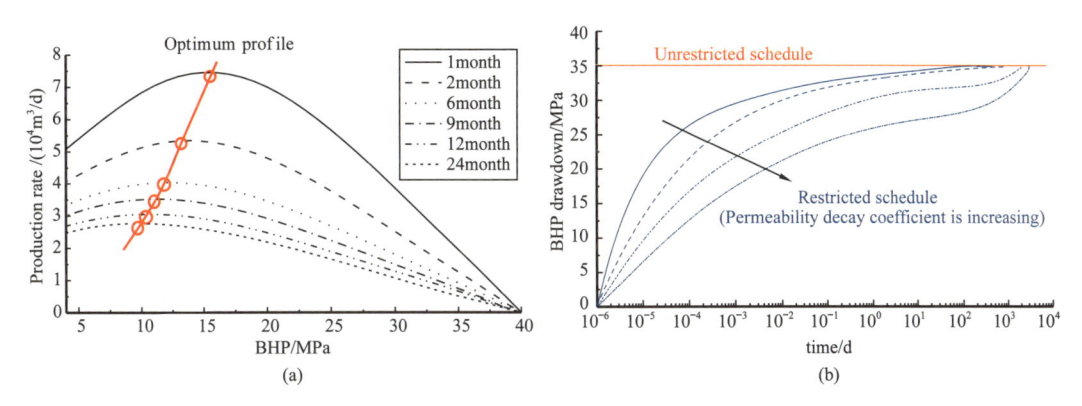

Fig.7 Optimum drawdown schedule verse time (a) and effect of the magnitude of stress dependency on optimum profile (b)

2 Field Case Study

Well performance is affected by many parameters, such as formation petrophysical properties, fluid properties, fracturing parameters, and operational condition. Numerous practices in the field demonstrate that when if petrophysical and fluid properties are fixed, the performance is consistent with engineering parameters including fractured horizontal length, number of fractures and volume of proppant.

Two horizontal wells from Zhaotong shale in China are selected to perform the analysis. Here, one well is controlled by use of unrestricted drawdown, and another well is controlled by use of optimum restricted drawdown. Other geology and engineering parameters are list in Tab.1. As might be analyzed from Tab.1, for unrestricted well, the fractured length and number of fractures are lower than restricted well, but the volume of proppant is higher than restricted well by 50%. Therefore, without taking the effect of operational condition into account, the overall productivity of unrestricted well approximates to restricted well.

Tab.1 Geology and engineering parameters for unrestricted and restricted wells

Well name	Operational condition	Horizontal length/m	Horizontal fractured length/m	Number of fractures	Volume of proppant/t
Unrestricted well	High drawdown	1510	1318	18	1673
Restricted well	Managed drawdown	1605	1568	22	1101

By analyzing the transient rate/pressure data in Fig.8, the decline rate of BHP for unrestricted well is 0.278MPa/d, but the value for restricted well is only 0.13MPa/d. Correspondingly, during the first 3 months, the average productivity index for restricted well is about $64.7 \times 10^4 m^3/MPa$, being higher than $37.8 \times 10^4 m^3/MPa$ for unrestricted well. The ratio of productivity index is about 1.7. During the first 12 month, the ratio is further increased by 2.0. The difference of productivity index is attributed to the effect of operational condition.

Based on the parameters presented in Tab.1, we established a conceptual reservoir model by use of semi-analytical modeling. The initial pressure is 38 MPa, and the finial line BHP is set to be 5MPa. Characteristic parameter a is $0.1108MPa^{-1}$, and characteristic parameter b is $0.00167MPa^{-2}$. Fracture conductivity and length are initially equal to 36mD and 90m respectively. The matrix porosity is 8%, and permeability initially equals to 0.0002mD. These parameters are all input into the semi-analytical model to perform history matching. BHP was used for simulation constraint and production rate is the history-matching variables. Fracture length, conductivity, and matrix permeability are mainly adjusted to achieve a good matching.

Next, according to the step presented in section 2.3, the optimum BHP schedule is achieved as shown in Fig.8 (a). Taking the effect of production history into account, Fig.8 (b) generates a practical transient IPR.

The optimum BHP-drawdown schedule presented in Fig.9 is then put into the restricted case to predict production performance, while line BHP-drawdown schedule is put into the unrestricted case. Fig.10 shows that the cumulative rate of unrestricted case is higher than the restricted case before 4.5 year. As time increases, it is more advantageous for restricted case to obtain more cumulative rate in long-time period. The 20-year cumulative rate for restricted case is about $0.835 \times 10^8 m^3$, while $0.636 \times 10^8 m^3$ for unrestricted case. Compared with unrestricted schedule, optimum restricted drawdown may cause an increase in EUR of up to 30%.

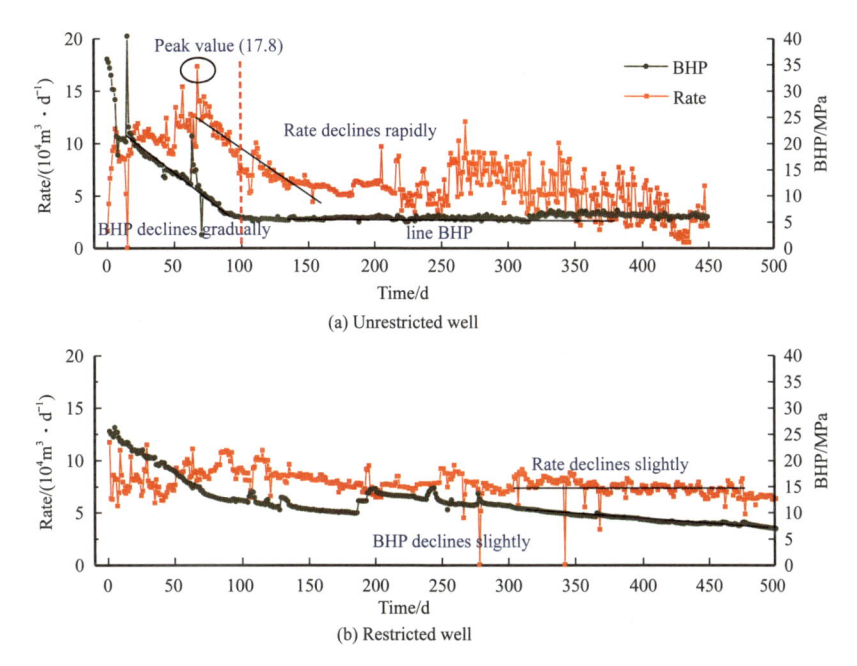

Fig.8 Comparison of production performance for restricted and unrestricted wells

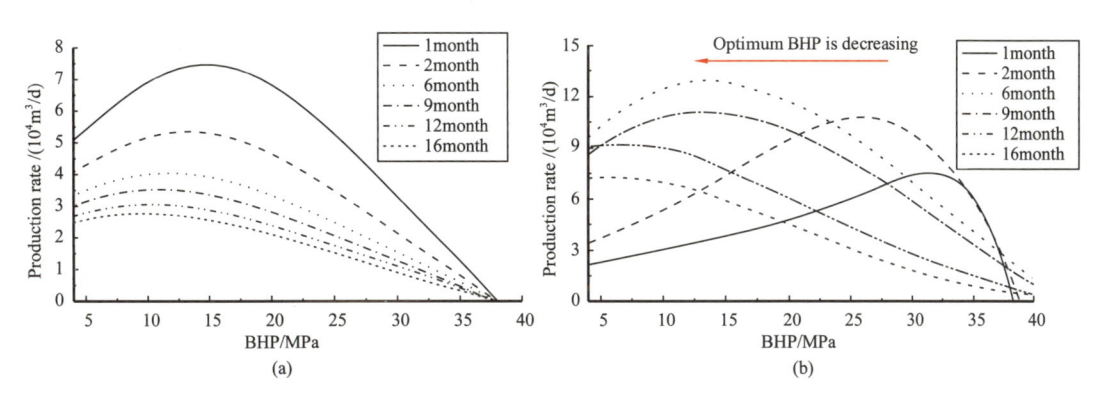

Fig.9 Optimum BHP drawdown for restricted well based on transient IPR without history (a) and with history (b)

3 Summary and Conclusions

In this study the effect of pressure-dependent conductivity on the performance of reservoir was investigated under different operational conditions. The relationship between production rate and BHP drawdown is quantified in the form of transient IPR. Afterward, an integrated approach was developed to give an efficient workflow of selecting an optimum BHP-drawdown schedule.

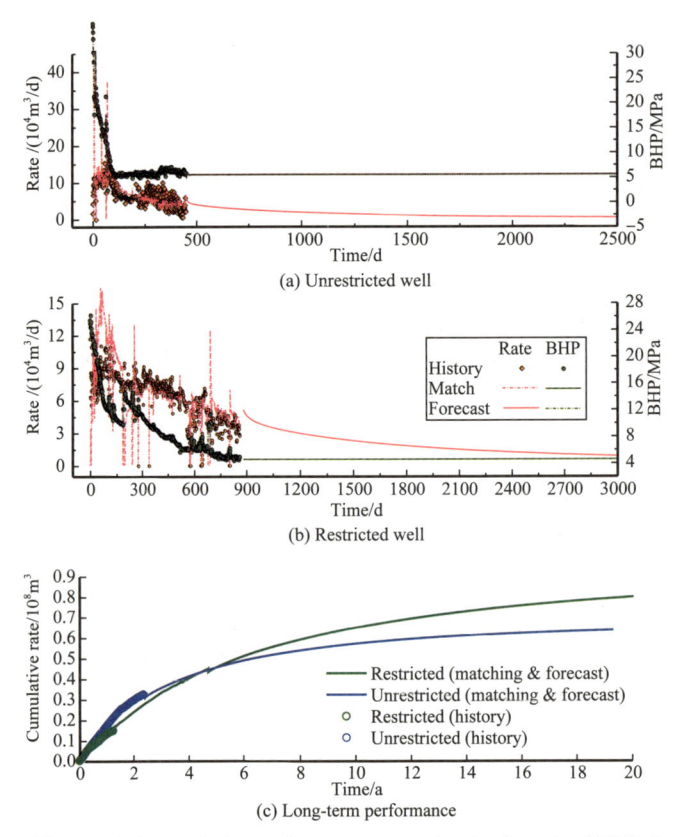

Fig.10 Simulation Zhaotong shale cumulative gas forecast for unrestricted and restricted BHP–drawdown schedules

Different from conventional IPR, in transient IPR the productivity index deteriorates with time. Meanwhile, due to the existence of pressure dependency of fracture conductivity, productivity index would deteriorate, which contributes to the production rate gain is lower than the predicted straight-line behavior. When the permeability-decay coefficient is a function of pressure drawdown, there exists a maximum value on transient IPR. Case study demonstrates that the resulting optimum BHP-drawdown schedule finds a reasonable tradeoff in which the fracture mains considerable conductive while maintaining a high enough drawdown to maximize the long-term performance.

Appendix A : dimensionless definitions used in semi-analytical model

For simplicity, the mathematical model is expressed in the dimensionless form, and the corresponding variables are defined as follows :

$$t_D = \frac{K_m t}{\varphi_m \mu c_t L_{ref}^2}, L_D = \frac{L}{L_{ref}}, C_{fD} = \frac{K_f(p_i) w_f}{K_m L_{ref}}, \gamma_{fD} = a_D p_{fD} + b_D \quad (A-1)$$

In the condition of constant-rate operation, the dimensionless definitions are given by

$$p_{fD} = \frac{2\pi K_{mi} h(p_i - p_f)}{q_w \mu B}, \gamma_{fD} = \frac{q_w \mu B}{2\pi K_m h}\gamma_f, q_{wfD} = \frac{q_{wf}}{q_w}, q_{fD} = \frac{2q_f L_{ref}}{q_w} \quad (A-2)$$

In the condition of constant-BHP operation, the dimensionless definitions are given by

$$p_{fD} = \frac{p_i - p_f}{p_i - p_w}, \gamma_{fD} = (p_i - p_w)\gamma_f, q_{wfD} = \frac{q_{wf}\mu B}{2\pi K_m h(p_i - p_w)}, q_{fD} = \frac{(2q_f L_{ref})\mu B}{2\pi K_{mi} h(p_i - p_w)} \quad (A-3)$$

References

[1] Cipolla C L, Williams M J, Weng X, et al. Hydraulic fracture monitoring to reservoir simulation: maximizing value [C] // Paper SPE 133877 presented at the SPE Annual Technical Conference and Exhibition, 2010, 19-22 September, Florence, Italy.

[2] Aybar U, Yu W, Eshkalak M O, et al. Evaluation of production losses from unconventional shale reservoirs [J].Journal of Natural Gas Science and Engineering, 2015, 23: 509-516.

[3] Britt L K, Smith M B, Klein H H, et al. Production benefits from complexity - effects of rock fabric, managed drawdown, and propped fracture conductivity [C] // Paper SPE-179159-MS presented at the SPE Hydraulic Fracturing Technology Conference, 2016, 9-11 Februray, Woodlands, Texas, USA.

[4] Stewart G. Integrated analysis of shale gas well production data [C] // Paper SPE-171420-MS presented at the SPE Asia Pacific oil & Gas Conference and Exhibition, 2014, 14-16 October, Adelaide, Australia.

[5] Rojas D, Lerza A. Horizontal well productivity enhancement through drawdown management approach in Vaca Muerta shale [C] // Paper SPE-189822-MS presented at the SPE Canada Unconventional Resources Conference, Calgary, Alberta, Canada, 2018, 13-14 March.

[6] Okouma V, Guillot F, Sarfare M, et al. Estimated ultimate recovery (EUR) as a function of production practices in the Haynesville shale [C] // Paper SPE 147623 presented at the SPE Annual Technical Conference and Exhibition [C] // 30 October-2 November, 2011, Denver, Colorado, USA.

[7] Clarkson C R, Qanbari N M, Heffner L. Incorporating geomechanical and dynamic hydraulic fracture property changes into rate-transient analysis: example form the Haynesville Shale [C] // Paper SPE-162526-MS presented at the SPE Canadian Unconventional Resources Conference, 2012, 30 October-1 November, Calgary, Alberta, Canada.

[8] Qanbari F, Clarkson C R. Analysis of transient linear flow in stress-sensitive formations [C] // Paper SPE-162741-MS presented at the SPE Canadian Unconventional Resource Conference, 2012, 30 October-1 November, Calgary, Canada.

[9] Wang S H, Ma M X, Ding W, et al. Approximate analytical-pressure studies on dual-porosity reservoirs with stress-sensitive permeability [J]. SPE Reservoir Evaluation & Engineering, 2015, 11: 523-533.

[10] Tabatabaie S H, Pooladi-Darvish M, Mattar L, et al. Analytical modeling of linear flow in pressure-sensitive formations [J]. SPE Reservoir Evaluation & Engineering, 2016, 20 (1): 215-227. SPE-181755-PA.

[11] Cho Y, Apaydin O G, Ozkan E. Pressure-dependent natural-fracture permeability in shale and its effect on shale-gas well production [J]. SPE Reservoir Evaluation & Engineering, 2012, 5: 216-228.

[12] Rahman K, Gui F, He W. Multistage hydraulic fracturing optimization for a shale oil field integrating geomechanics and production modeling [C] // Paper SPE-172097-MS presented at the Abu Dhabi International Petroleum Exhibiton and Conference, 2014, 10-13 November, Abu Dhabi, UAE.

[13] Mirani A, Marongiu-Porcu M, Wang H Y, et al. Production pressure drawdown management for fractured horizotnal wells in shale gas formations [C] // Paper SPE-181365-MS presented the SPE Annual Technical Conference and Exhibition, Dubai, UAE, 2016, 26-28 September.

[14] Wilson K. Analysis of drawdown sensitivity in shale reservoirs using coupled-geomechanics models [C] // Paper SPE-175029-MS presented at the SPE Annual Technical Conference and Exhibition, 2015, 28-29 September, Houston, Texas, USA.

[15] Wilson K. Efficient stress characterization for real-time drawdown managment [C] // Paper URTeC-221192-MS present at the Unconventional Resources Technology Conference, 2017, 24-26 July 2015, Austin, Texas, USA.

[16] Yong R, Wu J F, Shi X W, et al. Developmemt strategy optimization of Ning201 block Longmaxi shale gas [C] // Paper SPE-191452-18IHFT-MS presented by the SPE International Hydraulic Fracturing Technology Conference and Exhibition, 2018, 16-18 October, Muscat, Oman.

[17] Cinco-Ley H, Zeng H-Z. Pressure transient analysis of wells with finite conductivity vertical fractures in double porosity reservoirs [C] // Paper SPE 18172 presented at the 63rd Annual Technical Conference and Exhibition, 1988, 2-5 October, Houston, Texas, USA.

[18] Ozkan E, Raghavan R. New solutions for well-test-analysis problems : part 1 – analytical considerations [J]. SPE Formation Evaluation, 1991, 9: 359-371. SPE-18615-PA.

[19] Luo W J, Tang C F. A semianalytical solution of a vertical fractured well with varying conductivity under non-Darcy-flow condition [J]. SPE Journal, 2015, 7: 1-13. SPE-178423-PA.

[20] Wang J L, Jia A L, Wei Y S, et al. Semi-analytical simulation of transient flow behavior for complex fracture network with stress-sensitive conductivity [J]. Journal of Petroleum Science and Engineering, 2018, 171: 1191-1210.

[21] Wang J L, Luo W J, Chen Z M. An integrated approach to optimize bottomhole-pressure-drawdown management for a hydraulically fractured well using a transient inflow performance relationship [J]. SPE Reservoir Evaluation & Engineering, 2019, 23 (1). SPE-195688-PA.

页岩气水平井组产量递减特征及动态监测

谢维扬[1,2]，刘旭宁[3]，吴建发[1,2]，张　鉴[1,2]，吴天鹏[1]，陈　满[4]

（1. 中国石油西南油气田公司页岩气研究院；
2. 页岩气评价与开采四川省重点实验室；
3. 中国石油西南油气田公司页岩气勘探开发部；
4. 四川长宁天然气开发有限责任公司）

摘　要：国内外页岩气田大量采用水平井组开发模式，在提高了生产效率的同时也降低了管理成本。基于页岩气藏特殊渗流机理及解吸扩散规律，建立了页岩气多级压裂水平井渗流数学模型，并分析了其不稳定产量递减规律，结合平台式水平井组生产方式，模拟了水平井组内多井生产情况并分析了多井干扰对于水平井组生产效果的影响，进一步明确了在井组内存在不同井数和井距情况下的生产动态特征。通过多种方法对比，认为精细化的数值模拟法适用于页岩气水平井组的产量递减分析，并在此基础上进一步探索性地提出了页岩气水平井组动态监测方案。本文的研究结果可以为页岩气水平井组开发技术政策制定及动态分析方法优选带来一定的启示。

关键词：页岩气藏；压裂水平井；平台式井组；多井干扰；动态监测

页岩气在页岩储层中流动所受到的阻力相比常规气藏要大得多，从而导致了页岩气井生产能力低或无自然生产能力，再加上其天然的超低渗透率超低孔隙度特性，开采难度大，国外目前均采用水平井技术和大型水力压裂来进行有效的开发。早期页岩气井产量主要来自游离气，初期产量较高，但产量很快降低并趋于稳定，稳定期气井产量主要来自基质孔隙里的吸附气，由于吸附气解吸气量有限，扩散速度缓慢，导致了稳产。期产量普遍偏低[1-5]。

目前页岩气压裂水平井常用的产量递减分析方法有Arps递减分析法、幂律指数法、扩展指数法、改进扩展指数法、Duong方法以及渗流解析模型法等，各类方法有其适用性，Arps递减难以用固定系数公式描述整个生产期产量递减规律，一般用于页岩气井生产后期的产量预测；幂律指数法在气井生产时间较长、产量递减趋势表现较为明显的情况下能有效降低公式参数计算的多解性时，分析结果较为可靠；扩展指数法的优点是需要计算确定的参数相对较少，有利于降低多解性，不足之处是普适性下降，对储层渗透性较好的页岩气井适用性相对较差；Duong方法有严格的理论基础，适用于裂缝线性流或双线性流始终占绝对主导地位的情况，一般情况下在页岩气井开发早期预测效果较好[6-10]。

由于受到复杂地形及开采成本的制约，且兼顾了高效管理和合理开发的缘故，平台式水平井组模式在页岩气藏的开采中大规模使用。一个丛式水平井组一般在其上下半支分别部署2～5口压裂水平井，通过统一的管理和混合计量，达到节约占地面积，高效动用储层的作用。目前基于水平井组的研究还相对较少，张小涛等[11]通过数值模拟手段模拟了页岩气井渗流特征，为编制页岩气开发方案提供了理论依据。李国峰[12]通过研究大牛地气田丛式水平井组压裂工艺试验的应用现状，探索了丛式井组的复合裂缝监测方法。郭建春[13]通过建立地层应力场分布模型，讨论了丛式井组人工裂缝周围水平应力场的分布情况，指导了页岩气水平井组拉链压裂的优化设计。吕玉民[14]通过灰色关联法，定量评价了各项参数对于煤层气丛式井组产能的影响程度，这对页岩气丛式井组产能评价起到了一定的启示作用。目前丛式井组研究多针对工程方面，较少涉及其生产动态分析的特征讨论。

本文基于长宁—威远页岩气田大规模使用的丛式压裂水平井组开发方式，建立了页岩气藏压裂水平井渗流模型，绘制并分析了其产量递减特征曲线，然后在此基础上，构建了数值模拟模型，分析了水平井组半支内分别存在不同井距的1口，2口，3口及4口压裂水平井情况下的生产动态特征，讨论了各类动态分析方法的适用性，优选了适合平台式井组的动态分析方法，提出了平台式水平井组动态监测手段。

1 页岩气多级压裂水平井产量递减模型

传统的压裂水平井分析理论是基于流体渗流遵循达西定律的基础上进行的，并未考虑页岩气吸附解吸及扩散运移特征的影响，利用此传统分析模型对页岩气井进行研究并不符合真实情况。假设双重介质页岩气藏压裂水平井通过射孔方式进行完井，流体的主要流通通道为水力裂缝，流体在裂缝中的渗流满足达西定律，其基本渗流模型如图1所示。

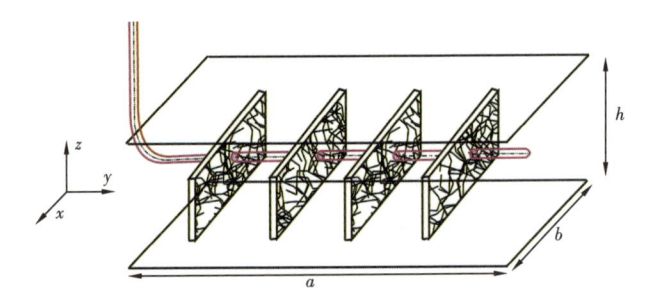

图1 多级压裂水平井渗流物理模型

在渗流区域中建立三维笛卡尔坐标系，裂缝和基质流动方程分别如下：

$$v_{\mathrm{f}} = -\frac{K_{\mathrm{f}}}{\mu_{\mathrm{g}}} \nabla p_{\mathrm{f}} \tag{1}$$

$$v_{\mathrm{m}} = -\frac{K_{\mathrm{m}}}{\mu_{\mathrm{g}}} \nabla p_{\mathrm{m}} \tag{2}$$

根据页岩气的解吸附特征，讨论解吸气拟稳态扩散情况，引入Langmuir等温吸附定律和Fick扩散定律，建立描述稳定扩散的表达式如下：

$$q_{\mathrm{m}} = -G\frac{\mathrm{d}c_{\mathrm{m}}}{\mathrm{d}t} = -G\rho_{\mathrm{sc}}\frac{6\pi^2 D}{R^2}(V_{\mathrm{a}} - V) \tag{3}$$

通过质量守恒定律，联系运动方程和状态方程，可得页岩气压裂水平井渗流微分方程如下：

$$\frac{\partial}{\partial x}\left(\frac{p_f}{\mu_g Z}\frac{\partial p_f}{\partial x}\right)+\frac{\partial}{\partial y}\left(\frac{p_f}{\mu_g Z}\frac{\partial p_f}{\partial y}\right)+\frac{K_{fv}}{K_{fh}}\frac{\partial}{\partial z}\left(\frac{p_f}{\mu_g Z}\frac{\partial p_f}{\partial z}\right)-\frac{p_{sc}T}{K_{fh}T_{sc}\rho_{sc}}q_m=\frac{\phi_f}{K_{fh}}\frac{\partial\left(\frac{p_f}{z}\right)}{\partial t} \tag{4}$$

引入气体压缩系数 c_g 和拟压力函数 m：

$$c_g = \frac{1}{p_f} - \frac{1}{Z}\frac{\partial Z}{\partial p_f} \tag{5}$$

$$m = \frac{\mu_i Z_i}{p_i}\int_{p_0}^{p}\frac{p_f}{\mu_g Z}\mathrm{d}p_f \tag{6}$$

定义无因次变量如表 1 所示。

表 1 无因次变量对照表

无因次变量	无因次定义表达式
无因次拟压力及无因次时间	$m_D = \frac{2\pi K_{fh}h}{q_{sc}B_{gi}\mu_i}(m_i - m)$, $\quad t_D = \frac{K_{fh}}{\sigma L^2}t$
储容比及窜流系数	$\omega = \frac{\phi_f\mu_g c_g}{\sigma}$, $\quad \lambda = \frac{\sigma L^2}{k_{fh}\gamma}$
无因次 x, y 坐标值及持续点源值	$x_D = \frac{x}{L}$, $\quad x_{wD} = \frac{x_w}{L}$, $\quad y_D = \frac{y}{L}$, $\quad y_{wD} = \frac{y_w}{L}$
无因次 z 坐标值及持续点源值	$z_D = \frac{z}{L}$, $\quad z_{wD} = \frac{z_w}{L}\sqrt{\frac{K_{fh}}{K_{fv}}}$
无因次裂缝微元段长度顶底值	$L_{fLDi} = \frac{L_{fLi}}{L}$, $\quad L_{fRDi} = \frac{L_{fRi}}{L}$
无因次基质块半径及无因次储层厚度	$r_D = \frac{r}{L}$, $\quad r_{wD} = \frac{r_w}{L}$, $\quad h_D = \frac{h}{L}$
无因次气体平衡浓度及裂缝气体平均浓度	$V_{aD} = V_a - V_i$, $\quad V_D = V - V_i$
无因次气体质量密度	$c_{mD} = (c_m - c_i)$
无因次井筒储集系数	$C_D = \frac{C}{2\pi h\phi_f c_g L^2}$
拟稳态扩散参数团	$\sigma_p = \phi_f\mu_g c_g + \frac{p_{sc}TZ_i}{T_{sc}p_i}\frac{2\pi K_{fh}h}{q_{sc}B_{gi}}$ $\gamma_p = \frac{R^2}{6\pi^2 D}$
非稳态扩散参数团	$\sigma_U = \phi_f\mu_g c_g + \frac{p_{sc}TZ_i}{T_{sc}p_i}\frac{6\pi K_{fh}h}{q_{sc}B_{gi}}$ $\gamma_U = \frac{R^2}{\pi^2 D}$

通过上述变换，页岩气压裂水平井渗流微分方程（4）可转化为如下形式：

$$\frac{\partial}{\partial x_D}\left(\frac{\partial m_D}{\partial x_D}\right)+\frac{\partial}{\partial y_D}\left(\frac{\partial m_D}{\partial y_D}\right)+\frac{\partial}{\partial z_D}\left(\frac{\partial m_D}{\partial z_D}\right)=\omega\frac{\partial m_D}{\partial t_D}+\frac{\sigma L^2}{K_{fh}\rho_{sc}}(1-\omega)q_m \qquad (7)$$

对上式进行关于无因次时间的 Laplace 变换：

$$\frac{\partial}{\partial x_D}\left(\frac{\partial \overline{m}_D}{\partial x_D}\right)+\frac{\partial}{\partial y_D}\left(\frac{\partial \overline{m}_D}{\partial y_D}\right)+\frac{\partial}{\partial z_D}\left(\frac{\partial \overline{m}_D}{\partial z_D}\right)=s\omega\overline{m}_D+\frac{\sigma L^2}{K_{fh}\rho_{sc}}(1-\omega)\overline{q}_m \qquad (8)$$

根据解吸气拟稳态扩散的表达形式，式（3）的右端第二项可以写为：

$$\frac{\sigma_p L^2}{K_{fh}\rho_{sc}}(1-\omega)\overline{q}_m=-\frac{\sigma_p L^2(1-\omega)}{K_{fh}\rho_{sc}}\left(G\rho_{sc}\frac{6\pi^2 D}{R^2}s\overline{c}_m\right) \qquad (9)$$

将上式化简并引入拟稳态扩散窜流系数，可以获得 Laplace 变换后的表达式：

$$\frac{\sigma_p L^2}{K_{fh}\rho_{sc}}(1-\omega)\overline{q}_m=-\lambda sG(1-\omega)\overline{c}_{mD} \qquad (10)$$

由于 c_{mD} 和 V_D 皆已无因次化，其在数值上是相等的，于是有 $\overline{c}_{mD}=\overline{V}_D$。通过移项，拟稳态扩散项则可以写成如下形式：

$$\overline{V}_D=\frac{\lambda}{\lambda+s}\overline{V}_{aD} \qquad (11)$$

为了将裂缝内拟压力与拟稳态扩散项联系起来，引入 Langmuir 等温吸附定律，其表达式如下：

$$V=V_m\frac{p}{p_L+p} \qquad (12)$$

将式（12）代入到 Laplace 空间下的表达式中，可以得到：

$$\overline{V}_{aD}=\overline{\left(V_m\frac{m_f}{m_L+m_f}-V_m\frac{m_i}{m_L+m_i}\right)}=-\overline{\left[\frac{V_m m_L(m_i-m_f)}{(m_L+m_f)(m_L+m_i)}\right]} \qquad (13)$$

代入无因次拟压力函数，上式可以变成：

$$\overline{V}_{aD}=-\overline{\left[\frac{V_m m_L(m_i-m_f)}{(m_L+m_f)(m_L+m_i)}\right]}=-\overline{\left[\frac{q_{sc}\mu_i}{2\pi K_{fh}h}\frac{V_m m_L}{(m_L+m_f)(m_L+m_i)}m_D\right]} \qquad (14)$$

引入解吸系数 α，将其代入上式中，可得：$\overline{V}_{aD}=-\alpha\overline{m}_D$，这其中，解吸系数定义为 $\alpha=\frac{q_{sc}\mu_i}{2\pi K_{fh}h}\frac{V_m m_L}{(m_L+m_f)(m_L+m_i)}$。

将化简后的 $\overline{V}_{aD}$ 及 $\overline{V}_D$ 表达式代入拟稳定扩散项表达式中，可得：

$$\frac{\sigma_p L^2}{K_{fh}\rho_{sc}}(1-\omega)\overline{q}_m=-\frac{\lambda s(1-\omega)}{\lambda+s}\alpha\overline{m}_D \qquad (15)$$

将上式代入式（8）并化简可得：

$$\frac{\partial}{\partial x_D}\left(\frac{\partial \overline{m}_D}{\partial x_D}\right)+\frac{\partial}{\partial y_D}\left(\frac{\partial \overline{m}_D}{\partial y_D}\right)+\frac{\partial}{\partial z_D}\left(\frac{\partial \overline{m}_D}{\partial z_D}\right)=f(s)\overline{m}_D \qquad (16)$$

引入球坐标转换和点源函数法便可以获得式（16）在顶底威封闭边界，横向上威无限大的连续点源解如下：

$$\overline{m}_D = \frac{\overline{q}}{\phi_f \mu_g c_g} \frac{p_{sc}T}{T_{sc}} \frac{\mu_i Z_i}{2\pi h_D L^3 p_i} \left\{ K_0\left[r_D\sqrt{f(s)}\right] + 2\sum_{n=1}^{\infty} K_0\left[r_D\sqrt{f(s)+\frac{n^2\pi^2}{h_D^2}}\right] \cos n\pi \frac{z_D}{h_D} \cos n\pi \frac{z_{wD}}{h_D} \right\}$$
（17）

基于假设，裂缝与水平井筒皆为无限导流，裂缝内任意微元段上的压力值等同于其他微元段及水平井井筒的压力值，但沿着人工裂缝方向上的流体流率不同。根据压降叠加原理，对裂缝段进行离散[10]，如图 2 所示。

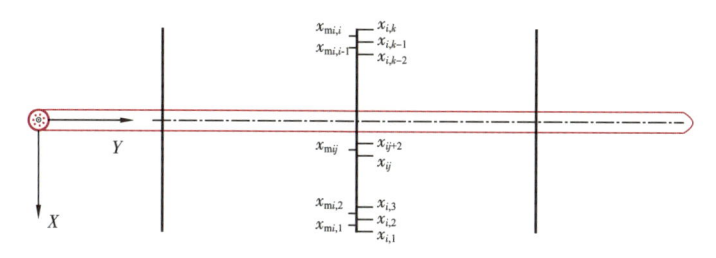

图 2 裂缝微元段离散示意图

引入无因次点源强度定义：

$$\overline{q}_D = L\left[\frac{\overline{q}}{\phi_f \mu_g c_g} \frac{p_{sc}T}{T_{sc}} \frac{\mu_i Z_i}{2\pi h_D L^3 p_i}\right] = \frac{1}{s}$$
（18）

构建裂缝微元段产量累加系数计算矩阵：

$$\begin{bmatrix} A_{1,1}, A_{1,2} \ldots\ldots A_{1,k} \ldots\ldots A_{1,2n\times n_f}, -1 \\ \ldots\ldots \\ A_{k,1}, A_{k,2} \ldots\ldots A_{k,k} \ldots\ldots A_{k,2n\times n_f}, -1 \\ \ldots\ldots \\ A_{2n\times n_f,1}, A_{2n\times n_f,2} \ldots\ldots A_{2n\times n_f,k} \ldots\ldots A_{2n\times n_f,2n\times n_f}, -1 \end{bmatrix} \times \begin{bmatrix} q_{D1} \\ q_{D2} \\ \ldots\ldots \\ q_{D2n\times n_f} \\ \overline{m}_{wD} \end{bmatrix} = \begin{bmatrix} 0 \\ 0 \\ \ldots \\ 0 \\ 1 \end{bmatrix}$$
（19）

由上述矩阵可以得到每个微元段的流率以及一个无因次拟压力值，引入 Duhamel 原理便可以获得生产过程中存在表皮效应和井筒储集效应的解析解如下：

$$\overline{m}_{wD} = \frac{s\overline{m}_D + S_K}{s + C_D s^2(s\overline{m}_D + S_K)}$$
（20）

根据 Everdingen 和 Hurst 的研究成果，在 Laplace 空间内，当气井以定井底压力生产时的无因次产量响应与定产量生产时的无因次拟压力响应在存在着一个倒数关系，因此顶底封闭水平方向上无限大的条件下的双重介质页岩气藏压裂水平井的无因次产量表达式如下：

$$\overline{q}_D = \frac{1 + sC_D\left\{\overline{q}_{Di,j}\int_{x_{Di,j}}^{x_{Di,j+1}} K_0\left[\sqrt{(x_D-\zeta)^2+(y_D-y_{Di,j})^2}\sqrt{f(s)}\right]d\zeta + S_K\right\}}{s\left\{s\overline{q}_{Di,j}\int_{x_{Di,j}}^{x_{Di,j+1}} K_0\left[\sqrt{(x_D-\zeta)^2+(y_D-y_{Di,j})^2}\sqrt{f(s)}\right]d\zeta + S_K\right\}}$$
（21）

对上式进行 Stehfest 数值反演[15]，可以获得双重介质页岩气藏压裂水平井产量递减特征曲线如图 3 所示。

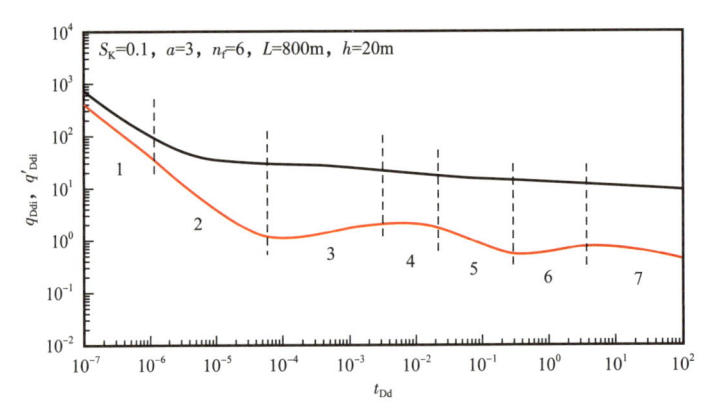

图 3 页岩气藏压裂水平井无因次产量积分及产量积分导数典型曲线

由图 3 可以看出，双重介质页岩气藏压裂水平井产量递减特征曲线大致可以分为 7 个流动阶段：井储晚期阶段，表皮效应影响流动阶段，裂缝线性流动阶段，裂缝径向流动阶段，基质向裂缝窜流早期流动阶段，基质向裂缝窜流晚期流动阶段，储层外边界控制流动阶段。页岩气井产量递减幅度最大的阶段主要为前 4 个流动阶段，后 3 个流动阶段为中后期受到吸附气解吸的影响，递减速度减缓。由于页岩气藏极致密的特性，若没有较长的生产历史，储层外边界控制流动阶段几乎是无法观测到的。

2 页岩气平台式水平井组产量递减特征及规律

2.1 数值模拟模型

平台式水平井组常见的布井模式是上半支和下半支分别部署 2～5 口压裂水平井，共同构成一个统一的水平井组开采模式，在连接处采用混合计量方式统一计量和管理采出的气量和水量。由于水平井组内存在多口压裂水平井共同生产，其井间干扰和井间连通情况会明显影响到生产。

长宁—威远国家级页岩气示范区大规模采用平台式水平井组进行开发，通过干扰试井测试，发现部分平台内由于井间间距过小存在井间连通的情况。以长宁—威远平台井组生产及工程参数为基础，考虑了优化参数的可能性，通过数值模拟建立了内含井间间距 200m、300m、400m、500m 及 600m 的 1 口、2 口、3 口及 4 口页岩气压裂水平井的水平井组模型，模型为存在基质、天然裂缝及水利裂缝三重介质的双孔双渗组分模型，考虑了甲烷吸附气的解吸及扩散，模拟水平段长度 1500m，模拟体积压裂效果 SRV 区域 12 段，并对该区域进行网格加密，最大网格数 107820，如图 4 所示。

2.2 水平井组内多井干扰分析

通过模拟分析发现，受到多井干扰的影响，相同井距情况下井数越多，储层动用程度越大，压裂过程中人工裂缝极易延伸至邻井压降波及范围内，地层压力消耗越厉害，单井产量递减速度及幅度越大，产量差值最高可达到 66%，相同井数情况下井距越大，发生干扰的时间越晚，干扰程度受储层致密性影响越弱，单井产量递减速度及幅度越小，如图5所示。

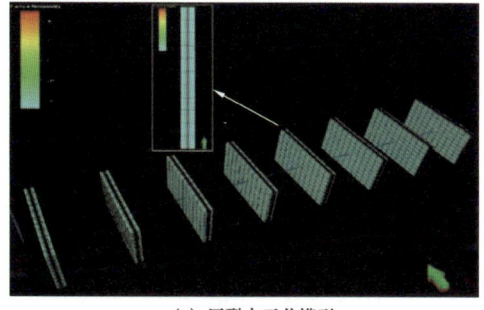

(a) 压裂水平井模型　　　　　　　　　(b) 平台内三井生产模型

图4　平台式压裂水平井组模型

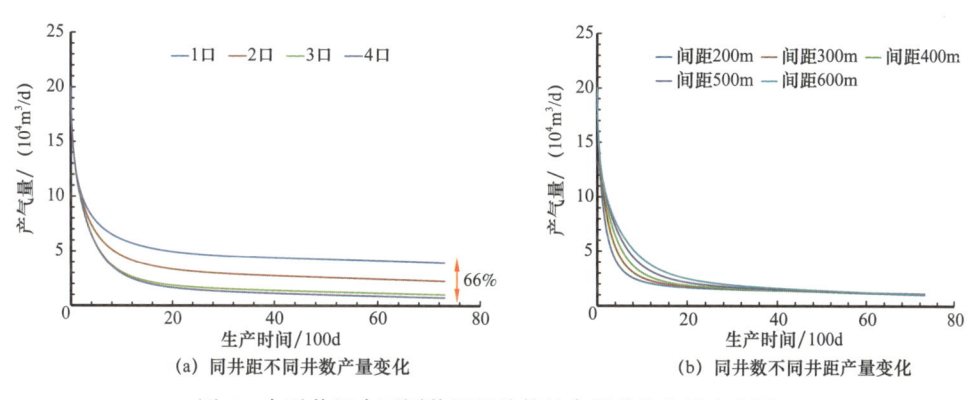

(a) 同井距不同井数产量变化　　　　　(b) 同井数不同井距产量变化

图5　水平井组内不同井距及井数的参照单井产量变化图

生产初期，水平井组内部各井之间由于储层致密的缘故，压力波及程度未能到达影响产量的级别，水平井组不存在明显干扰，等同于多个单井互不干扰的生产。随着生产时间的增加，压降波及范围扩大到邻井周围，井间干扰开始出现明显影响，最先受到影响的是靠近中部的井，其裂缝线性流动阶段持续时间明显缩短，更早地进入了裂缝径向流动阶段，产量递减速度及幅度远高于靠近两侧的井，如图6所示。

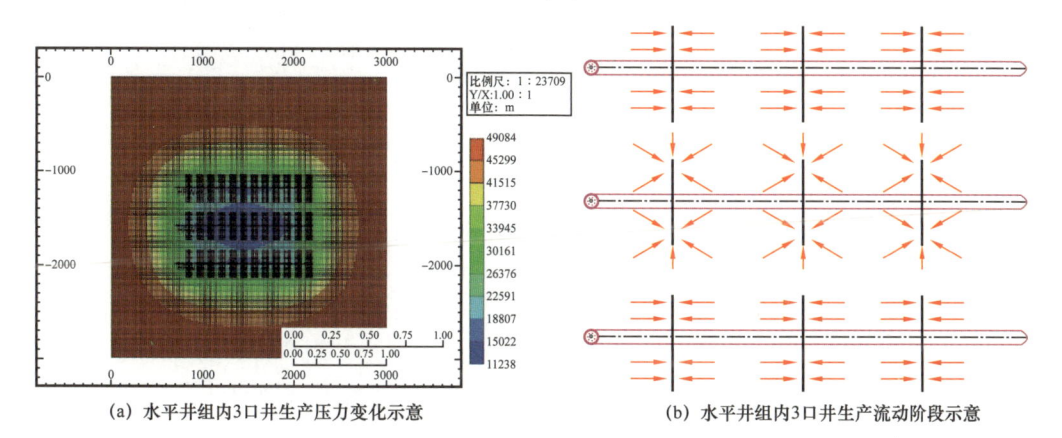

(a) 水平井组内3口井生产压力变化示意　　　(b) 水平井组内3口井生产流动阶段示意

图6　水平井组内不同井距及井数的参照单井产量变化图

水平井组内井间干扰发生时间受井距及井数的影响，井距越小，包含井数越多，则水平井组发生明显井间干扰的时间越早，如表2所示。

表2 水平井组内不同井间间距及井数的井间干扰发生时间变化表

	明显干扰发生时间 /d				
	井间间距 200m	井间间距 300m	井间间距 400m	井间间距 500m	井间间距 600m
水平井组内2口井	40	80	180	250	350
水平井组内3口井	30	60	150	200	300
水平井组内4口井	20	40	120	180	280
平均值	30	60	150	210	310

根据上表可以看出，不同井距及不同井数生产会导致在开井后几十甚至数百天内发生明显井间干扰，井距小而井数多的水平井组会在极短生产时间内发生明显井间干扰影响生产，在这种情况下，目前广泛用于单井产量递减预测的方法将不再适用，若水平井组内井距大而井数少，则发生干扰的时间会大大延后，单井产量递减预测方法则能满足部分情况的预测。基于目前国内页岩气井生产情况，生产数据还相对较少，受多井干扰的影响，使用单井预测方法仅通过前期的产量递减趋势来预测水平井组产量会导致水平井组最终产量被高估。

根据数值模拟结果，多种井距及井数方案所获得的水平井组EUR（生产井最终可采储量）存在着较大的差异，如表3所示。

表3 存在不同井间间距及井数的水平井组最终可采储量变化表

	水平井组 EUR/$10^8 m^3$					
	井间间距 200m	井间间距 300m	井间间距 400m	井间间距 500m	井间间距 600m	平均值
水平井组内1口井	3.69	3.69	3.69	3.69	3.69	3.69
水平井组内2口井	4.68	4.91	5.12	5.31	5.48	5.06
水平井组内3口井	5.39	5.86	6.27	6.66	7.00	6.23
水平井组内4口井	6.02	6.70	7.27	7.71	8.13	7.16
平均值	4.95	5.29	5.59	5.84	6.07	

从表中可以看到，随着水平井组内井数的增加及井间间距的加大，水平井组EUR呈现一个增长的趋势，但增长的幅度逐渐降低。井间间距大小及井数多少直接关系到开发成本，在实际的开发方案编制过程中应顾及开采效率。井距过大到会导致水平井组占地较多且水平井组内布井数量减少，不适合于地面条件较差的水平井组布井；若井距过小则井网密度相对较大，单个水平井组成本较高，严重的井间干扰会影响到水平井组实际收益，通常建议水平井组内压裂水平井井间间距为300～500m，水平井组内单支部署井3～4口。

2.3 平台式水平井组产量递减分析方法评价及优选

由于水平井组内多井生产的井间干扰影响，适用于页岩气单井的产量递减预测方法存在时间上的适用性，早期生产过程中井间干扰未发生或者影响不明显，单井预测方法基本适用于水平井组预测，但随着生产的进行，井间干扰对产量的影响越来越明显，此时若继续使用单井预测方法来预测水平井组生产则会导致高估水平井组产量。

根据不同组合的水平井组初始参数所对应的不同数值模拟结果将水平井组内井间干扰特征分为了3大类：（1）几乎无干扰的水平井组（井组内2口水平井，间距500m），（2）存在微弱干扰的水平井组（井组半支井数3口，井间间距300m），（3）存在明显干扰的水平井组（井组半支井数大于4口，井间间距200m），针对上述3类分别对比了幂律指数法、扩展指数法以及Duong方法的拟合效果，如图7至图9所示。

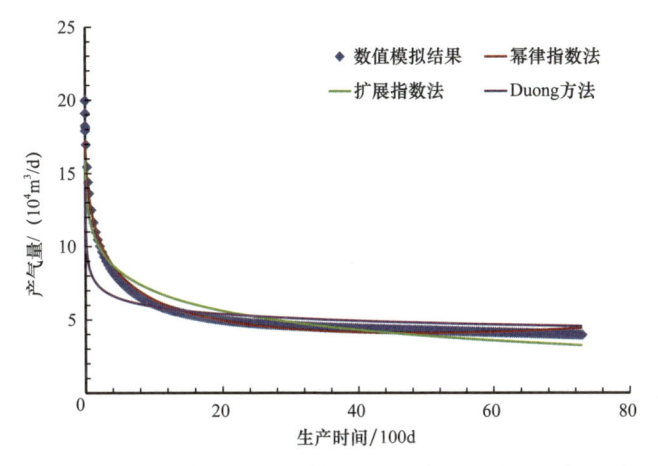

图7 水平井组内不存在井间干扰情况下的产量递减预测方法对比图

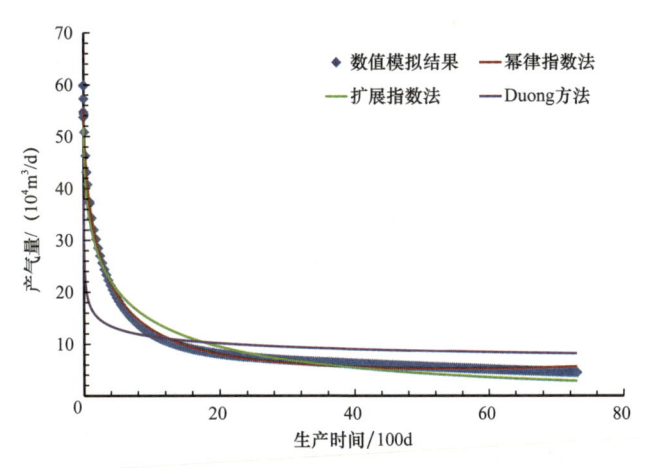

图8 平台内存在明显井间干扰情况下的产量递减预测方法对比图

数值模拟结果是基于气井放压生产，过程中没有发生生产制度改变以及增压操作，由于在生产早期没有出现明显井间干扰，幂律指数法及扩展指数法拟合效果较好，生产中后期由于井间干扰明显加剧，这两种方法已不能对生产动态起到良好的预测作用，Duong方法则拟合效果较差。

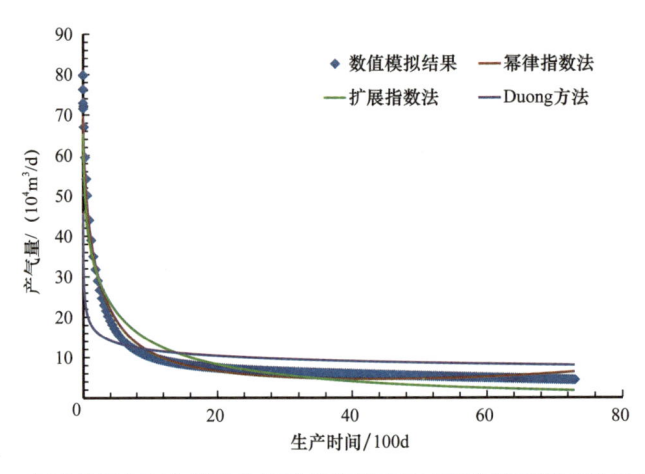

图 9 水平井组内存在严重井间干扰情况下的产量递减预测方法对比图

在确定了储层及工程参数并结合微地震监测资料后，构建精细化的数值模拟模型，能够较为准确的预测平台式水平井组的产量递减趋势，幂律指数法及扩展指数法在生产早期能够较好地完成预测工作，但后期的预测效果与实际效果相差较大。

3 页岩气平台式水平井组动态监测方法探索

气井动态监测的是科学管理气井的重要技术手段，通过对气井在生产过程中的产量、压力、流体物性的变化，以及井下、地面工程的变化等监测，及时有效地指导其合理开采。动态监测的内容主要包括了压力、温度、产量、产出流体理化性质以及工程参数变化等。

目前针对页岩气单井的动态监测方法基本成形，如表 4 所示，但由于该方法基于单井，尚未考虑对平台的支持，加之大部分水平井组在计量上都选用混合计量方式，从平台总体产气产液量上要比单井计量更为准确，水平井组内水平井数逐渐增多，井网密度逐渐加大，多井干扰明显，需要对现有的单井动态监测手段进行基于平台式井组的改进。

表 4 页岩气井动态监测项目表

序号	监测项目	所取资料	监测目的
1	压力恢复试井	井筒静止压力梯度/流动压力梯度、井筒静止温度梯度/流动温度梯度、压力恢复试井数据井底压力数据	分析储层渗流参数，评估气井压裂效果；掌握气井井筒积液状况，为工艺措施选择提供依据
2	产能试井	井筒静止压力梯度/流动压力梯度、井筒静止温度梯度/流动温度梯度、产能试井井井底压力数据	计算气井产能；掌握气井井筒积液状况，为工艺措施选择提供依据
3	干扰试井	压力恢复试井井底压力数据、激动井生产数据和井底流压数据、观测井井底压力数据	分析储层渗流参数，评估气井压裂效果；计算激动井产能；分析气井的井间干扰；掌握气井井筒积液状况，为工艺措施选择提供依据
4	生产测井	产层段流体剖面和流动压力剖面	掌握气井流体剖面，定量分析压裂段的压裂效果，分析气井产能
5	流体性质监测	井口气、水样全分析数据	分析气井产出流体性质，实施掌握压裂液返排状况

基于前两节的研究成果，明确了页岩气水平井组生产动态特征会因井间干扰而发生明显变化，而根据数值模拟结果显示，井间干扰程度主要受井距、水平井组内井数、压裂规模等影响，再根据平台式水平井组的布井特点，应考虑将监测重点放在多井干扰分析及人造裂缝延伸上，因此在平台式井组动态监测方案的设计上加强了干扰试井的测试力度，增加井下微地震监测、返排液矿化度监测、化学示踪剂监测及生产测井。干扰试井一般需要选取水平井组的上半支进行测试，如果条件允许，对下半支也就是下倾井也应该进行测试。微地震监测能够全面系统的描述页岩储层压裂后裂缝走向及缝网连通情况，也是目前最有效的评估储层改造效果的手段[16]。返排液矿化度监测、化学示踪剂监测及生产测井能够在井组内水平井之间精细化的描述井间连通情况，并可以间接的确定目标层位微裂缝的发育情况。平台式井组动态监测方案如表5所示。

表5 页岩气平台式水平井组动态监测改进方案表

监测类型	获得数据	生产指导
压力恢复试井	储层渗透、地层压力、人工裂缝延伸距离、表皮系数	1. 摸清区块地质特征，确定气井基础参数 2. 间接判断气井压裂规模和效果 3. 确定单井生产状况，判断是否需要进行生产制度优化（控压、下油管、柱塞气举等）
注入—停注试井	原始地层参数、压裂缝长与缝宽	
返排液矿化度监测	返排液不同时期矿化度变化值	
井筒压力梯度监测	气井井筒流体分布情况，是否积液	
生产测井	各个压裂段产气产液量	1. 判断是否存在井间干扰 2. 综合SRV体积和裂缝延伸形态确定对邻井是否形成窜通 3. 结合各压裂段产气产液剖面和邻井裂缝延伸形态判断窜通对生产效果的影响 4. 根据上述分析结果确定该地区同样地质条件下的合理井距和压裂规模，提高单井产量
微地震监测	裂缝延伸形态、宽度和纵深、气井压裂改造后的SRV大小	
干扰试井	井间是否连通，连通强弱	
化学示踪剂监测	示踪剂流动位置和深度	
压力/温度监测	邻井生产或关井中，观测井压力温度变化值	

根据一个页岩气水平井组的井数和间距不同综合考虑，建议将其分为3类：（1）几乎无干扰或干扰出现时间相对较晚的水平井组（井组半支井数少于3口，井间间距大于500m），主要开展常规气井动态监测，包括井筒压力/温度监测、压力恢复试井，适当弱化井间干扰和井距监测相关的试验以减少支出；（2）存在微弱干扰的水平井组（井组半支井数3~4口，井间间距300~400m），需主要开展井筒压力/温度监测、压力恢复试井、微地震监测、生产测井及干扰试井，通过井间干扰监测，确定其对生产的影响以便制定下一步优化方案；（3）存在明显干扰或井间存在特别发育的微裂缝的水平井组（井组半支井数大于5口，井间间距小于300m），需主要开展表5所示的所有动态监测项目，并建议进一步开展井距试验，摸清目标储层的微裂缝发育情况，确定合理的压裂规模，避免因井间窜通而导致的气井废气。

4 结论

本文建立了并分析了页岩气多级压裂水平井不稳定产量递减规律，在此基础上进而建立了页岩气平台式压裂水平井组生产模型，描述了其内部存在的多井干扰特征及生产特征，通过对比优选，提出了页岩气水平井组产量递减分析方法，最后通过对常规页岩气单井动态监测方法的讨论，提出了页岩气水平井组动态监测改进方案。通过本文的研究，可以得到如下结论：

（1）双重介质页岩气藏压裂水平井产量递减特征曲线大致可以分为 7 个流动阶段，页岩气井产量递减幅度最大的阶段主要为前 4 个流动阶段，后 3 个流动阶段为中后期受到吸附气解吸的影响，递减速度减缓。

（2）建立了水平井组多井数值模拟模型，分析了多井干扰对产量递减的影响，多井生产会导致产量递减速度及幅度增大，但能有效增大水平井组 EUR。

（3）单井产量递减预测方法若用于平台式井组将会造成平台最终产量被高估，精细化的数值模拟分析法较适合存在多井干扰的水平井组动态分析研究，目前相对合理的水平井组内井距为 300～500m，水平井组单支部署井 3～4 口。

（4）单井动态监测存在局限性，提出了平台式井组动态监测改进方案，该方案加强了井间间距的相关监测和试验。

符号注释：

dt— 时间增量，h；dp/dr—压力梯度，MPa/m；D—基质块扩散系数，m^2/s；G—球形基质块几何因子；h—储层厚度，m；K—视渗透率，μm^2；K_f—裂缝渗透率，μm^2；L—水平井长度，m；M— 气体摩尔质量，kg/kmol；m—拟压力函数，MPa；n_f— 裂缝条数；p—气体压力，MPa；p_L—Langmuir 压力，MPa；p_0—原始地层压力，MPa；$\bar{p}_0$—平均压力，等于渗流单元流入流出端压力的平均值，MPa；q_m—页岩气藏基质扩散流速，kg/($m^3 \cdot s$)；R—页岩气藏球形基质块半径，m；r—孔隙半径，m；r_m—球形基质系统径向半径，m；s—Laplace 变量；T— 气体绝对温度，K；V_m—Langmuir 等温吸附常数，m^3/t；V_L—Langmuir 体积，反映页岩的最大吸附能力，$V_L=V_m$，m^3/t；V_a—拟稳态扩散中的气体平衡浓度，m^3/m^3；V—水力裂缝中的气体平均浓度，m^3/m^3；V_m—Langmuir 吸附常数；ρ—气体密度，kg/m^3；ϕ—孔隙度；μ—流体黏度，$mPa \cdot s$；μ_i—初始气体黏度，$mPa \cdot s$；ρ_{sc}—标准状况下的天然气密度，kg/m^3。

参 考 文 献

[1] 张卫东，郭敏，杨延辉. 页岩气钻采技术综述[J]. 中外能源，2010，15（6）：35–40.

[2] Arps J J. Analysis of decline curves[J]. Trans. AIME, 1945（160）：228–247.

[3] Guo J J, Zhang L H, Wang H T, et al. Pressure transient analysis for multi-stage fractured horizontal well in shale gas reservoirs[J]. Transport in Porous Media, 2012, 106（3）：635–653.

[4] Eric S C, James C M. Devonian shale gas production: mechanisms and simple models[C] // Paper SPE 19311 presented at the 1989 SPE Eastern Regional Meeting held in Morgantown, WV, USA, 21–23 October, 1989.

[5] Medeiros F, Kurtoglu B, Ozkan E. Analysis of production data from hydraulically fractured horizontal wells in shale reservoirs[C]// Paper SPE 110848 presented at the SPE Annual Technical Conference and Exhibition held in Anaheim, California, USA, 11–14 November, 2012.

[6] 苗和平, 王鸿勋. 水平井压后产量预测及裂缝数优选[J]. 石油钻采工艺, 1992, 14(6): 51-56.

[7] 郎兆新, 张丽华. 压裂水平井产能研究[J]. 石油大学学报(自然科学版), 1994, 18(2): 43-46.

[8] 杨龙, 王晓东, 韩永新. 垂直裂缝井产量递减曲线研究[J]. 天然气工业, 2003, 12(3): 76-79.

[9] 李建秋, 曹建红, 段永刚, 等. 页岩气井渗流机理及产能递减分析[J]. 天然气勘探与开发, 2011, 34(2): 34-37.

[10] 张荻萩, 李治平, 苏皓. 页岩气产量递减规律研究[J]. 岩性油气藏, 2015, 27(6): 138-144.

[11] 张小涛, 吴建发, 冯曦, 等. 页岩气藏水平井分段压裂渗流特征数值模拟[J]. 天然气工业, 2013, 33(3): 47-52.

[12] 李国锋. 丛式水平井组整体压裂裂缝监测方案优选及应用[J]. 重庆科技学院学报(自然科学版), 2014, 16(3): 23-28.

[13] 郭建春, 周鑫浩, 邓燕. 页岩气水平井组拉链压裂过程中地应力的分布规律[J]. 天然气工业, 2015, 35(7): 44-48.

[14] 吕玉民, 柳迎红, 曲英杰, 等. 煤层气井组产能差异的影响因素评价[J]. 煤炭科学技术, 2015, 43(12): 80-84.

[15] Stehfest H. Numerical inversion of Laplace transform[J]. Communications of the ACM, 1970, 13(1): 47-49.

[16] 刘海鑫. 微地震在页岩气水平井压裂中的作用[J]. 江汉石油职工大学学报, 2016, 29(2): 17-21.

四、生产应用类

地质—工程—经济一体化页岩气水平井井距优化
——以国家级页岩气开发示范区宁 209 井区为例

雍 锐[1]，常 程[2]，张德良[2]，吴建发[2]，黄浩勇[2]，敬代骄[3]，郑 健[4]

（1. 中国石油西南油气田公司；
2. 中国石油西南油气田公司页岩气研究院；
3. 中国石油西南油气田公司天然气经济研究所；
4. 四川长宁天然气开发有限责任公司）

摘　要：为最大限度地提高资源动用率，通常采用一次性井网整体部署的方式开发页岩气，水平井井距设计是页岩气井网部署的关键。在确定最优井距时，既需要掌握地质特征和钻井压裂工艺，也必须考虑气价、成本等经济因素的影响，目前国内外均没有形成较为可靠的页岩气水平井井距设计方法。为此，首次建立了一种基于地质—工程—经济一体化的页岩气水平井井距分析方法，通过地质建模、数值模拟、现金流分析 3 种技术手段，使用最终可采储量（EUR）、采收率和内部收益率（IRR）等 3 项指标对长宁区块宁 209 井区页岩气开发井距进行综合评价。研究结果表明：宁 209 井区在当前的地质、工程、经济条件下，井距大于 240m 可以确保页岩气平台开发的内部收益率大于 8%；井距控制在 330～380m 时，可以同时兼顾单井 EUR、平台采收率和经济效益。该研究支撑了宁 209 井区页岩气开发技术政策的制定，为该井区实现规模效益开发奠定了基础。

关键词：页岩气；合理井距；井间压窜；地质建模；数值模拟；一体化；工作流；经济效益；宁 209 井区

页岩气藏只有通过"水平井+体积压裂"的方式才能实现有效开发。水平段压裂后会在井筒周围形成复杂的人工裂缝网络，目前由于微地震、示踪剂等监测技术在定量刻画裂缝几何形态与缝内支撑剂分布方面仍在存在挑战，很难准确判断两口水平井之间应该采用多大的井距较为合适。井距过大会导致井间的储量没有充分动用，造成资源浪费；井距过小会发生井间干扰，甚至严重影响气井的生产效果。国内在页岩气井距方面的研究相对较少，北美页岩气经过长期的开发总结了一些经验：Cakici D 等[1]在 Marcellus 页岩气藏开展井组内变井距动态监测试验，确定出了裂缝的有效延伸距离；Lalehrokh 等[2]通过引入经济模型进一步研究了 Eagle Ford 页岩气藏井距与净现值（NPV）之间的关系；Kim 等[3]将压力特征曲线与数值模拟相结合，通过井距敏感性分析判断出了井间干扰发生的时间；Orozco 等[4]运用修正物质平衡方程，在假设水平井泄气面积为矩形的条件下计算了井距；Pankaj 等[5]采用地质工程一体化建模和数模的方法，对 Marcellus 页岩气井距进行了评价，认为在当前压裂工艺条件下最优井距在 300m 左右，并提出井距与压裂规模密切相关。但究竟如何判断最合理的井距，国内外至今没有形成统一的认识，有必要开展针对性的研究。

1 长宁区块井间压窜现状

长宁—威远国家级页岩气示范区长宁区块自建成以来累计投产井接近200口,气井生产效果不断提升,主力产层为上奥陶统五峰组—下志留统龙马溪组页岩[6]。随着开发井距的不断调整,在不同批次水平井压裂投产过程中,井间压窜和生产干扰现象普遍存在,给新井和老井的开发效果造成了一定程度的影响。如图1所示,2014—2019年期间,长宁区块页岩气水平井井距从500~600m缩小到300~400m,井间压窜发生的概率逐渐增加,尤其是2017年主体井距缩小到400m以内,井间干扰更加明显。

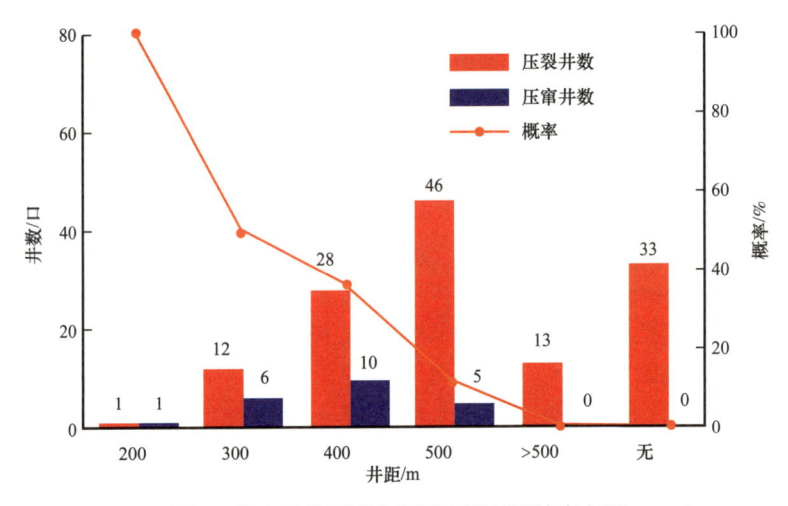

图1 长宁区块不同井距压窜情况统计直方图

根据北美页岩气的开发经验,井间压窜是由于两口井之间的水力裂缝相互连通造成的。引起水力裂缝窜通的原因很多,客观原因包括储层非均质性、地应力特征、天然裂缝发育情况,主观原因包括水平井钻遇层位、压裂规模、投产先后顺序等。长宁区块的井间压窜主要包括压裂—生产、压裂—压裂、压裂—关井等3种情况。压裂—生产是指页岩气井压裂时对先期投产的邻井造成了产量和压力的干扰,主要原因是邻井生产后在其周边形成了压降漏斗,导致压裂缝更易向低压区延伸[7];压裂—压裂和压裂—关井均是在邻井没有明显先期压降的情况下,压裂规模太大导致压裂液直接进入邻井压裂缝内引起的窜通。

现场三维地震蚂蚁体追踪和井口压力监测表明,井间压窜主要以某一段或某几段的压窜为主,并未有井间大范围的窜通(图2、图3)。虽然在压力、产量数据中能看到一些明显的扰动,但更应该研究这种局部的压窜是否会对气井的EUR和井组的整体采收率造成影响,这才是设计井距的关键,也是目前国内外页岩气开发关注的焦点。

2 地质工程一体化工作流

作为一种非常规资源,页岩气必须采用地质工程一体化的思路才能实现规模效益开发,总体要求是以地质研究为中心,多学科多专业融合,多工程技术协同,其核心就是要打造精细的三维地质模型[8]。

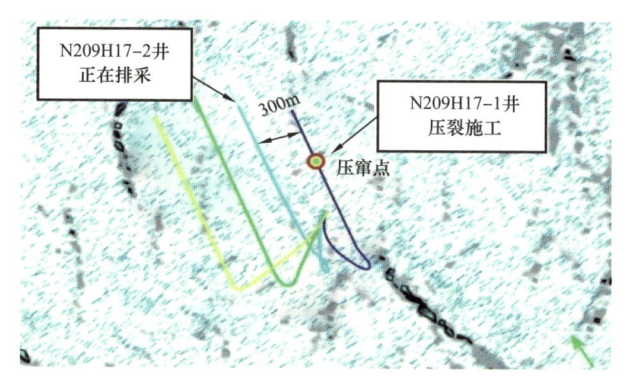

图 2　宁 209 井区 H17 平台蚂蚁体预测天然裂缝模型图

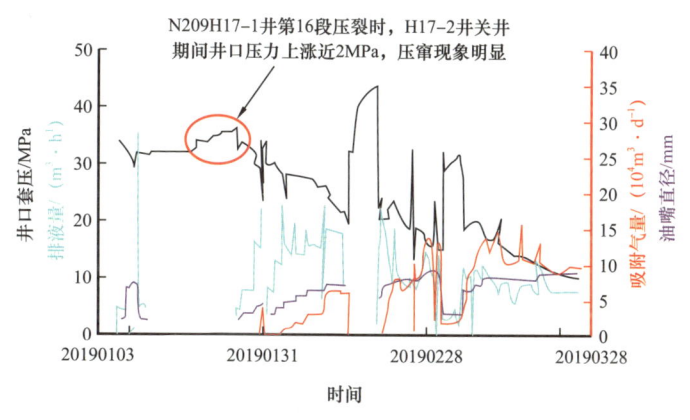

图 3　宁 209 井区 H17-2 井排采曲线图

为了准确评价合理井距，需要先建立能够客观反映工区地质工程特征的三维模型，在此基础上运用实际的压裂工艺参数对人工缝网进行模拟，最后建立页岩气多级压裂水平井数值模型。目前业界普遍采用斯伦贝谢公司的建模和数模软件，该类软件能够综合考虑地质、地球物理、岩石力学、气藏工程等多种信息源的耦合，通过在三维空间还原气井生产时压力场、应力场的变化模拟井间干扰[9-10]。在北美已经有学者采用较为固定的工作流研究合理井距，包括建模、压裂模拟、产能预测、应力场更新、子井压裂模拟及产能预测 5 个步骤（图 4），对 Haynesville 气田的井距进行优化研究，如可以考虑在不同井距条件下，母井投产后地层先期压降对子井压裂缝网扩展的影响[11-12]。

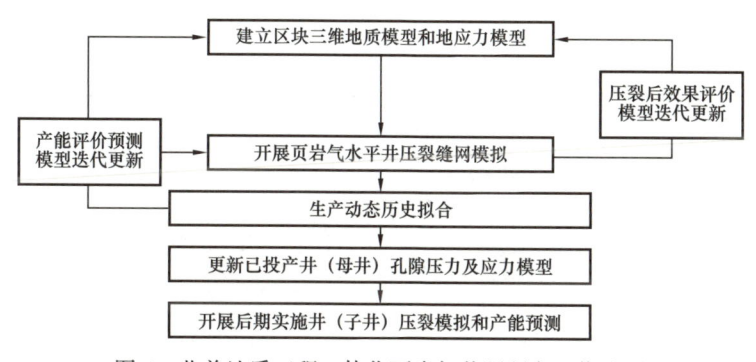

图 4　北美地质工程一体化页岩气井距研究工作流图

3 经济效益评价

3.1 考虑经济效益评价的必要性

与常规气相比，页岩气的产量递减快、开采成本高，最终 EUR 的预测存在不确定性。有学者已经从经济可行性的角度对页岩气开发进行了相关研究，提出了最大程度降低成本是页岩气可持续发展的首要任务[13-14]。因此有必要结合经济分析判断合理井距，降低页岩气资源的开发风险。

众所周知，在北美从事页岩气开发的中小型油公司较多，它们主要采用初期大井距后期加密的"滚动开发"模式[15]。北美页岩气的开发理念是"一切以效益为王"，追求单井 EUR 最大化的同时更注重经济效益最大化。因此北美的油公司会根据已购买或者租赁区域的面积进行井网调整，开发早期在面积较大的区域用大井距进行整体控制，母井生产 1~4 年后再根据气价变化决定加密井距，只要有效益就会不断进行加密。然而国内长宁区块的页岩储层地质工程特征与北美存在差异[16-17]，又主要采用一次性井网整体部署的开发模式，这些客观条件要求作业者必须因地制宜，采用地质—工程—经济一体化的工作思路，运用多种方法手段综合论证合理井距，才能确保兼顾产量、效益和采收率。

3.2 经济评价参数

净现值（NPV）和内部收益率（IRR）是国内企业评价项目投资是否具有经济效果的普遍标准，两者通过考虑资金的时间价值来判断项目的资金状况，能够反映投资的有效性和质量高低[18-19]。笔者主要采用内部收益率评价不同井距条件下气井生产的经济效益。内部收益率的表达式为：

$$\sum_{t=0}^{n}(CI-CO)_t(1+IRR)^{-t}=0 \qquad (1)$$

IRR 表示内部收益率，若 IRR＞基准折现率，则项目具有经济效益；CI 表示评价期内页岩气井的现金流入，万元；CO 表示评价期内页岩气井的现金流出，万元；t 表示评价期，本文取 t=20 年（与单井 EUR 计算截止时间一致）。

现金流入主要为页岩气销售收入，现金流出主要包括投资（钻完井及地面工程费用）、生产成本（操作成本和设备折旧费等）以及相应的税费。本文使用的经济评价基础参数如表 1 所示。

表 1 宁 209 井区页岩气开发经济评价基础参数表

经济参数	取值
地面建设工程费用 / 万元	1000
钻井成本 / 万元	1800
压裂成本 / 万元	2500
页岩气操作成本 / [（元 /10^3m^3）]	200

续表

经济参数	取值
天然气价格 /[(元/10^3m^3)]	1275
增值税税率 /%	9
基准折现率 /%	8

4 案例应用

4.1 基本情况

宁209井区是目前长宁区块的主要建产区，目的层五峰组—龙马溪组的Ⅰ+Ⅱ类储层厚度介于32～36m，埋藏深度介于3000～3500m，压力系数介于1.8～2.0，属于超压气藏，最小水平主应力介于71～73MPa、水平应力差平均为16.7MPa，天然裂缝较为发育。已实施水平井的平均压裂长度为1500m，主体压裂工艺参数如表2所示。

表2 宁209井区主要地质、工程特征参数表

类别	属性参数	取值
地质工程参数	孔隙度 /%	5.1～6.3
	含气量 /($m^3·t^{-1}$)	5.3～6.2
	TOC/%	3.2～3.6
	Ⅰ+Ⅱ类储层厚度 /m	32～36
	压力系数	1.8～2.0
	埋藏深度 /m	3000～3500
	最小水平主应力 /MPa	71～73
	水平应力差 /MPa	16.7
压裂工艺参数	压裂水平段长 /m	1500
	段间距 /m	60
	单段簇数 /簇	3
	簇间距 /m	20
	加砂强度 /($t·m^{-1}$)	2.0～2.5
	施工排量 /($m^3·min^{-1}$)	15～16

4.2 平台井组建模

按照"一体化研究工作流"建立宁209井区平台三维模型，大小为1700m×1400m×30m，储量丰度为$5.17×10^8m^3/km^2$。分别设计2、3、4、5、6口井共5种井组方案，模拟

井距在 200～600m 时的井间干扰影响（图 5）。水平井压裂缝网模拟均采用宁 209 井区的平均压裂参数，所有井全部完成压裂后同时投产。

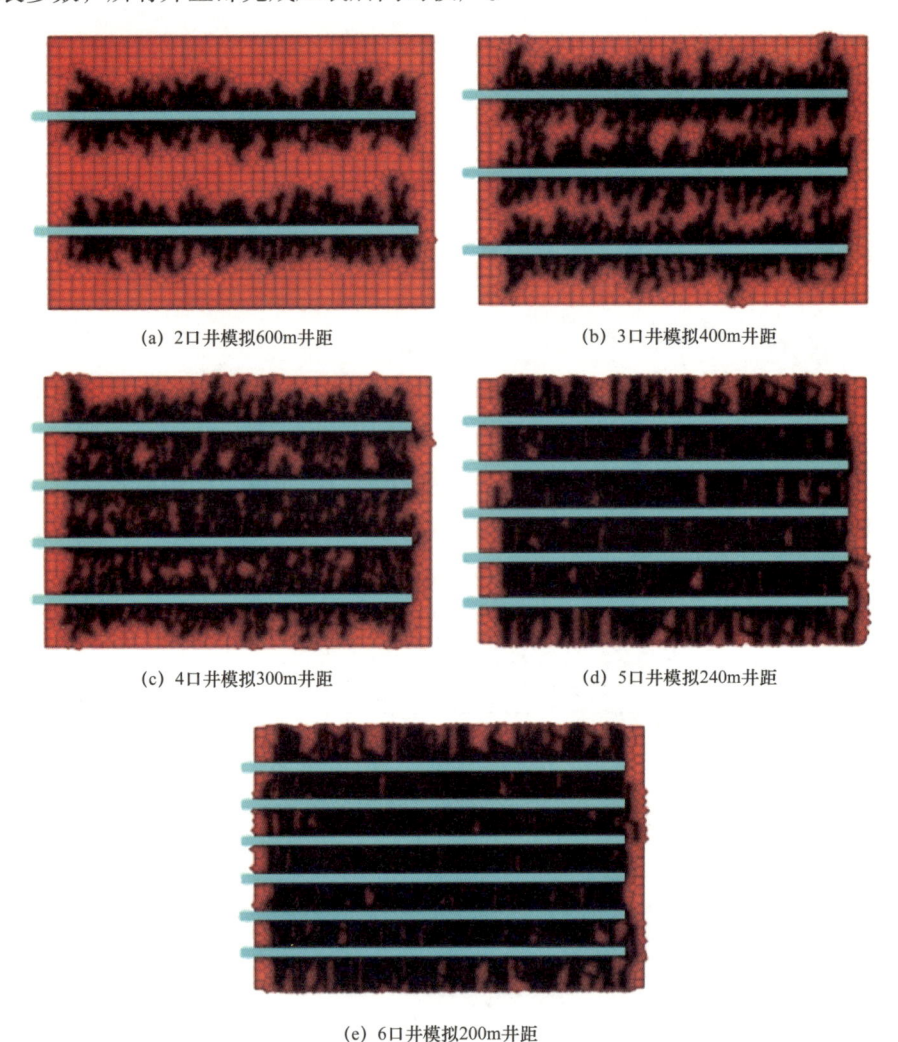

(a) 2口井模拟600m井距

(b) 3口井模拟400m井距

(c) 4口井模拟300m井距

(d) 5口井模拟240m井距

(e) 6口井模拟200m井距

图 5　2～6 口井地质工程一体化数值模型图

4.3　结果分析

图 6 是 5 种方案模拟得到的井均日产气和 EUR 的结果。随着井距的缩小，井间干扰程度加重，页岩气井的日产气量逐渐变差，EUR 不断下降。由于 200m 井距的平台模型中有 5 口井，采收率仍然比较高，但更加严重的井间干扰导致 200m 井距 6 口井的平台采收率比 240m 井距 5 口井的平台更小（图 7）。

图 8 是不同井距预测 20 年后的地层孔隙压力分布情况。当井距大于 400m 时，井间仍有剩余储量没有采出；井距为 300m 时，大部分人工裂缝已经连通，但仍然存在储量没有充分采出的区域；井距进一步缩小到 240m 以下，井间干扰变得十分严重，井控范围内的地层压力明显下降，表明井控储量已基本被采出。

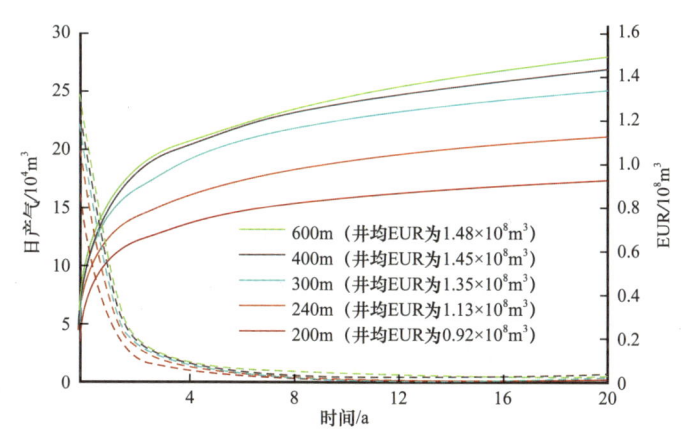

图 6 不同井距下单井平均日产气与 EUR 曲线图

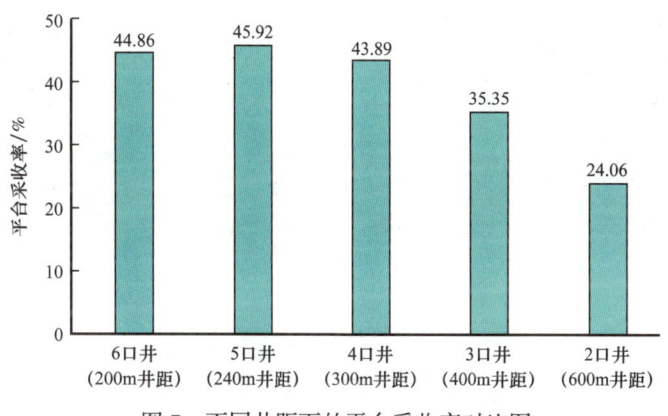

图 7 不同井距下的平台采收率对比图

将 5 种方案模拟计算的 EUR 代入公式（1），通过现金流分析就能够获得各方案生产 20 年对应的内部收益率，计算结果如图 9 所示。随着井距的增加，同一平台内需要开钻的井数减少，钻完井投资也相应减少，平台生产获得的内部收益率就越大。所以说 2 口井的平台相比于 6 口井的平台，虽然采收率低，但投资也相对减少，经济效益反而更好。从经济效益的角度分析，井距也并非越大越好，而是存在一个临界值，井距一旦超过该临界值，内部收益率的增加就不明显了。本案例计算得到的临界井距在 380m，对应的内部收益率为 16.9%。

将 EUR、采收率和内部收益率随井距的变化绘制在同一张图上（图 10）。当井距大于 240m 时内部收益率能够达到 8%，整个平台气井生产是有经济效益的，但此时 EUR 仍有进一步提升的空间；把井距增大到 380m 时内部收益率达到 16.9%，表明经济效益接近最大，但此时平台采收率已经下降至 42%。若要兼顾单井 EUR、平台采收率和经济效益，井距控制在 330~380m 是较为合理的。即最小井距不能小于 EUR 和采收率的交点，最大井距不超过临界经济井距的上限。

笔者提出的基于"地质—工程—经济一体化"的页岩气井距研究方法表明，任何一个页岩气藏都不存在"唯一的最佳井距"。地质特征发生变化、工艺技术不断优化、钻完井成本不断下降均会导致最佳井距发生改变。根据当前油公司掌握的最新地质认识、工艺技术水平和经济参数取值，采用地质—工程—经济一体化的思路（图11）可以论证出页岩气井距的可接受范围，根据该井距范围进行页岩气水平井平台部署才是较为合理的。

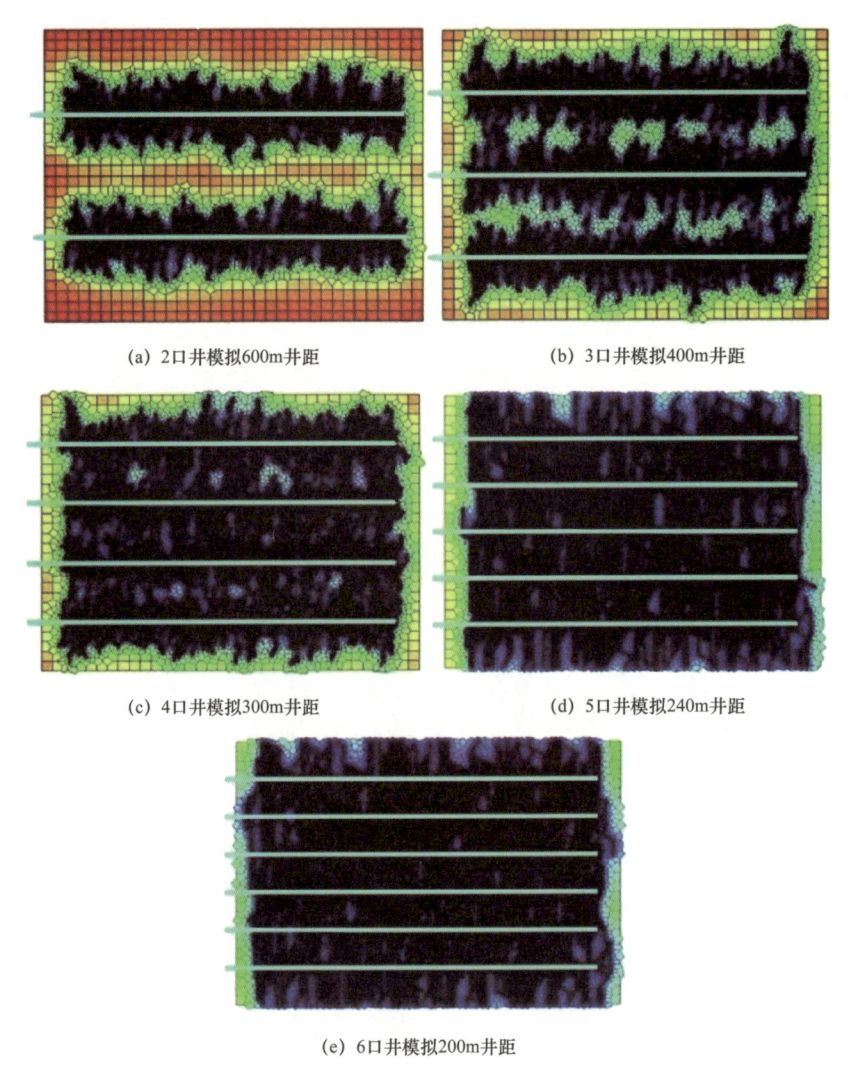

(a) 2口井模拟600m井距

(b) 3口井模拟400m井距

(c) 4口井模拟300m井距

(d) 5口井模拟240m井距

(e) 6口井模拟200m井距

图8 预测20年页岩气平台地层孔隙压力分布图

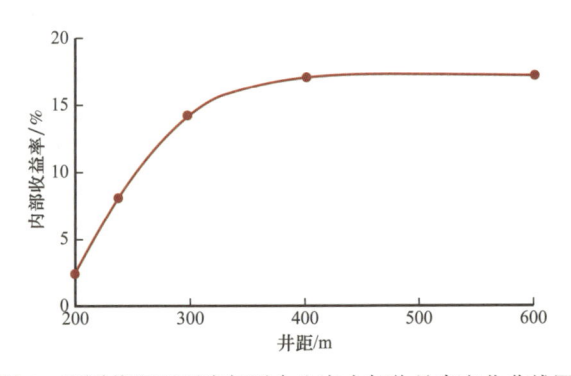

图9 不同井距下页岩气平台生产内部收益率变化曲线图

- 194 -

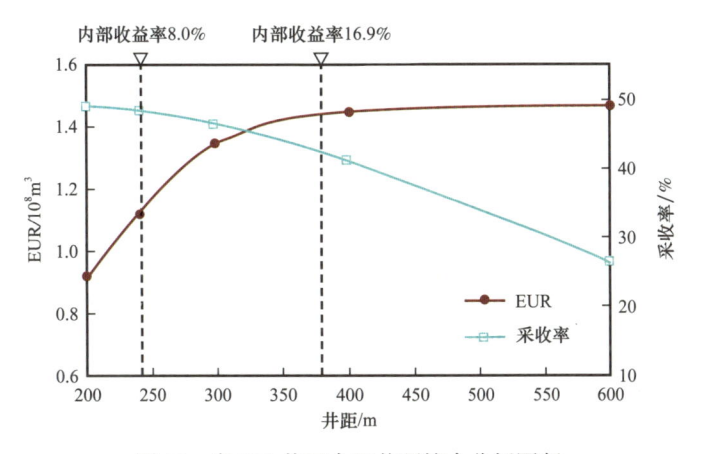

图10　宁209井区合理井距综合分析图版

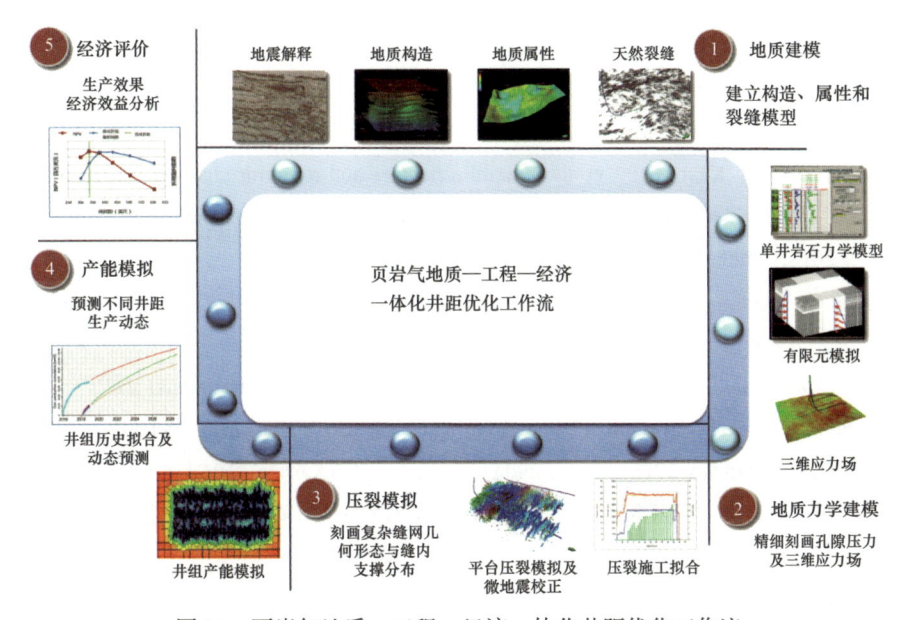

图11　页岩气地质—工程—经济一体化井距优化工作流

5　结论

水平井井距设计对于页岩气开发至关重要，但世界上没有任何两个页岩气藏是完全相同的，地质特征、工程技术存在差异，甚至是气价调整，都会使合理井距发生变化，因此有必要运用地质—工程—经济一体化的思路进行综合研究。

（1）井距从500~600m缩小至300~400m后，长宁区块的井间压窜概率逐渐增加，但主要表现为某一段或者某几段压窜，并没有大面积井间窜通，不能仅靠现场压力监测和产量数据变化判断井距是否合适。

（2）在北美地质工程一体化工作流的基础上，结合国内经济评价方法综合开展页岩气井距论证，不仅能真实模拟人工压裂缝网窜通后的气井生产效果，又能同时考虑方案的净盈利和平台井数设计是否合理。

（3）宁209井区在当前地质认识和工程技术条件下，兼顾单井EUR、平台采收率以及开发经济效益的合理井距范围为330～380m。

参 考 文 献

[1] Cakici D, Dick C, Mookerjee A, et al. Marcellus well spacing optimization-pilot data integration and dynamic modeling study [C] //SPE/AAPG/SEG Unconventional Resources Technology Conference, 12-14 August 2013, Denver, Colorado, USA. DOI: 10.1190/urtec2013-130.

[2] Lalehrokh F, Bouma J. Well spacing optimization in Eagle Ford [C] //SPE/CSUR Unconventional Resources Conference-Canada, 30 September-2 October 2014. Calgary, Alberta, Canada. DOI: 10.2118/171640-MS.

[3] Kim J, Chun M, Jung W, et al. Optimum design of multi-stage hydraulically fractured multi-horizontal shale gas well using flow regime analysis [J]. Geosciences Journal, 2015, 19（3）: 481-487.

[4] Orozco D, Aguilera R. Use of dynamic data and a new material-balance equation for estimating average reservoir pressure, original gas in place, and optimal well spacing in shale gas reservoirs [J]. SPE Reservoir Evaluation & Engineering, 2018, 21（4）: 1035-1044.

[5] Pankaj P, Shukla P, Kavousi P, et al. Hydraulic fracture and reservoir characterization for determining optimal well spacing in the Marcellus shale [C] //SPE Liquids-Rich Basins Conference-North America, 5-6 September 2018, Midland, Texas, USA. DOI: 10.2118/191802-MS.

[6] 谢军, 鲜成钢, 吴建发. 长宁国家级页岩气示范区地质工程一体化最优化关键要素实践与认识 [J]. 中国石油勘探, 2019, 24（2）: 174-185.

[7] Lindsay G, Miller G, Xu T, et al. Production performance of infill horizontal wells vs. pre-existing wells in the major US unconventional basins [C] //SPE Hydraulic Fracturing Technology Conference and Exhibition, 23-25 January 2018, The Woodlands, Texas, USA. DOI: 10.2118/189875-MS.

[8] 胡文瑞. 地质工程一体化是实现复杂油气藏效益勘探开发的必由之路 [J]. 中国石油勘探, 2017, 22（1）: 1-5.

[9] Yong R, Zhuang X, Wu J, et al. An integrated modelling workflow to optimiseinitial production rate and well spacing for Longmaxishale gas play [C] //International Petroleum Technology Conference, 26-28 March 2019, Beijing, China. DOI: 10.2523/IPTC-19104-MS.

[10] 杨兆中, 李扬, 李小刚, 等. 页岩气水平井重复压裂关键技术进展及启示 [J]. 西南石油大学学报（自然科学版）, 2019, 41（6）: 75-86.

[11] 段永刚, 张泰来, 魏明强, 等. 页岩气藏"井工厂"模式下水平井裂缝分布优化 [J]. 油气藏评价与开发, 2019, 9（6）: 78-84.

[12] Zheng W, Xu T, Baihly J, et al. Advanced modeling of production induced pressure depletion impact on infill well using cloud computation in the Haynesville [C] //International Petroleum Technology Conference, 26-28 March 2019, Beijing, China. DOI: 10.2523/IPTC-19460-MS.

[13] 高世葵, 朱文丽, 殷诚. 页岩气资源的经济性分析——以Marcellus页岩气区带为例 [J]. 天然气工业, 2014, 34（6）: 141-148.

[14] 杨济源, 李海涛, 张劲. 四川盆地川南页岩气立体开发经济可行性研究 [J]. 天然气勘探与开发,

2019，42（2）：95-99.

[15] 位云生，王军磊，齐亚东.页岩气井网井距优化[J].天然气工业，2018，38（4）：129-137.

[16] 蒋裕强，董大忠，漆麟.页岩气储层的基本特征及其评价[J].天然气工业，2010，30（10）：7-12.

[17] 王兰生，廖仕孟，陈更生.中国页岩气勘探开发面临的问题与对策[J].天然气工业，2011，31（12）：119-122.

[18] 赵金洲，任岚，沈骋.页岩气储层缝网压裂理论与技术研究新进展[J].天然气工业，2018，38（3）：1-14.

[19] 李庆辉，陈勉，金衍.压裂参数对水平页岩气井经济效果的影响[J].特种油气藏，2013，20（1）：146-150.

基于多井模型的压裂参数—开发井距系统优化

王军磊，贾爱林，位云生，贾成业，齐亚东，袁 贺，金亦秋

（中国石油勘探开发研究院）

摘 要：以变缝宽导流裂缝为基本流动单元，建立考虑井（缝）间干扰的多水平井渗流数学模型，采用半解析方法求解模型，模拟多井开发平台全生命周期生产动态，分析裂缝维数（裂缝长度、导流能力）、井距、缝距等因素对生产效果的影响。以此为基础，提出了以压裂规模为内部约束条件、以经济效益为外部约束条件的"嵌套式"全局优化方法，以优化水平井—裂缝等钻完井参数。研究表明：当不考虑约束条件时，增加裂缝与地层接触面积、降低缝/井间干扰强度、平衡裂缝与地层流入流出关系均能有效提高平台开发效果，不存在着最优钻完井参数；当仅考虑内部约束条件时，裂缝维数内存在最优匹配关系，但不存在最优井距、缝距；当同时考虑内外部约束条件时，裂缝维数、井距、缝距等因素间产生关联，出现参数优化空间，在压裂规模较小情况下宜采用小井距、宽缝距、短裂缝的水平井部署模式，在压裂规模较大情况下则宜采用大井距、窄缝距、长裂缝的部署模式。本文旨在提供一种联合生产动态预测模型、压裂设计模型和经济评价模型的多参数系统优化方法，为开发平台的开发技术政策制定提供理论依据。

关键词：多段压裂；水平井；裂缝导流能力；裂缝长度；井距；缝距；参数优化

水平井钻井及分段压裂技术广泛应用于非常规油气开发，研究表明影响井产能的主控因素众多，主要分为地层参数、流体参数及钻完井参数[1]。其中地层参数和流体参数是不可控因素，钻完井参数由工程设计确定，是人为可控因素。从油藏工程角度看，水平井钻完井的目的是在尽可能增加裂缝与地层接触面积前提下，提高裂缝内部的有效导流能力，降低流动过程中的渗流阻力，改善生产效果。对于包括多口压裂水平井的开发平台而言，最主要的设计指标为井距、段距、支撑剂用量、裂缝长度及导流能力等[2-4]。

对开发平台设计指标（或称为主控因素、影响参数）优化时的难点主要体现在：一是合理模拟能够充分体现缝间干扰、井间干扰以及与地层耦合的裂缝导流影响的生产动态特征，二是厘清主控因素与生产动态的对应关系及主控因素间的关联性。在生产动态模拟方面，在 Chen 和 Raghavan[5]研究基础上王晓冬等[6]采用均匀流量裂缝＋裂缝导流因子的方法获得了不稳态渗流解析解，解决裂缝间干扰问题，较好地模拟了单口压裂水平井压力动态特征；Chen 等[7]建立了考虑有限导流裂缝空间分布复杂性的地层—裂缝耦合流动模型，并给出了相应的边界元求解方法，有效解决了复杂形态裂缝的非稳态渗流问题；方文超等[8]建立了考虑复杂裂缝跨尺度特性的离散裂缝数值模型，突破了单井尺度的规模限制，模拟了包含 3 口长体积压裂水平井的开发平台生产动态；Yu 等[9]采用半解析方法描述了不同井间连通条件下的多口压裂水平井间响应机制，重点研究了开发井距对邻井压力干扰响应的影响。在主控因素影响分析方面，目前多数研究采用单因素控制法逐个分析

特定因素影响，该方法忽略了各因素间关联性，分析结果局限性大，难以考虑其他因素随之产生的影响[10-13]；正交试验、灰色关联等分析手段虽然改进了单因素控制分析法的局限性，但仍属于多方案设计对比分析方法，存在着难以全部覆盖最优解空间的弊端[14-16]；以输出结果为导向的遗传算法、神经网络等智能算法，计算代价大，尤其是当井数和压裂段数增加时，计算量呈几何倍数增加，不便推广应用[17-18]。

综上所述，水平井—压裂设计参数的同步优化是一项复杂的多参数优化技术，既是非线性规划的数学问题，又是涉及多学科交叉的油藏工程问题。本文以变导流裂缝为基本流动单元，建立多井系统全生命周期生产动态模型，并获得产量、累计产量的半解析解；以此为基础，将经典支撑剂指数法（UFD）改进应用至多井/缝的非稳态生产阶段，同时结合压裂设计和经济评价模型，约束影响参数的优化空间，最终形成一套具有油藏工程意义的多参数全局优化方法。

1 多井非稳态产能评价模型

1.1 模型说明

本文以"井工厂"平台的半支为基本研究单元，为了减小非均质性影响，平台采用水平井均匀部署、裂缝平行排列的模式，同时裂缝属性相同（图1）。其中，平台横向宽度为 x_{ef}，有 n_w 口水平井排列，平台纵向长度为 y_{ef}，每口水平井有 n_f 条裂缝，水平井以常压生产，井底压力为 p_{wf}。假设储层为渗透率 K_m 的均质地层，裂缝高度等于储层厚度，裂缝内产出微可压缩流体。气体在致密性介质内遵循非线性渗流规律，根据 Ertekin 模型，基质内流动速度可分解为经典达西流和滑脱流[19]：

$$v_m = \frac{K_m}{\mu_g}\nabla p_m + \frac{M_g D_g}{\rho_g R_g T}\nabla\left(\frac{p_m}{Z_g}\right) = \frac{K_m}{\mu_g}\left(1 + \underbrace{\frac{D_g \mu_g c_g p_m}{K_m}}_{b_{am}}\frac{1}{p_m}\right)\nabla p_m \quad (1)$$

其中，∇p_m 为压力梯度；b_{am} 为滑脱因子；M_g 为分子质量；D_g 为基质扩散系数；ρ_g 为气体密度；μ_g 为气体黏度；T 为温度。

式（1）中的滑脱速度是由基质内浓度差引起的，在基质与水力裂缝交界面上浓度差可忽略不计[20]，滑脱效应趋近于0，即：

$$b_{am}(x\ y) = \begin{cases} 0 & \text{交界面} \\ b_{am} & \text{非交界面} \end{cases} \quad (2)$$

对气体流动问题通常引入拟函数变量对非线性流动进行拟线性化处理。其中裂缝内拟压力为 $m_f = \frac{\mu_{gi}Z_{gi}}{p_i}\int_{p_i}^{p}\frac{\xi}{\mu_g(\xi)Z_g(\xi)}d\xi$，基质内拟压力为 $m_m = \frac{\mu_{gi}Z_{gi}}{p_i}\int_{p_i}^{p}\frac{\gamma_m(\xi)\xi}{\mu_g(\xi)Z_g(\xi)}d\xi$，这里非线性流动修正因子为 $\gamma_m(p_m) = 1 + \frac{b_{am}}{p_m}$。与之对应的拟时间定义为

$$t_a = \int_0^t \frac{\gamma_m(\tau)\mu_{gi}c_{ti}}{\mu_g(\tau)c_t(\tau)}d\tau = t\beta(t) \quad (3)$$

其中黏度—压缩比值 $\lambda(t)=\dfrac{\gamma_{\mathrm{m}}(\tau)}{\left[\mu_{\mathrm{g}}(\tau)/\mu_{\mathrm{gi}}\right]\left[c_{t}(\tau)/c_{\mathrm{ti}}\right]}$，非线性影响因子 $\beta(t)=\dfrac{1}{t}\int_{0}^{t}\lambda(\tau)\mathrm{d}\tau$。处理后可以将非线性气体流动在形式上等效转换为线性化液体流动。

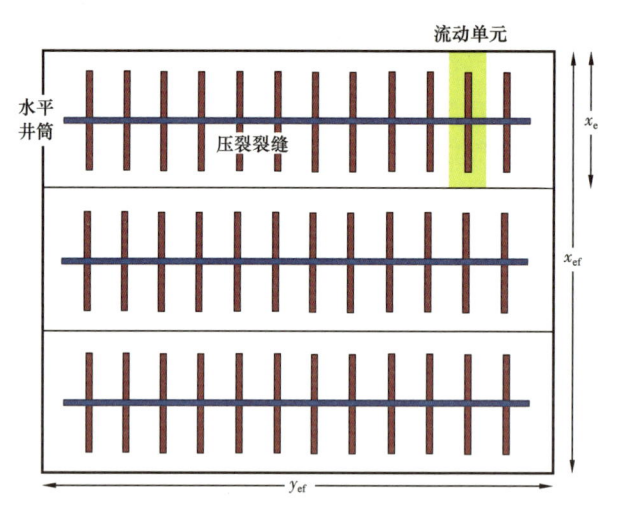

图 1　平台半分支多口压裂水平井部署示意图

由于整个渗流系统处于同一压力系统，流动过程可分解为连续的裂缝内和地层内流动两部分。对流动全过程进行建模时，裂缝和地层采用两套独立空间坐标，通过将两部分流动在裂缝面进行压力和流量耦合，可得到不同时刻地层任一点的压力和沿裂缝流量分布。

考虑不同井距条件下水力压裂对裂缝开启的影响，本文采用变缝宽裂缝模型。当裂缝延展方向平行于最大主应力方向时，越远离井筒井间应力阴影越显著，导致裂缝开启所需净压力增加、支撑剂运移受阻，裂缝宽度自井筒向外呈现递减趋势[21]，形式为：

$$w(x_{\mathrm{hf}})=\dfrac{4(1-v^{2})p_{\mathrm{net}}}{E}\left(-x_{\mathrm{hf}}^{2}+2x_{\mathrm{f}}x_{\mathrm{hf}}\right) \qquad (4)$$

式（1）可重新表述为：

$$w(x_{\mathrm{hf}})=w_{\max}\left[-(x_{\mathrm{hf}}/x_{\mathrm{f}})^{2}+2(x_{\mathrm{hf}}/x_{\mathrm{f}})\right] \qquad (5)$$

其中，x_{hf} 为裂缝坐标；$w_{\max}$ 为井筒处最大裂缝宽度；v 为泊松比；E 为杨氏模量；p_{net} 为净压力；x_{f} 为裂缝半长。

为了便于数学建模，本文定义无量纲压力、无量纲产量、无量纲累计产量：

$$p_{\mathrm{mD}}=\dfrac{m_{\mathrm{m}}(p_{i})-m_{\mathrm{m}}(p_{\mathrm{m}})}{m_{\mathrm{m}}(p_{i})-m_{\mathrm{m}}(p_{\mathrm{wf}})},\quad p_{\mathrm{fD}}=\dfrac{m_{\mathrm{f}}(p_{i})-m_{\mathrm{f}}(p_{\mathrm{f}})}{m_{\mathrm{m}}(p_{i})-m_{\mathrm{m}}(p_{\mathrm{wf}})},\quad q_{\mathrm{wD}}=\dfrac{\alpha q_{\mathrm{w}}\mu_{\mathrm{gi}}B_{\mathrm{gi}}}{k_{\mathrm{m}}h\left[m_{\mathrm{m}}(p_{i})-m_{\mathrm{m}}(p_{\mathrm{wf}})\right]},$$

$$G_{\mathrm{pD}}=\dfrac{\alpha B_{\mathrm{gi}}G_{\mathrm{p}}}{h\left[m_{\mathrm{m}}(p_{i})-m_{\mathrm{m}}(p_{\mathrm{wf}})\right]\varphi_{\mathrm{m}}c_{\mathrm{ti}}L_{\mathrm{ref}}^{2}}$$

其余无量纲时间、无量纲长度、无量纲导流能力及无量纲裂缝流量密度为：

$$t_D = \frac{\beta K_m t_a}{\varphi_m \mu_{gi} c_{ti} L_{ref}^2}, \quad \xi_D = \frac{\xi}{L_{ref}}, \quad C_{fD} = \frac{K_f w_f}{K_m L_{ref}}, \quad q_{fD}(u,t) = \frac{\alpha q_f(u,t) L_{ref} \mu_{gi} B_{gi}}{k_m h \left[m_m(p_i) - m_m(p_{wf}) \right]}$$

其中 α、β 为单位转换系数。当采用 SI 单位制时，$\alpha=0.5/\pi$，$\beta=1$；当采用矿场单位制时，$\alpha=1.842$，$\beta=0.0036$。

1.2 地层流动模型

根据模型假设，平台内多口水平井处于相同压力系统，将地层内气体流动控制方程变量转化为拟压力、拟时间，控制方程符合线性流动规律，因此可以采用压力叠加原理解决缝间、井间的相互干扰问题。压力干扰效应可以将平台流动系统分解为一系列以单裂缝为基本单元的子流动系统，如图 2 所示。根据物质守恒原理，气体拟压力控制方程满足：

$$\frac{\partial^2 m_m}{\partial x^2} + \frac{\partial^2 m_m}{\partial y^2} + \frac{\mu_{gi} B_{gi}}{k_m h} S_f(x,y,t) - \frac{\mu_{gi}}{k_m} q_{de}(t) = \frac{\phi_m \mu_{gi} c_{ti}}{k_m} \frac{\partial m_m}{\partial t_a} \quad (6)$$

其中，$S_f(x, y, t)$ 为水力裂缝流动引起的扰动函数：

$$S_f = \int_0^t \int_{x_{of}}^{x_{of}+L_f} q_f(u,t-\tau) \delta(x-x_{of}-u,\tau) \delta(y-y_{of},\tau) \mathrm{d}u \mathrm{d}\tau \quad (7)$$

当考虑页岩解吸气影响时，$q_{de}(t)$ 为解吸气提供的气体供给函数[7]：

$$q_{de}(t) = \frac{Z_{gi} p_{sc} T}{T_{sc} p_i} \frac{\partial V}{\partial t} = \frac{Z_{gi} p_{sc} T}{T_{sc} p_i} \frac{6 D_g \pi^2}{R_m^2} (V_E - V) \quad (8)$$

将裂缝视为一个汇集中分布的连续函数 $\tilde{q}_{fD}(s)$，将额外气体供给处理为双孔介质模型 $f(s)$，同时无量纲化处理式（6）并进行 Laplace 时间变换，可以得到无量纲拟压力控制方程：

$$\frac{\partial^2 \tilde{p}_{mD}}{\partial x_D^2} + \frac{\partial^2 \tilde{p}_{mD}}{\partial y_D^2} + \pi \int_{x_{ofD}}^{x_{ofD}+L_{fD}} \tilde{q}_{fD}(s) \delta(x_D - x_{ofD} - u_D) \delta(y_D - y_{ofD}) \mathrm{d}u_D = s f(s) \tilde{p}_{mD} \quad (9)$$

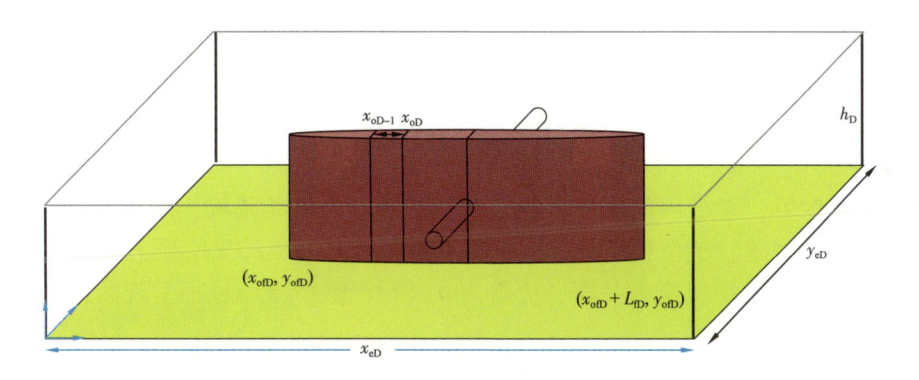

图 2 单裂缝引起的压力扰动示意图

从储层角度看，可以将裂缝进一步分解为 N 个带有不同流量强度的微元体，同时假设每个微元体内流量分布均匀，流量为 q_{fDi} 且长度为 Δx_{fDi}。根据压力叠加原理，可以得到多裂缝在地层内任一点的压力：

$$\tilde{p}_{mD}(x_{Dj}, y_{Dj}) = \sum_{i=1}^{N} \tilde{q}_{Di} \cdot \tilde{p}_{uDj,i}(\beta_{Dj}, \beta_{wDi}, x_{fDi}, x_{eD}, y_{eD}; s) \tag{10}$$

其中，$\tilde{p}_{uDj,i}$ 代表第 i 个微元在第 j 个微元处引起的压力扰动，利用 Green 函数和 Newman 乘积法[22]，同时结合 Laplace 变换，计算裂缝微元段引起的不稳定压力分布：

$$\tilde{p}_{uDj,i}(\beta_{Dj}, \beta_{wDi}) = \frac{2\pi \Delta x_{fDi}}{x_{eD}} \tilde{H}_0 + \tilde{F}_{ji} + 4\sum_{n=1}^{\infty} \frac{\tilde{H}_n - 1}{n\varepsilon_n} \cos\frac{n\pi x_{Dj}}{x_{eD}} \sin\frac{n\pi x_{wDi}}{x_{eD}} \cos\frac{np\Delta x_{fDi}}{x_{eD}} \tag{11}$$

这里

$$\begin{cases}\tilde{H}_n = \dfrac{\cos h\left[\varepsilon_n(y_{eD} - |y_{Dj} - y_{wDi}|)\right] + \cos h\left[\varepsilon_n(y_{eD} - |y_{Dj} + y_{wDi}|)\right]}{\sinh(\varepsilon_n y_{eD})} \\ \tilde{F}_{ji} = \displaystyle\int_{-\Delta x_{fDi}}^{\Delta x_{fDi}}\left(\frac{2\pi}{x_{eD}}\sum_{n=1}^{\infty}\frac{1}{\varepsilon_n}\cos\frac{n\pi x_{Dj}}{x_{eD}}\cos\frac{n\pi u_D}{x_{eD}}\right)du_D\end{cases} \tag{12}$$

上式中上标 "~" 为 Laplace 空间下的解析函数式，$\beta_{Dj} = (x_{Dj}, y_{Dj})$，$\beta_{wDi} = (x_{wDi}, y_{wDi})$，$\varepsilon_n = \sqrt{sf(s) + n^2\pi^2/x_{eD}^2}$，$f(s) = \omega + \dfrac{aD_D(1-\omega)}{s + D_D}$。其中 s 为 Laplace 变量，D_D 为无量纲扩散系数，a 为气体解吸系数，ω 为基质系统储容比，以上变量均为无量纲量，具体表达式可参考文献[7]。

特殊地，可以获得沿着裂缝面的地层压力分布。为了便于计算，我们将式（10）改写成矩阵形式，

$$\begin{pmatrix}\tilde{p}_{mD1} \\ \vdots \\ \tilde{p}_{mDj} \\ \vdots \\ \tilde{p}_{mDN}\end{pmatrix} = \begin{pmatrix}\tilde{p}_{uD1,1}(\beta_{wD1},\beta_{wD1}) & \cdots & \tilde{p}_{uD1,i}(\beta_{wD1},\beta_{wDi}) & \cdots & \tilde{p}_{uD1,N}(\beta_{wD1},\beta_{wDN}) \\ \vdots & & \vdots & & \vdots \\ \tilde{p}_{uDj,1}(\beta_{wDj},\beta_{wD1}) & \cdots & \tilde{p}_{uDj,i}(\beta_{wDj},\beta_{wDi}) & \cdots & \tilde{p}_{uDj,N}(\beta_{wD1},\beta_{wDN}) \\ \vdots & & \vdots & & \vdots \\ \tilde{p}_{uDN,1}(\beta_{wDN},\beta_{wD1}) & \cdots & \tilde{p}_{uDN,i}(\beta_{wDN},\beta_{wDj}) & \cdots & \tilde{p}_{uDN,N}(\beta_{wDN},\beta_{wDN})\end{pmatrix} \cdot \begin{pmatrix}\tilde{q}_{D1} \\ \vdots \\ \tilde{q}_{Di} \\ \vdots \\ \tilde{q}_{DN}\end{pmatrix} \tag{13}$$

1.3 变导流裂缝流动模型

从裂缝角度看，裂缝内的流动可视为有源汇的一维流动区域，其中源指的是有流体不断从地层流入裂缝。单位长度裂缝流量即流量密度为 $q_f(x, t)$，而在裂缝与井筒交汇处存在着汇，流体从裂缝流入井筒，流量为 $q_w(t)$，整个流动过程呈现典型的变质量流（图 3a）。

本文建立了变缝宽的裂缝模型（图 3b），裂缝内的一维流动规律可使用如下无量纲数学模型进行描述：

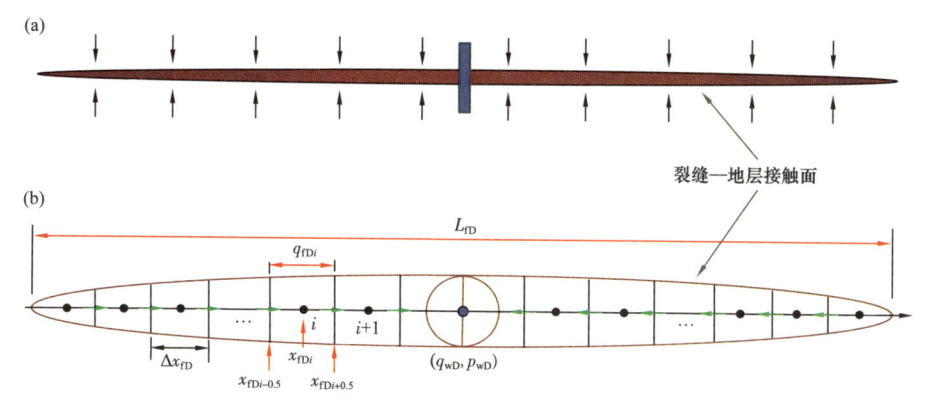

图 3 变质量裂缝流动示意图

$$\frac{\partial}{\partial x_{\text{Dhf}}}\left(C_{\text{fD}}(x_{\text{Dhf}})\frac{\partial \tilde{p}_{\text{fD}}}{\partial x_{\text{Dhf}}}\right) - 2\pi \tilde{q}_{\text{fD}}(x_{\text{Dhf}}) + 2\pi\left[\tilde{q}_{\text{wD}}(t_{\text{D}})\delta(x_{\text{Dhf}}, x_{\text{Dwhf}})\right] = 0 \quad (14)$$

其中，$\delta(\ )$ 为 Dirac 函数，无量纲变导流能力函数为 $C_{\text{fD}}(x_{\text{Dhf}}) = C_{\text{fDmax}}\left(-x_{\text{Dhf}}^2 + 2x_{\text{Dhf}}\right)$。式（14）为非常系数线性微分方程，无法直接求解，引入坐标变换方法对其进行常系数化处理。其中，坐标转换方程为

$$\xi_{\text{D}}(x_{\text{Dhf}}) = \hat{C}_{\text{fD}} \int_0^{x_{\text{Dhf}}} C_{\text{fD}}^{-1}\left[p_{\text{fD}}(x_{\text{D}})\right] dx_{\text{D}}, \hat{C}_{\text{fD}} = L_{\text{fD}} \Big/ \int_0^{L_{\text{fD}}} C_{\text{fD}}^{-1}\left[p_{\text{fD}}(x_{\text{D}})\right] dx_{\text{D}} \quad (15)$$

使用式（15）对式（14）进行处理，可以在新坐标 ξ_{D} 下获得常系数线性微分方程，在此基础上使用边界元方法进行双重积分处理[23]，得到裂缝内的无量纲压力分布：

$$\tilde{p}_{\text{wD}} - \tilde{p}_{\text{fD}}(\xi_{\text{D}}) = \frac{2\pi}{\hat{C}_{\text{fD}}}\tilde{q}_{\text{wD}}G(\xi_{\text{D}}, \xi_{\text{Dwhf}}) - \frac{2\pi}{\hat{C}_{\text{fD}}}\left[I(\xi_{\text{D}}) - I(\xi_{\text{wD}})\right] \quad (16)$$

这里 Fredholm 积分式为 $I(\xi_{\text{D}}) = \int_0^{\xi_{\text{D}}} d\zeta \int_0^{\zeta} \tilde{q}_{\text{fD}}(\varsigma) d\varsigma$。为了考虑裂缝与井筒相交点周围的聚流效应，引入聚流表皮因子[24]，修正裂缝内压力降：

$$\tilde{p}_{\text{wD}} - \tilde{p}_{\text{fD}}(\xi_{\text{D}}) = \frac{2\pi}{\hat{C}_{\text{fD}}}\tilde{q}_{\text{wD}}G(\xi_{\text{D}}, \xi_{\text{Dwhf}}) - \frac{2\pi}{\hat{C}_{\text{fD}}}\left[I(\xi_{\text{D}}) - I(\xi_{\text{wD}})\right] + \tilde{q}_{\text{wD}}\underbrace{\frac{2h_{\text{D}}}{\hat{C}_{\text{fD}}L_{\text{fD}}}\left[\ln\left(\frac{h_{\text{D}}}{2r_{\text{wD}}}\right) - \frac{\pi}{2}\right]}_{S_{\text{c}}} \quad (17)$$

同样地，可以将式（17）改写为矩阵形式：

$$\begin{pmatrix}\tilde{p}_{\text{wD}}\\ \vdots\\ \tilde{p}_{\text{wD}}\\ \vdots\\ \tilde{p}_{\text{wD}}\end{pmatrix} - \begin{pmatrix}\tilde{p}_{\text{fD1}}\\ \vdots\\ \tilde{p}_{\text{fD}i}\\ \vdots\\ \tilde{p}_{\text{fD}N}\end{pmatrix} = \frac{2\pi}{\hat{C}_{\text{fD}}}\begin{pmatrix}\tilde{q}_{\text{wD}}\gamma_1\\ \vdots\\ \tilde{q}_{\text{wD}}\gamma_i\\ \vdots\\ \tilde{q}_{\text{wD}}\gamma_N\end{pmatrix} - \frac{2\pi}{\hat{C}_{\text{fD}}}\begin{pmatrix}\alpha_{1,1} & \cdots & \alpha_{1,i} & \cdots & \alpha_{1,N}\\ \vdots & & \vdots & & \vdots\\ \alpha_{j,1} & \cdots & \alpha_{j,i} & \cdots & \alpha_{j,N}\\ \vdots & & \vdots & & \vdots\\ \alpha_{N,1} & \cdots & \alpha_{N,i} & \cdots & \alpha_{N,N}\end{pmatrix} \cdot \begin{pmatrix}\tilde{q}_{\text{fD1}}\\ \vdots\\ \tilde{q}_{\text{fD}i}\\ \vdots\\ \tilde{q}_{\text{fD}N}\end{pmatrix} + S_{\text{c}}\begin{pmatrix}\tilde{q}_{\text{wD}}\\ \vdots\\ \tilde{q}_{\text{wD}}\\ \vdots\\ \tilde{q}_{\text{wD}}\end{pmatrix} \quad (18)$$

上式中 Ξ 为井点所处裂缝微元编号，行列式内元素为 $\gamma_i = G(\xi_{\text{D}i}, \xi_{\text{Dwhf}})$，其中 $G(\ )$

— 203 —

为 Heaviside 阶跃函数的积分函数 $G(x,x_0) = \begin{cases} x-x_0, x \geq x_0 \\ 0, x < x_0 \end{cases}$；矩阵内元素 $\alpha_{j,i} = \text{RS}_i + \text{RT}_i$，其中

$$\text{RS}_i = \begin{cases} \xi_{Dj}\Delta\xi_{Dj} - 0.5(\xi_{oDi}^2 - \xi_{oDi-1}^2), i < j \\ \xi_{Dj}(\xi_{Dj} - \xi_{oDj-1}) - 0.5(\xi_{Dj}^2 - \xi_{oDj-1}^2), i = j \end{cases}, \quad \text{RT}_i = \begin{cases} (\xi_{\text{Dwhf}} - \xi_{Di})\Delta\xi_{Dj}, i < 1+\Xi \\ \left(\xi_{\text{Dwhf}} - \sum_{i=1}^{\Xi}\Delta\xi_{Di}\right)\left(0.5\xi_{\text{Dwhf}} + \sum_{i=1}^{\Xi}\Delta\xi_{Di}\right), i = 1+\Xi \end{cases}$$

2 模型求解及验证

地层流动模型与裂缝流动模型在裂缝面流量、压力耦合，满足如下公式：

$$\tilde{p}_{\text{fD}}(x_{\text{Dhf}}) = \tilde{p}_{\text{mD}}(x_{\text{ofD}} + x_{\text{Dhf}}, y_{\text{ofD}}), \tilde{q}_{\text{fD}} = \tilde{q}_{\text{D}} \quad (19)$$

值得注意的是，经过坐标转换后的裂缝流量 q_{wD} 与裂缝流量密度 $q_{\text{fD}j}$ 的关系为：

$$\tilde{q}_{\text{wD}} = \sum_{j=1}^{N} \tilde{q}_{\text{fD}j} \underbrace{\Delta\xi_{\text{D}j}/\Delta x_{\text{D}j}}_{f_{cj}} \quad (20)$$

这样式（13）中矩阵元素修正为 $f_{cj} \times \tilde{p}_{\text{uD}j,i}(\beta_{\text{wD}j}, \beta_{\text{wD}i})$。

利用压降叠加原理，将多井平台渗流系统分解为单级裂缝，以单裂缝为基本评价单元，对于每条裂缝均联立式（13）和式（18）辅以相应的耦合条件式（19）~式（20），通过求解耦合矩阵可以计算沿裂缝的流量分布，进而计算产量，同时根据累计产量与产量间在 Laplace 空间下的转换关系 [$\tilde{G}_{\text{pD}} = \tilde{q}_{\text{wD}}/s$]，获得累计产量。上述求解方程为未知量可表述为如下行列式：

$$X^{\text{T}} = \begin{pmatrix} \underbrace{\tilde{q}_{\text{fD}1,1}^{(1)}, \tilde{q}_{\text{fD}1,2}^{(1)}, \cdots \tilde{q}_{\text{fD}1,N}^{(1)}}_{N}, \cdots \underbrace{\tilde{q}_{\text{fD}N_f,1}^{(1)}, \tilde{q}_{\text{fD}N_f,2}^{(1)}, \cdots \tilde{q}_{\text{fD}N_f,N}^{(1)}}_{N}, \tilde{q}_{\text{wD}}^{(1)}, \tilde{G}_{\text{pD}}^{(1)}; \cdots \\ \cdots \underbrace{\tilde{q}_{\text{fD}1,1}^{(N_w)}, \tilde{q}_{\text{fD}1,2}^{(N_w)}, \cdots \tilde{q}_{\text{fD}1,N}^{(N_w)}}_{N}, \cdots \underbrace{\tilde{q}_{\text{fD}N_f,1}^{(N_w)}, \tilde{q}_{\text{fD}N_f,2}^{(N_w)}, \cdots \tilde{q}_{\text{fD}N_f,N}^{(N_w)}}_{N}, \tilde{q}_{\text{wD}}^{(N_w)}, \tilde{G}_{\text{pD}}^{(N_w)} \end{pmatrix} \quad (21)$$

其中，$\tilde{q}_{\text{fD}n,i}^{(w)}$ 代表第 w 口井第 n 条裂缝的第 i 个裂缝微元。需要指明的是，拟时间是关于时间和空间的积分函数，可将拟时间近似处理为平均地层压力关于时间的单重积分函数 $t_\alpha = t\beta[p_{\text{avg}}(t)]$，需要结合物质平衡方程 $\dfrac{p_{\text{avg}}(t)}{Z_{\text{avg}}(t)} = \dfrac{p_i}{Z_i}\left(1 - \dfrac{G_p}{\text{OGIP}}\right)$ 进行计算，具体算法可参考文献 [25]。

为了验证模型可靠性，分别使用解析模型和数值模型进行验证。Poe 等将裂缝单翼分成 4 个不同流动区域，建立了非均质裂缝定产量模型，并获得了经典无量纲压力图版[26]。本文将聚流表皮因子去除后计算定产量条件下的无量纲压力，计算结果与 Poe 图版进行对比。如图 4 所示（实离散点为本文结果），两种模型计算结果完全一致，从而验证了本文算法在单条变缝宽裂缝压力动态模拟条件下的适用性。

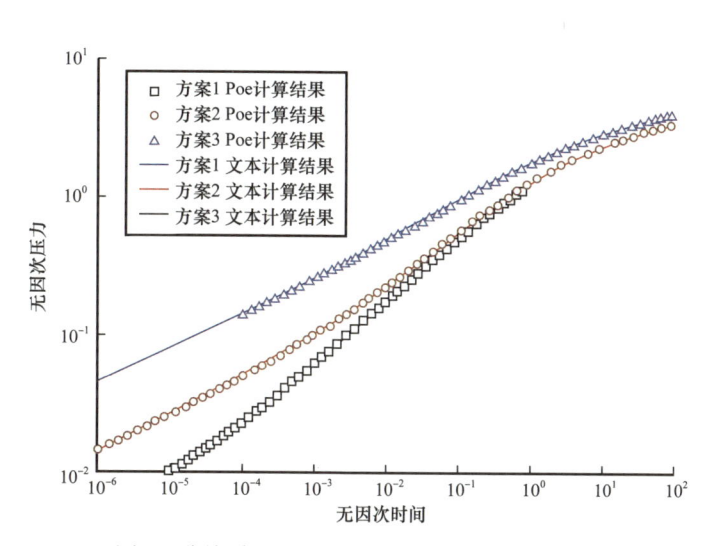

图4 非均质导流能力垂直裂缝无量纲压力对比

为进一步验证，借助ECL数值模拟器建立多口分段压裂水平井模型。模型所用参数如下：原始地层压力45.0MPa，井底压力6.5MPa，有效裂缝高度20m，地层有效孔隙度8%，含油饱和度85%，综合压缩系数4.35×10^{-4}MPa^{-1}，有效泄流区域内地层平均渗透率0.01mD，井底压力6.5MPa。原油黏度1mPa·s，体积系数1.0m^3/sm^3，井筒半径0.019m。裂缝段数30段，地层维数为1500m×1600m，裂缝有效半长114.95m，裂缝宽度为0.0127m，孔隙度为35%。为精确模拟生产动态裂缝采用网格对数加密描述（图5），在裂缝导流能力相同条件下，数值模型中裂缝等效宽度0.1m，等效孔隙度$\phi_e = \phi_e w_f / w_e = 0.044$，等效裂缝最大渗透率为$12 \times [(K_f w_f)/(K_m x_f)]$mD，沿裂缝渗透率分布满足公式（2）（图5）。

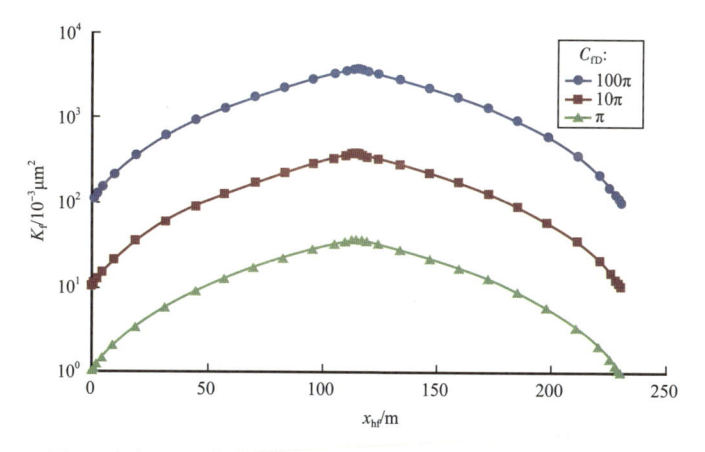

图5 包括3口井的开发平台数值模型及压力分布模拟结果

由于采用均匀布井，三口井生产动态一致（图5）。对比本文算法与数值模拟获得的无量纲生产动态数据（图6），结果表明两种模型获得的单井产量与累计产量高度吻合，进一步验证了本文算法在多井、多裂缝条件下的适用性。

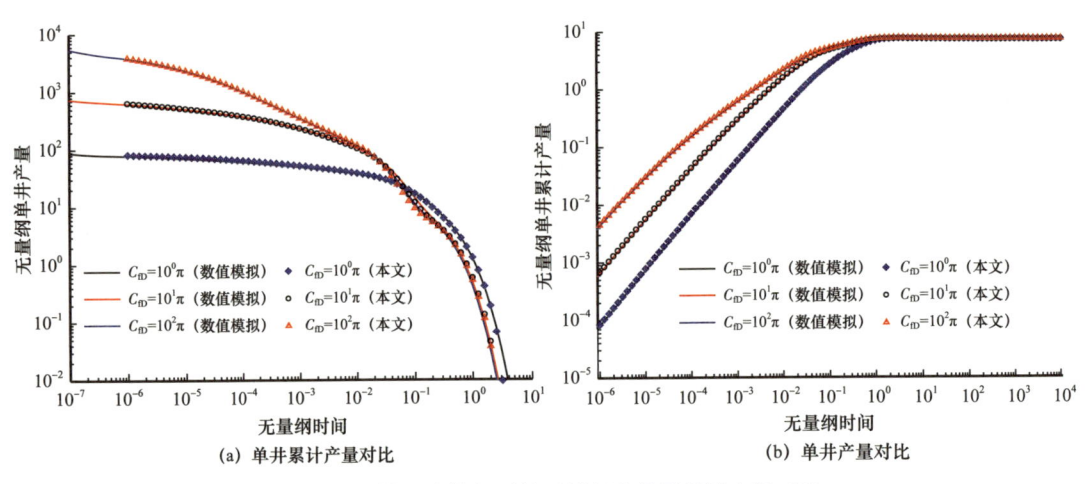

图 6　两种模型计算得到的无量纲产量及累计产量对比

3　裂缝与井距参数优化

影响开发平台生产动态的主要影响因素包括裂缝条数、裂缝长度、导流能力、水平井数、压裂段长度以及相应的位置分布等。由于影响因素较多，为便于讨论，把握主要因素间内在关联，本文假设一种理想模式（裂缝与水平井均匀分布，缝与缝、井与井之间参数一致）。具有相同属性的裂缝占有相同的子泄流面积（子泄流面积等于总泄流面积的 $1/(n_w \times n_f)$），这样影响多井平台生产特征因素可归结为两大类：裂缝总长度和裂缝导流能力。

在给定支撑剂量条件下裂缝总长度和裂缝导流能力增加会共同争夺支撑剂体积，当两者间达到平衡状态时，生产井产能指数达到最大，即 Economides 等提出的支撑剂指数法（UFD）。需要强调的是，传统 UFD 理论仅适用于单条均匀导流裂缝的拟稳态假设条件，以拟稳态产能指数为优化目标，与时间无关[27-28]。

本文以多井系统为研究对象，以总支撑剂体积（与压裂规模相关）为约束条件，以动态累积产量为优化目标，建立全生命周期的压裂—水平井参数动态优化方法。其中，改进的支撑剂指数 N_{prop} 定义为：

$$N_{prop} = \frac{2K_f V_{prop}}{K_m V_{res}} = \frac{2K_f}{K_m x_{ef} y_{ef}} \sum_{m=1}^{n_w} \sum_{n=1}^{n_f} \int_0^{L_{fm,n}} w_{fm,n}(\zeta) d\zeta = \left(\frac{4x_{ef}}{3y_{ef}}\right) \times \left(\frac{n_f C_{fDmax} I_x^2}{n_w}\right) \quad (22)$$

支撑剂体积 $V_{prop}=M_{prop}/[\rho_{prop}(1-\phi_{prop})]$，无量纲导流能力 $C_{fDmax}=[(K_{f,max}w_f)/(K_m x_f)]$。其中，$M_{prop}$ 为支撑剂质量；ρ_{prop} 为支撑剂密度；ϕ_{prop} 为支撑裂缝孔隙度。此外，为方便标记下文中提及的无量纲导流能力即为式（14）中的 C_{fDmax}。

3.1　裂缝维数优化

裂缝维数主要包括裂缝导流能力、长度，是平衡裂缝与地层流入流出关系的关键指标。这里以单裂缝为研究对象，聚焦裂缝维数优化过程，地层维数假设为 400m×50m×20m，使用裂缝穿透率表征裂缝长度（$I_x=L_f/x_e$），无量纲导流能力表征裂缝导流能力。

裂缝维数随支撑剂指数变化的优化结果见图8,其中蓝色曲线代表不考虑支撑剂体积约束的模拟结果,红色曲线代表考虑支撑剂体积约束的模拟结果。图8表明,当压裂规模不受约束时,无量纲累计产量随裂缝穿透比和导流能力增加而单调递增,但递增幅度逐渐减小,直到极限值,此时裂缝完全贯穿地层且达到无限导流能力;而考虑约束时,累计产量与裂缝维数存在最优值,即图中离散点,最大无量纲累计产量($G_{pD,max}$)随着支撑剂指数增加而增加。对比图7a和图7b可以看出,不同时刻下最优裂缝维数结果不同。

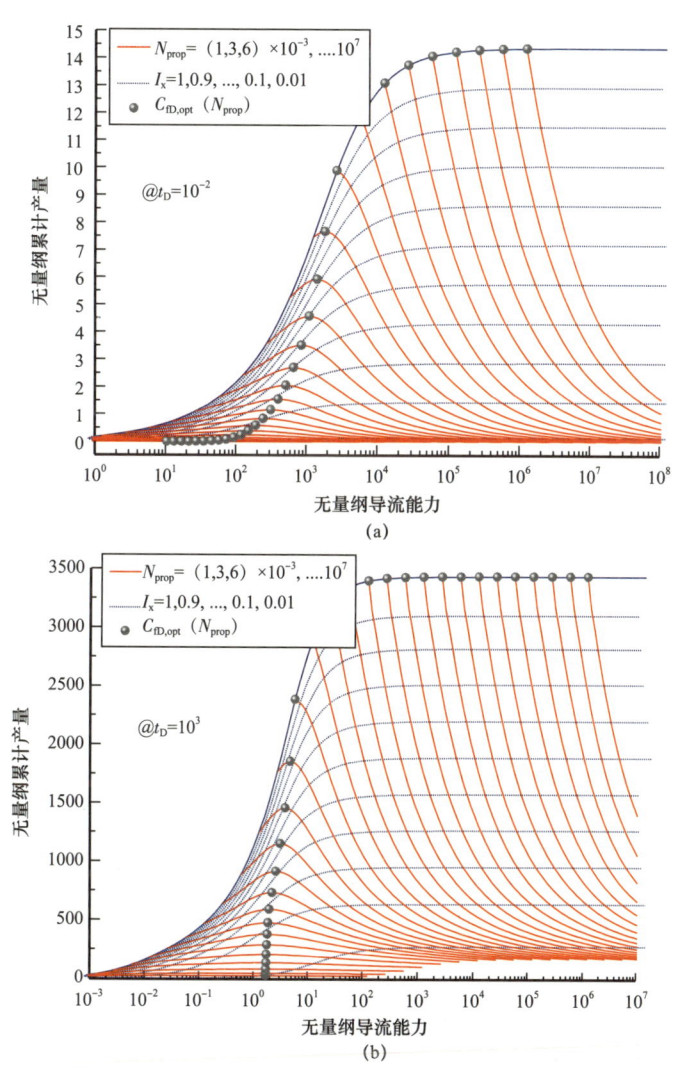

图7 不同时刻下有限导流裂缝维数优化图版

为厘清生产时间对优化结果的影响,重新计算获得裂缝维数随时间变化的优化结果(图8)。以图8a为例,最大无量纲累计产量随着生产时间增加而增加,而最优裂缝无量纲导流能力($C_{fD,opt}$)却逐渐递减且趋近于常数。对比图8a—d可以看出:支撑剂指数越大,在相同时刻下所对应的最大无量纲累计产量越大;支撑剂指数越大,在相同时间间隔内最优无量纲导流能力值的变化区间越大，且所趋近的常数也越大。具体地，图8b（$N_{prop}=1$）在 $t_D=0.01$~1000 时间间隔内对应的最优无量纲导流能力变化区间为 $C_{fD,opt} \in$（1，50），趋近值为 $\lim\limits_{t_D \to \infty} C_{fD,opt}=1.62$；图9d（$N_{prop}=100$）在相同的时间间隔内 $C_{fD,opt} \in$（15，400），$\lim\limits_{t_D \to \infty} C_{fD,opt}=15$。

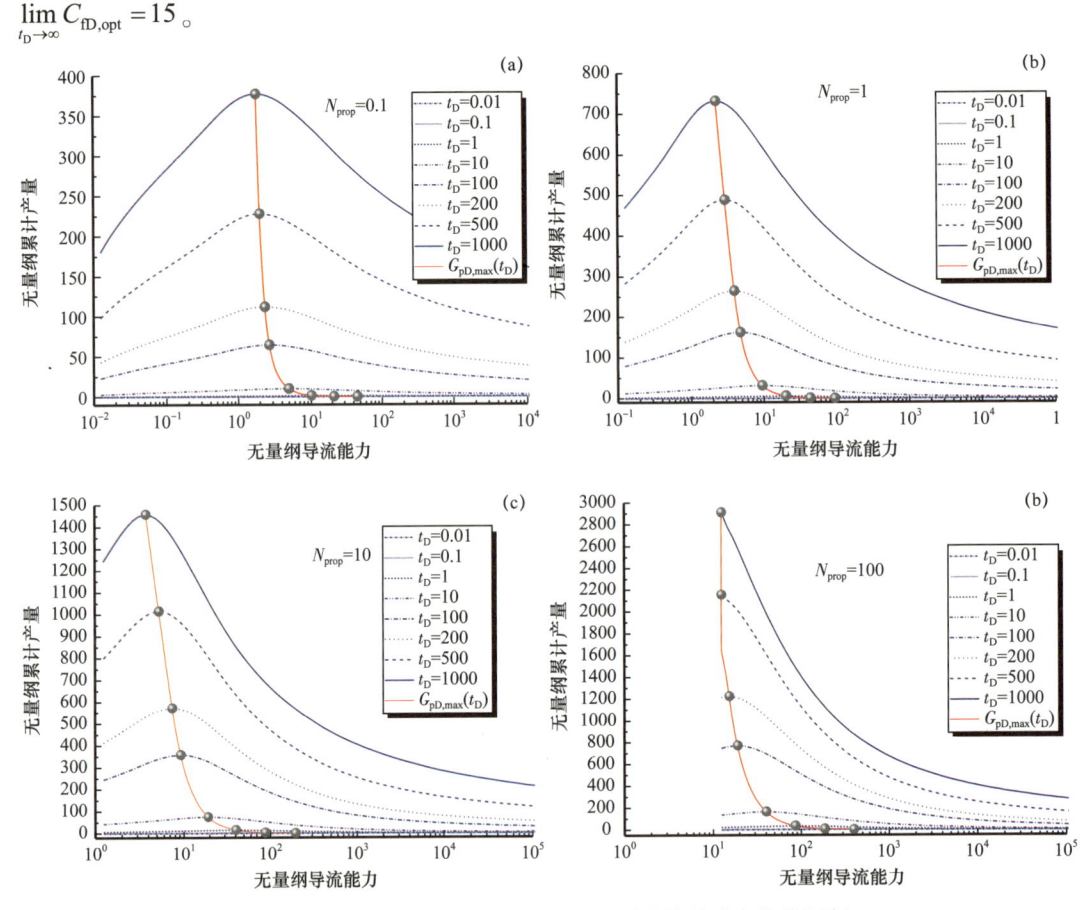

图8 不同支撑剂指数下的有限导流裂缝维数动态优化图版

图9总结了不同时刻、不同支撑剂指数下的最优裂缝维数、最大累计产量的变化规律。图9a表明当支撑剂指数较小时，随着支撑剂指数的减小最优无量纲导流能力逐渐递减且趋近于常量；无量纲时间越大，最优无量纲导流维持常量所对应的支撑剂指数区间越大，如 $t_D=1000$ 时 $C_{fD,opt}$ 在 $N_{prop} \in$（10^{-8}，10^{-1}）范围均为常量，$t_D=0.01$ 时对应区间为 $N_{prop} \in$（10^{-8}，10^{-7}），这与图7a、b中离散点分布特征一致。随着支撑剂指数增加，$C_{fD,opt}$ 和 $I_{x,opt}$ 均增加，当裂缝完全贯穿地层时（$I_x=1$），$C_{fD,opt}$ 与 N_{prop} 呈线性关系。图10b表明当支撑剂指数一定时，随着生产时间的增加，最优无量纲导流不断减小而趋近于常量，且常量值与 N_{prop} 呈正相关关系，如 $N_{prop}=10^{-2}$ 时对应 $C_{fD,opt}=1.62$，$N_{prop}=10^3$ 时对应 $C_{fD,opt}=175$，这与图8a—d反映的特征一致。总的来看，当无量纲生产时间（生产周期）较短时，可压裂长度较短且导流能力较高的裂缝；当无量纲生产时间较长时，可压裂长度较长且导流能力较低的裂缝。需要强调的是，所趋近的1.62特征值与Volko和Economides优化结果基本一致[24]，也证明了本文优化方法的可靠性。

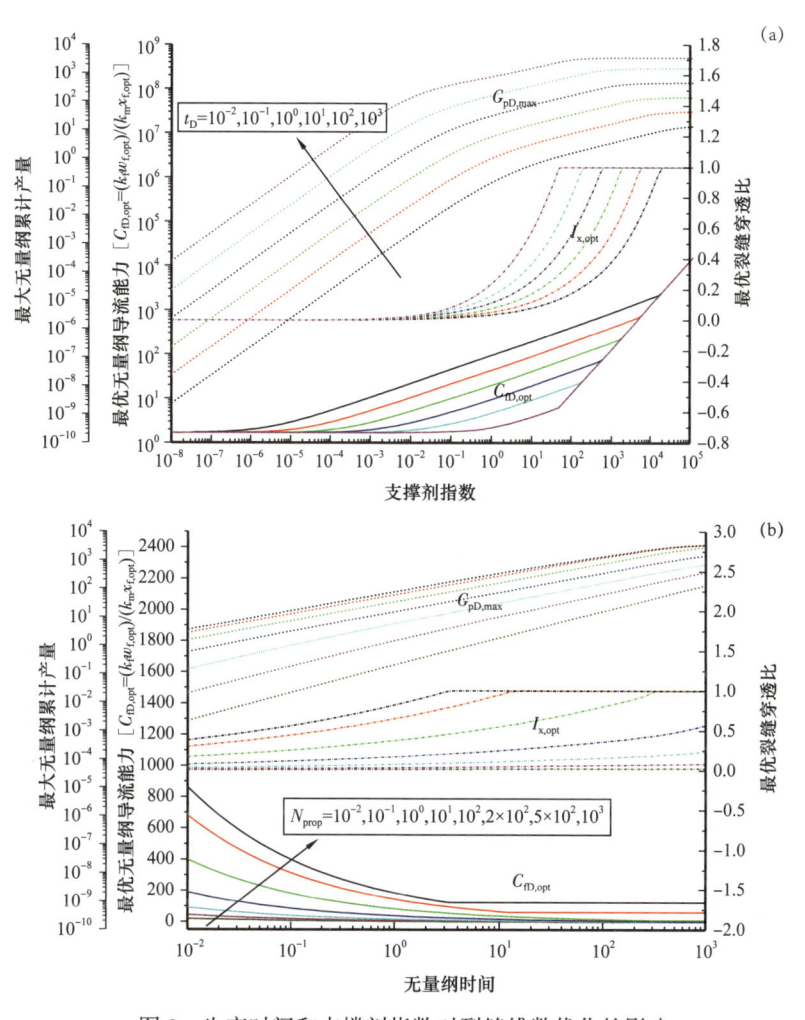

图 9 生产时间和支撑剂指数对裂缝维数优化的影响

3.2 缝距—井距优化

在泄流面积不变条件下（$A_{rea}=x_e \times y_e$），裂缝泄流面积形状会影响裂缝维数的优化结果，本文使用缝距与井距比值（$\lambda=y_e/x_e$）表征泄流面积形状，该值由裂缝条数和水平井井数确定（$y_e=y_{ef}/n_f$，$x_e=x_{ef}/n_w$）。图 10 反映了 λ 值对裂缝生产动态的影响，其中裂缝假设为完全贯穿（$I_x=1$）、无限导流（$C_{fD}=\infty$）。注意产量与累计产量间的对应关系，当产量递减至近似为 0 时累计产量趋近于极限值（$\lim G_{pD}=0.5 x_{eD} y_{eD}/\pi$），由于泄流面积相等，不同 λ 值下的极限累计产量相同。当 λ 值较小时，对应泄流面积呈长条状，裂缝与地层接触面积较大，渗流阻力较小，裂缝早期（累计）产量较高，能在较短时间内达到极限累计产量。同时，λ 较小值所对应的无量纲累计产量在整个生产周期内始终高于 λ 较大值。

考虑支撑剂指数约束后，λ 值对最大无量纲累计产量和最优裂缝维数均会产生影响（图 11）。如图 11a 所示，随着支撑剂指数增加最大无量纲累计产量趋近于最高值，而且 λ 值越小（窄缝距、大井距模式）对应的最高值越大，此时裂缝达到无限导流能力且完全贯穿地层（$I_x=1$，$C_{fD}>300$，图 11b）。当支撑剂指数较小时（$N_{prop}<10^3$），较大 λ 值对应的

最优无量纲导流能力相同，但最优裂缝穿透率较大（图11b），导致较小的渗流阻力，所以相应的最大无量纲累计产量值较大（图11a）。需要强调的是，图版所反映的特征与生产时间取值有关，当时间足够大时所有λ值对应的最高值均相等，而当时间足够小时在低N_{prop}情况下λ值对图版的影响很小。

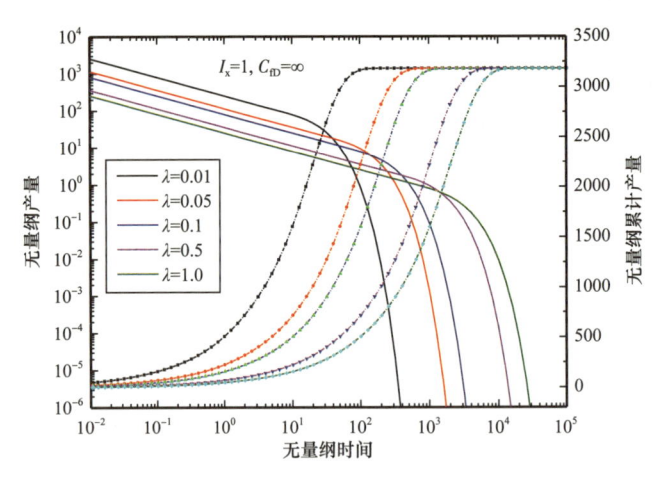

图10　缝距/井距比对生产动态特征的影响图版

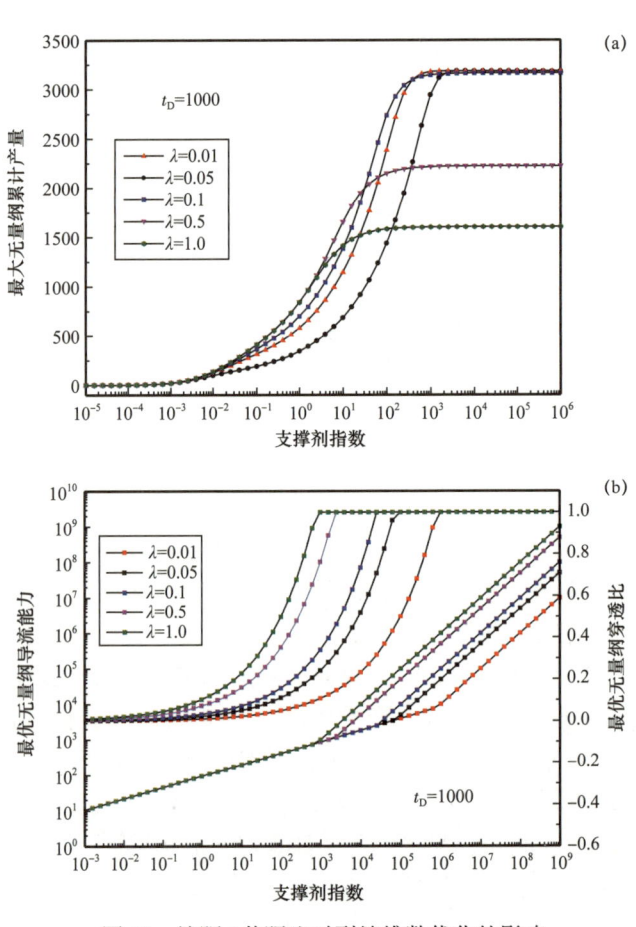

图11　缝距/井距比对裂缝维数优化的影响

3.3 多参数系统优化

多井平台下压裂水平井的开发效果优化具有明确的油藏工程意义，主要是通过增加裂缝与地层接触面积、降低井间干扰、缝间干扰、平衡裂缝与地层的流入流出关系实现，当4种渗流关系达到平衡时生产效果最佳。本文假定半支平台几何尺寸为1500m×1500m×20m，以支撑剂体积（或称压裂规模）为约束条件，以总NPV值（或称经济效益）为目标函数，采用嵌套式方法进行多参数优化（图12）。优化流程主要分为以下步骤：

步骤1：定义输入基本参量，包括地层参数、流体参数、支撑剂参数和生产周期；

步骤2：定义3种待优化变量，包括平台内井数（n_w）、单井压裂段数（n_f）和支撑剂体积（V_p）；

步骤3：根据3.1～3.2节所提供的UFD方法计算不同井数、段数和支撑剂体积条件下的最优裂缝维数及对应的最大累计产量（$G_{pD,\,max}$）；

步骤4：计算相应的NPV值；

步骤5：基于多元函数Powell全局优化算法重复步骤2，直到NPV值最大，此时对应的水平井—压裂参数即为最优设计参数。

其中NPV计算模型为

$$\mathrm{NPV} = \sum_{j=1}^{n} \frac{\left(G_{p,j} - G_{p,j-1}\right)}{\left(1+i_r\right)^j} - \left[FC + \sum_{k=1}^{n_w}\left(C_{\mathrm{well}} + \sum_{kk=1}^{n_f} C_{\mathrm{fracture}}\right)\right] \tag{23}$$

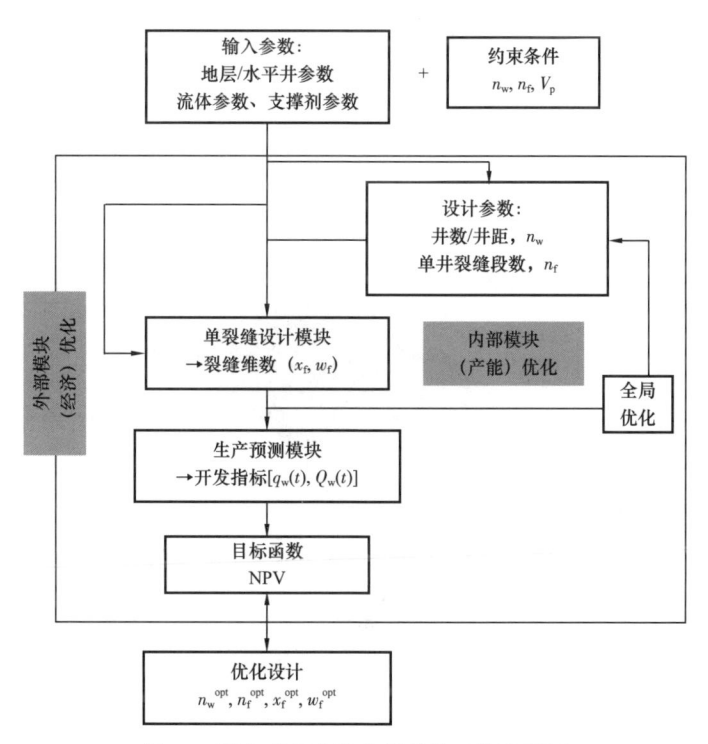

图12 嵌入式多参数全局优化工作流程图

其中，$G_{p,j}$为第j年累计产量；FC为固定总投资；C_{well}为单井钻井成本；$C_{fracture}$为单段裂缝压裂成本；n_w为水平井井数；n_f为单口井压裂段数；n为生产年限；i_r为年利率。相应的经济参数参考文献[16]。本文设定无量纲时间1000为生产周期，考虑到实际压裂规模受工程条件限制，将平台总支撑剂体积设定为固定的约束条件，使用图解法演绎多参数优化流程。

图13为不同水平井井数、单井压裂段数条件下的平台最大累计产量的变化规律。由图13可知，平台最大累计产量（$G_{pD,\,max}$）随井数和单井压裂段数的增加而单调增加，当支撑剂指数较小时平台最大累计产量随井数和段数的增加幅度基本一致（图13a）；当支撑剂指数较大时平台最大累计产量增加幅度相对减小，在$n_f>30$和$n_w>5$的区域内平台最大累计产量几乎不再增加（图13b）。因此，以平台最大累计产量为目标函数时，在仅以压裂规模为约束条件的情况下，平台内部仅存在着最优裂缝维数，但并不存在最优井距（井数）、缝距（缝数）。

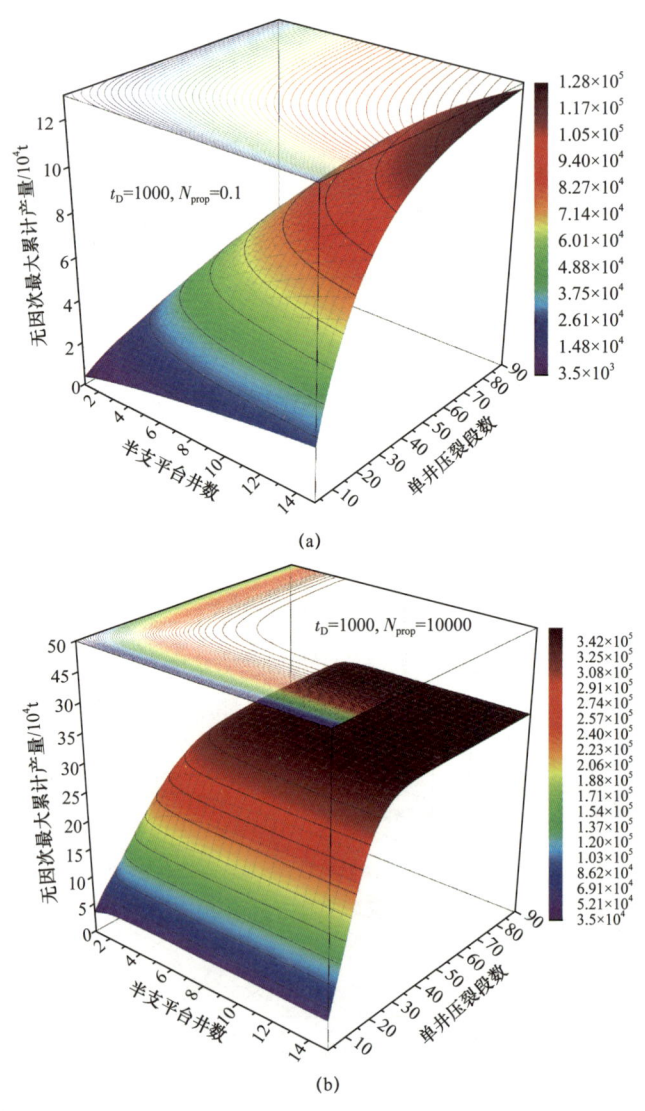

图13 不同支撑剂指数条件半支平台最大无量纲累计产量值

将开发指标计算结果代入经济评价模型［式（15）］，以净现值（NPV）为目标函数重新进行优化，结果见图 14。与图 13 不同的是，图 14 中明显出现了极值点，说明存在着最优井距、缝距。同样地，对应的裂缝维数优化结果见图 15。这是由于随着压裂段数和井数的增加，虽然提高了平台的开发效果，但投资成本随之增加，当开发效果增加幅度小于投资增长幅度时经济效益变差。因此在压裂规模和经济效益双重约束下多井平台内存在最优井距、缝距及裂缝维数，这为开发技术政策的制定提供了优化空间。

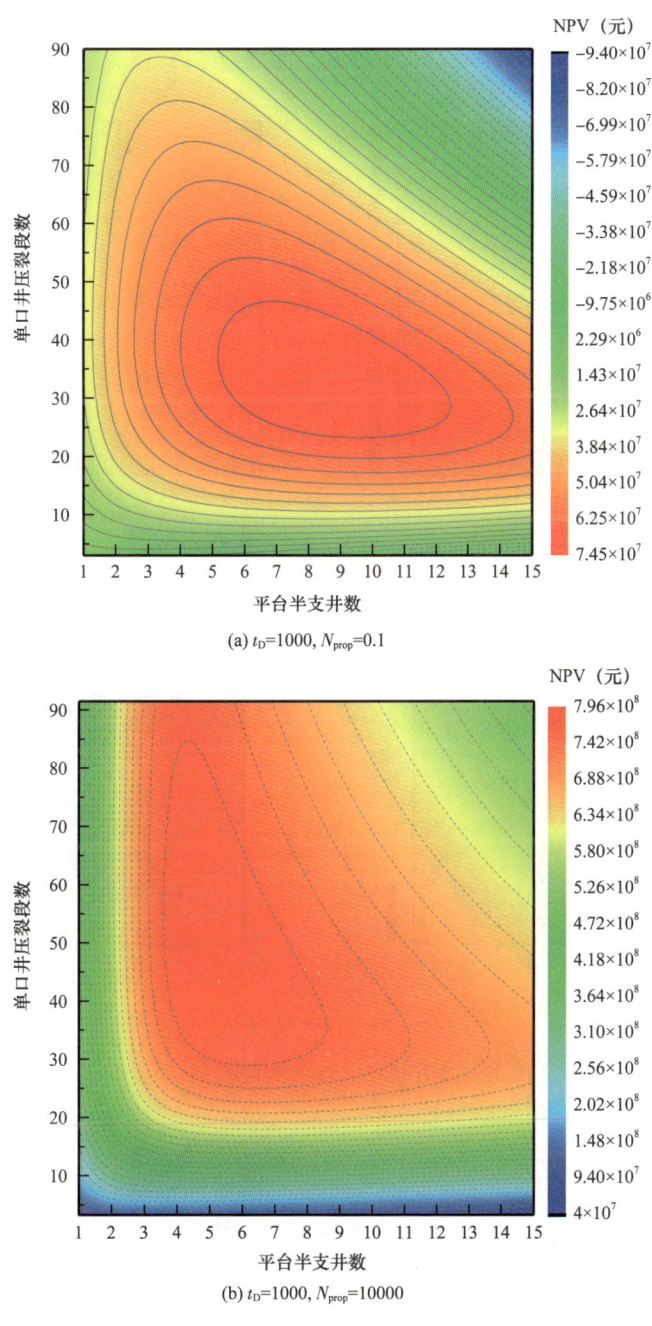

(a) $t_D=1000$, $N_{prop}=0.1$

(b) $t_D=1000$, $N_{prop}=10000$

图 14　最优井距—缝距值及平台最大 NPV 值

由图15可知，在本文算例中，当压裂规模较小时（如N_{prop}=0.1），最优井距较小（$n_{w,opt} \approx 8$）、最优缝距较宽（$n_{f,opt} \approx 35$）、裂缝穿透率较小（$I_{x,opt} \approx 0.13$）、裂缝导流能力较低（$C_{fD,opt} \approx 2.37$）。当压裂规模较大时（如$N_{prop}=10^4$），相比于小压裂规模情况，最优井距增加（$n_{w,opt} \approx 6$）、最优缝距减小（$n_{f,opt} \approx 50$），裂缝穿透率大幅度提高（$I_{x,opt} \approx 0.84$）、但裂缝导流能力仍处于较低水平（$C_{fD,opt} \approx 3.54$）。

需要指出的是，开发平台水平井压裂参数优化结果与设定的参数背景条件相关，如平台几何尺寸、开发评价周期、经济参数、裂缝形态复杂程度以及压裂规模约束类型（固定单缝、单井或平台支撑剂体积）等，限于篇幅，本文不再赘述。

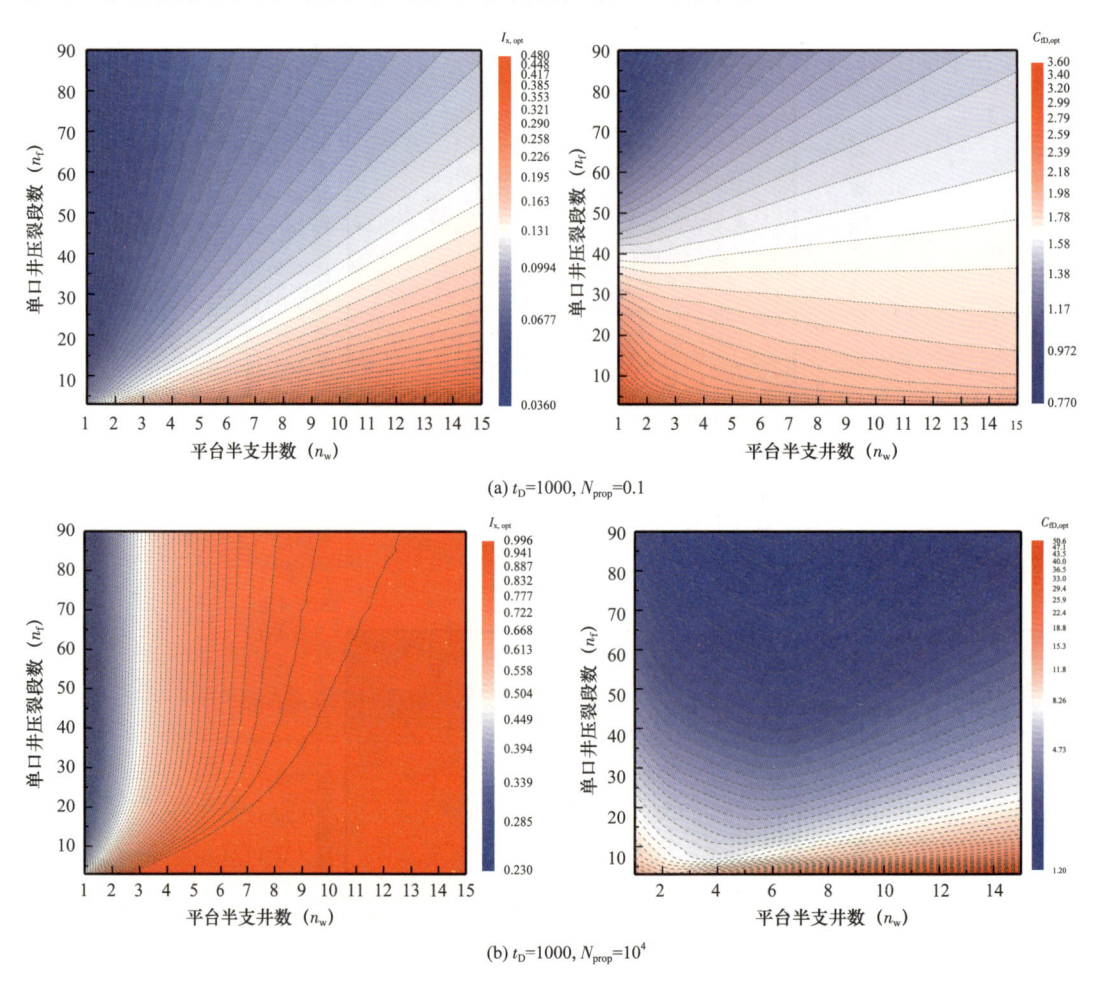

(a) t_D=1000, N_{prop}=0.1

(b) t_D=1000, $N_{prop}=10^4$

图15 多井平台下的裂缝维数优化图版

4 结论与建议

（1）针对开发平台内多口带有变缝宽裂缝的水平井，利用压力叠加原理，给出了一种新的变导流裂缝与储层耦合流动的半解析模型。通过坐标变换将变导流裂缝转换为常导流裂缝，可以处理空间位置及导流能力分布复杂的裂缝流动，并给出灵活的地层—裂缝耦合流动矩阵方便模型快速求解。

（2）考虑压裂规模约束时，以累计产量为目标函数，裂缝维数间存在最佳匹配关系，最大累计产量随着支撑剂指数和生产时间增加而增加，与之对应的最优裂缝穿透率增加，而随着支撑剂指数减小、生产时间增加，最优无量纲导流能力值不断减小且趋近于经典特征值（1.62）。缝距与井距比值也将影响裂缝维数的优化结果，在高支撑剂指数情况下，大井距、窄缝距的部署模式优势更明显。

（3）当不考虑经济效益时，开发平台的生产效果随着单井裂缝条数、水平井井数的增加而单调增加，不存在最优井数、段数，但增加幅度随着支撑剂指数的增加而降低，当井数与单井压裂段数同时超过特定值时，平台累计产量几乎不再增加。

（4）以净现值为优化目标函数，累计产量提高导致的正现金流与井数、段数增加产生的负现金流间相互冲抵，出现参数优化空间，存在最优裂缝维数、井数、段数，在压裂规模较小情况下宜采用小井距、宽缝距、短裂缝的水平井部署模式，在压裂规模较大情况下则宜采用大井距、窄缝距、长裂缝的部署模式。

符号注释：

x、y——地层系统坐标，m；x_{hf}——裂缝系统坐标；w_f——裂缝宽度，m；x_f——裂缝半长，m；L_f——裂缝全长，m；p——压力，Pa；p_i——原始地层压力，Pa；p_{wf}——井底压力，Pa；K_m——地层渗透率，m^2；K_f——裂缝渗透率，m^2；h——地层厚度，m；B_g——气体体积系数，m^3/m^3；ϕ_m——储层孔隙度；t——时间，s；c_t——地层综合压缩系数，Pa^{-1}；L_{ref}——参考长度，m；q_w——瞬时产量，m^3/s；q_{ref}——参考产量，m^3/s；G_p——累计产量，m^3；q_f——裂缝流量密度，m^2/s；x_{of}、y_{of}——裂缝端点坐标，m；x_{whf}——裂缝与井筒交叉位置，m；x_{ef}——储层横向宽度，m；y_{ef}——储层纵向长度，m；x_e——子泄流区域横向宽度，m；y_e——子泄流区域纵向长度，m；r_w——井筒半径，m；Z_g——气体偏差系数，无量纲；V_{prop}——支撑剂体积，m^3；V_E——球形基质中心处气体浓度，m^3/m^3；V——气体浓度，m^3/m^3；M_{prop}——支撑剂质量，g；R_m——解吸气球形基质半径，m；ρ_g——气体密度，g/m^2；μ_g为气体黏度，Pa·s；T——温度，K；b_{am}——滑脱因子，无量纲量；M_g——分子质量，g/mol；D_g——基质扩散系数，m^2/s；ρ_{prop}——支撑剂密度，g/m^3；ϕ_{prop}——支撑剂填充的裂缝孔隙度；C_{fD}——无量纲导流能力；ξ_D——无量纲裂缝转换坐标；n_f——单井裂缝条数；n_w——开发平台内井数。下标：D——无量纲；i、j——裂缝微元编号；m、n——裂缝编号；opt——最优值；max——最大值；sc——标准状况。上标：〈 〉——井编号

参 考 文 献

[1] Al-Rbeawi S. Productivity-index behavoir for hydraulically fractured reservoirs depleted by constant production rate considering transient-state and semisteady-state conditions [J]. SPE Production & Operations, 2018, 8: 11-31.

[2] 姚军, 孙海, 黄朝琴, 等. 页岩气藏开发中的关键力学问题 [J]. 中国科学E辑, 2013, 43（12）: 1527-1547.

[3] Sahai V, Jackson G, Rai R. Effect of non-uniform fracture spacing and fracture half-length on well spacing for unconventional gas reservoirs [R]. SPE 164927, 2013.

[4] 位云生, 王军磊, 齐亚东, 等. 页岩气井网井距优化 [J]. 天然气工业, 2018, 38（4）: 129-137.

[5] Chen C C, Raghavan R. A multiple-fractured horizontal well in a rectangular drainage region [J]. SPE

Journal, 1997, 2: 455-465.

[6] 王晓冬, 罗万静, 侯晓春, 等. 矩形油藏多段压裂水平井不稳态压力分析[J]. 石油勘探与开发, 2014, 37(11): 43-52.

[7] Chen Z M, Liao X W, Zhao X L. A practical methodology for production-data analysis of single-phase unconventional wells with complex fracture geometry[J]. SPE Reservoir Evaluation & Engineering, 2018, 2: 1-21.

[8] 方文超, 姜汉桥, 李俊键, 等. 致密储集层跨尺度耦合渗流数值模拟模型[J]. 石油勘探与开发, 2017, 44(3): 415-422.

[9] Yu W, Wu K, Zuo L H, et al. Physical models for inter-well interference in shale reservoirs: relative impacts of fracture hits and matrix permeability[R]. URTeC: 2457663, 2016.

[10] 李龙龙, 姚军, 李阳, 等. 分段多簇压裂水平井产能计算及其分布规律[J]. 石油勘探与开发, 2014, 41(4): 457-461.

[11] 孙贺东, 欧阳伟平, 张冕, 等. 考虑裂缝变导流能力的致密气井现代产量递减分析[J]. 石油勘探与开发, 2018, 45(3): 455-463.

[12] 李海涛, 王俊超, 李颖, 等. 基于体积源的分段压裂水平井产能评价方法[J]. 天然气工业, 2015, 35(9): 1-9.

[13] He Y W, Cheng S Q, Li S, et al. A semianalytical methodology to diagnose the location of underperforming hydraulic fractures through pressure-transient analysis in tight gas reservoir[J]. SPE Journal, 2017, 7: 924-939.

[14] 梁涛, 常毓文, 郭晓飞, 等. 巴肯致密油藏单井产能参数影响程度排序[J]. 石油勘探与开发, 2013, 40(3): 357-362.

[15] 李波, 贾爱林, 何东博, 等. 低渗致密气藏压裂水平井产能分析与完井设计[J]. 中南大学学报(自然科学报), 2016, 47(11): 3775-3783.

[16] Yu W, Sepehrnoori K. An efficient reservoir-simulation approach to design and optimize unconventional gas production[J]. Journal of Canadian Petroleum Technology, 2014, 3: 109-121.

[17] 姜瑞忠, 刘明明, 徐建春, 等. 遗传算法在苏里格气田井位优化中的应用[J]. 天然气地球科学, 2014, 25(10): 1603-1609.

[18] Adibifard M, Tabatabaei-Nejad S, Khodapanah E. Artificial neural network (ANN) to estimate reservoir parameters in naturally fractured reservoirs using well test data[J]. Journal of Petroleum Science and Engineering, 2014, 122: 585-594.

[19] Ertekin T, King G R, Schwerer F C. Dynamic gas slippage: a unique dual-mechanism approach to the flow of gas in tight formation[J]. SPE Formation Evaluation. 1986, 2: 43-52.

[20] Ozkan E, Raghavan R, Apaydin O G. Modeling of fluid transfer from shale matrix to fracture network[R]. SPE 134830, 2010.

[21] Xiao L, Zhao G. Study of 2-D and 3-D hydraulic fractures with non-uniform conductivity and geometry using source and sink function methods[R]. SPE 162542, 2012.

[22] Gringarten A C, Ramey H J. The use of source and Green's functions in solving unsteady-flow problems in reservoirs[R]. SPE 3818, 1973.

[23] 王晓冬, 张义堂, 刘慈群. 垂直裂缝井产能及导流能力优化研究[J]. 石油勘探与开发, 2004, 31(6):

78-81.

[24] Wang X D, Li G H, Wang F. Productivity analysis of horizontal wells intercepted by multiple finite-conductivity fractures [J]. Petroleum Science, 2010, 7: 367-371.

[25] Ye P, Ayala L F. A density-diffusivity approach for the unsteady state analysis of natural gas reservoirs [J]. Journal of Natural Gas Science and Engineering, 2012, 7 (1): 22-34.

[26] Poe B, Shan P, Elbel J. Pressure transient behavior of a finite-conductivity fractured well with spatially varying fracture properties [R]. SPE-24707-MS, 1992.

[27] Valko P P, Economides M J. Heavy crude production from shallow formations: long horizontal wells versus horizontal fractures [R]. SPE 50421, 1998.

[28] Bhattacharya S, Nikolaou M, Economides M J. Unified Fracture Design for very low permeability reservoirs [J]. Journal of Natural Gas Science and Engineering, 2012, 9: 184-195.

Optimization workflow for stimulation-well spacing design in a multiwell pad

Wang Junlei, Jia Ailin, Wei Yunsheng, Jia Chengye, Qi Yadong, Yuan He, Jin Yiqiu

(Research Institute of Petroleum Exploration and Development, PetroChina)

Abstract: A flow mathematical model with multiple horizontal wells considering interference between wells and fractures was established by taking the variable width conductivity fractures as basic flow units. Then a semi-analytical approach was proposed to model the production performance of full-life cycle in well pad and to investigate the effect of fracture length, flow capacity, well spacing and fracture spacing on estimated ultimate recovery (EUR). Finally, an integrated workflow is developed to optimize drilling and completion parameters of the horizontal wells by incorporating the productivity prediction and economic evaluation. It is defined as nested optimization which consists of outer-optimization shell (i.e., economic profit as outer constraint) and inner-optimization shell (i.e., fracturing scale as inner constraint). The results show that, when the constraint conditions aren't considered, the performance of the well pad can be improved by increasing contact area between fracture and formation, reducing interference between fractures/wells, balancing inflow and outflow between fracture and formation, but there is no best compromise between drilling and completion parameters. When only the inner constraint condition is considered, there only exists the optimal fracture conductivity and fracture length. When considering both inner and outer constraints, the optimization decisions including fracture conductivity and fracture length, well spacing, fracture spacing are achieved and correlated. When the fracturing scale is small, small well spacing, wide fracture spacing and short fracture should be adopted. When the fracturing scale is large, big well spacing, small fracture spacing and long fracture should be used.

Key words: horizontal well; multi-staged fracturing; fracture flow capacity; fracture length; well space; fracture space; parameters optimization

Horizontal drilling and multistage-hydraulic fracturing have been widely applied in the development of unconventional oil and gas resources. It is well known that many factors determines the productivity of horizontal wells, including formation petrophysical properties, fluid properties, and completion and fracturing parameters[1]. Among them, formation petrophysical properties and fluid properties are uncontrollable factors, and completion and fracturing parameters are controllable. From the perspective of reservoir engineering, the aim of horizontal well drilling is to increase contact area between fractures and reservoir, conductivity in propped hydraulic fractures, and reduce flow resistance to enhance well productivity. In the design of development well-pad with multiple fracturing wells, the main indexes need to be considered are well spacing, fracturing stage spacing, amount of proppant per fracture and fracture dimensions (length, width and height)[2-4].

The difficulty in optimization of well-pad design is to make sure that the simulation can reflect the characteristics of pro-duction performance under the effects of inter-fracture interference, interwell interference, and fracture conductivity, so the relationships between main factors and production performance and the correlation between the main factors can be made clear. In the terms of performance simulation of MFHW, several analytical and numerical methods have been developed. On the basis of instantaneous functions presented by Chen and Raghavan[6], Wang et al.[5] obtained the transient response of single MFHW by integrating the analytical solution for uniform-flux fracture with an impact function of fracture conductivity. Chen et al.[7] proposed a semi-analytical method derived from an analytical reservoir solution and a numerical fracture solution to forecast the full life-cycle performance of MFHW, which has the flexibility of capturing the geometry complexity of fracture network and the interplay between reservoir and fractures. Fang et al.[8] established a numerical model considering the multi-scale flow in discrete fractures, and simulated the production performance of a well pad with three volume fractured horizontal wells with long lateral section. Considering the well interference through both matrix permeability and hydraulic fracture hits, Yu et al.[9] developed a numerical, compositional model in combination with an embedded discrete fracture model (EDFM) to simulate both the response to interwell interference, to find out the impact of well spacing on pressure interference in the adjacent wells.

In the aspect of analyzing the impact of main control factors, single factor control method is commonly used to examine the factors one by one, which ignores the correlations between the factors and thus could lead to wrong results[10-13]. Although orthogonal test and grey correlation methods improve in limitedness than the single factor control analysis method, they are still multiple-scheme comparison methods, unable to cover all possible optimal solutions[14-16]. Automated optimization algorithms such as artificial neural network and genetic algorithm can provide globally optimal design, but huge in computation, they are time-consuming and can't be used widely, especially when the number of wells and fracturing stages are large[17-18].

Systematic optimization of completion parameters of MFHWs is a complicated nonlinear problem in the field of reservoir engineering and mathematical optimization. On the basis of a semi-analytical modeling, this paper developed a comprehensive optimization workflow to maximize NPV by integrating extended UFD and globally optimal algorithm. The principle objective of this work is to provide a quantitative assessment of optimal decisions such as number of wells, number of fractures per well, mass of proppant and fracture dimensions.

1　Performance modeling of multi-well pad

1.1　Model description

The configuration of the MHFW system used in this study is shown in Fig. 1. All MFHWs are assumed to be evenly located and completed in an isotropic, homogeneous and horizontal formation in rectangular shape. The formation fully penetrated by hydraulic fractures has uniform thickness, permeability and porosity, all of which are constant. All transverse fractures are of identical properties. Besides, all MFHWs are produced at constant bottomhole pressure (BHP).

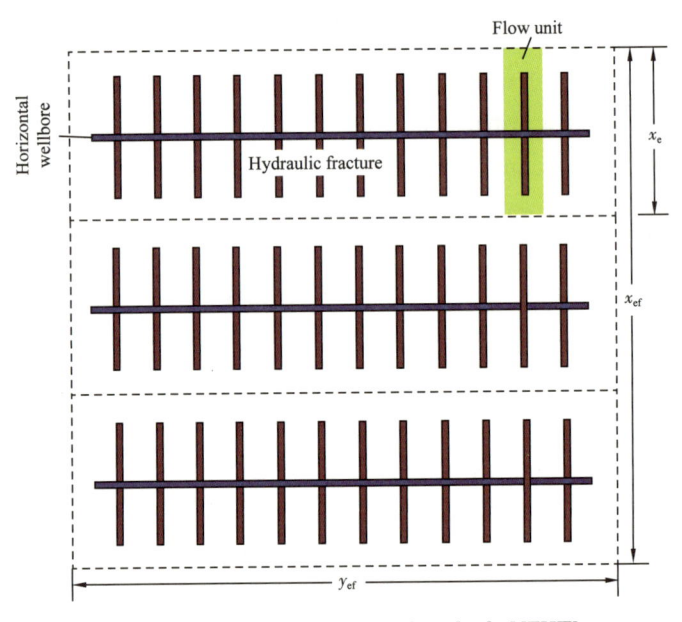

Fig. 1　Sketch of the well pad with multiple MFHWs

The fluid flow in the reservoir with low permeability is governed by non-Darcy's law. Based on Ertekin model [19], the flow velocity in the reservoir is defined as the sum of the component of the Darcy flow velocity and the slip flow velocity:

$$v_m = \frac{K_m}{\mu_g}(1+b_{am})\nabla p_m \tag{1}$$

where the apparent gas slippage term is not a constant but a variable given by
$b_{am} = \frac{D_g \mu_g c_g p_m}{K_m} \frac{1}{p_m}$.

The slip velocity is presented by modifying Fick's law, which is caused by the concentration gradient in the matrix. At matrix-fracture interface, the concentration gradient is negligible, and the slippage effect approaches zero, that is:

$$b_{am}(x,y) = \begin{cases} 0 & \text{at interface} \\ b_{am} & \text{within matrix} \end{cases} \tag{2}$$

To linearize the gas-flow governing equations, the pseudo-function approach is defined. Pseudo-pressures in the fracture and reservoir are respectively defined by

$$m_f = \frac{\mu_{gi} Z_{gi}}{p_i} \int_{p_i}^{p} \frac{\xi}{\mu_g(\xi) Z_g(\xi)} d\xi \tag{3}$$

$$m_m = \frac{\mu_{gi} Z_{gi}}{p_i} \int_{p_i}^{p} \frac{\gamma_m(\xi) \xi}{\mu_g(\xi) Z_g(\xi)} d\xi \tag{4}$$

And the nonlinear factor, γ_m, is introduced to account for the influence of gas slippage, which is expressed as

$$\gamma_m(p_m) = 1 + b_{am} \tag{5}$$

Correspondingly, the concept of pseudo-time in the reservoir is defined as

$$t_a = \int_0^t \frac{\gamma_m(\tau) \mu_{gi} c_{ti}}{\mu_g(\tau) c_t(\tau)} d\tau = t \beta_m(t) \tag{6}$$

where β_m is a dimensionless rescaling factor which describes the behavior the fluid viscosity and compressibility during depletion, and λ_m is a dimensionless viscosity-compressibility ratio, which are defined as follows:

$$\beta_m(t) = \frac{1}{t} \int_0^t \lambda_m(\tau) d\tau, \lambda_m(t) = \frac{\gamma_m(t)}{\dfrac{\mu_g(t)}{\mu_{gi}} \dfrac{c_t(t)}{c_{ti}}}$$

Substituting pseudo-pressure and pseudo-time into the gas governing equation renders gas diffusivity equation amenable to the liquid diffusivity equation.

Since the fluid flow caused by different MFHWs is confined to an identical pressure system, the fluid flow can be decomposed into continuous flow in fractures and flow in reservoirs. In the modeling of the fluid flow, two sets of separate coordinates were adopted for the fracture and reservoir, and the pressure and influx distribution along fracture face can be achieved by coupling the reservoir flow model and fracture model on the basis of continuity condition that the reservoir pressure equals to the fracture pressure on the fracture-reservoir interface.

To reflect the effect of interwell fracturing interference on the opening and propagation of hydraulic fractures, a fracture model with variable width is adopted in this study. To our knowledge, when the propagation direction of hydraulic fracture is parallel to the maximum stress direction, the farther from the wellbore, the stronger the interwell stress will be, thus the net pressure to open up fracture will increase, the migration of proppant will be blocked, and the fracture width will decrease gradually from the wellbore outward [21]. That is expressed as:

$$w_f(x_{hf}) = w_{f,\max} \left[-(x_{hf}/x_f)^2 + 2(x_{hf}/x_f) \right] \tag{7}$$

For simplicity, dimensionless pseudo-pressure in reservoir, dimensionless pseudo-pressure in fracture, dimensionless production rate and dimensionless cumulative production are respectively defined by

$$p_{mD} = \frac{m_m(p_i) - m_m(p_m)}{m_m(p_i) - m_m(p_{wf})} \tag{8}$$

$$p_{fD} = \frac{m_f(p_i) - m_f(p_f)}{m_m(p_i) - m_m(p_{wf})} \tag{9}$$

$$q_{wD} = \frac{\chi q_w \mu_{gi} B_{gi}}{K_m h [m_m(p_i) - m_m(p_{wf})]} \tag{10}$$

$$G_{pD} = \frac{\chi B_{gi} G_p}{h[m_m(p_i) - m_m(p_{wf})] \phi_m c_{ti} L_{ref}^2} \tag{11}$$

and dimensionless time, dimensionless length, dimensionless conductivity and dimensionless influx density are defined as follows:

$$t_D = \psi \frac{K_m t_a}{\phi_m \mu_{gi} c_{ti} L_{ref}^2} \tag{12}$$

$$\zeta_D = \frac{\zeta}{L_{ref}} \tag{13}$$

$$C_{fD} = \frac{K_f w_f}{K_m L_{ref}} \tag{14}$$

$$q_{fD} = \chi \frac{q_f L_{ref} \mu_{gi} B_{gi}}{K_m h [m_m(p_i) - m_m(p_{wf})]} \tag{15}$$

where χ and ψ are scaling factors. In SI unit, $\chi=0.5/\pi$ and $\psi=1$; in Darcy unit, $\chi=1.842$ and $\psi=0.0036$.

1.2 Analytical solution of fluid flow in the reservoir

Since all MFHWs are located in a pressure system, the drainage area can be divided into a set of sub-drainage areas according to the principle of pressure superposition. Fig. 2 shows the equivalent sub-drainage area drained by single fracture (sub-drainage area equals to $1/n_w n_f$ of total drainage area). Using pseudo-function definitions, the equation governing gas flow in the reservoir is given by

$$\frac{\partial^2 m_m}{\partial x^2} + \frac{\partial^2 m_m}{\partial y^2} + \frac{\mu_{gi} B_{gi}}{K_m h} S_f(x,y,t) - \frac{\mu_{gi}}{K_m} q_{de}(t) = \frac{\phi_m \mu_{gi} c_{ti}}{K_m} \frac{\partial m_m}{\partial t_a} \tag{16}$$

where $S_f(x, y, t)$ is the source function for influx distribution of hydraulic fracture with the length of L_f:

$$S_{f} = \int_{0}^{t}\int_{x_{o}}^{x_{o}+L_{f}} q_{f}(u,t-\tau)\delta(x-x_{o}-u,\tau)\delta(y-y_{o},\tau)\mathrm{d}u\mathrm{d}\tau \tag{17}$$

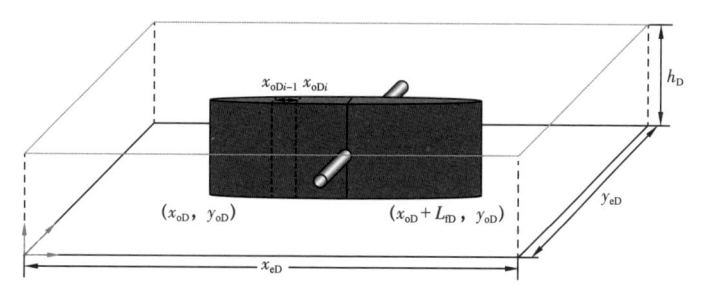

Fig.2　Schematic of pressure disturbance caused by a fracture

To account for the effect of gas desorption, $q_{de}(t)$ is defined to describe the supply of gas desorption[7]:

$$q_{de}(t) = \frac{Z_{gi}p_{sc}T}{T_{sc}p_{i}} \frac{6D_{g}\pi^{2}}{R_{m}^{2}}(V_{E}-V) \tag{18}$$

The hydraulic fracture can be regarded as the Green's function for a line source along which flux density is satisfied as $q_{fD}(s)$. The gas desorption is considered in the dual-porosity model of $f(s)$. After using the Laplace transformation, Eq. (16) can be written as the following dimensionless form,

$$\frac{\partial^{2}\tilde{p}_{mD}}{\partial x_{D}^{2}} + \frac{\partial^{2}\tilde{p}_{mD}}{\partial y_{D}^{2}} + \pi\int_{x_{oD}}^{x_{oD}+L_{fD}} \tilde{q}_{fD}(s)\tilde{\delta}(x_{D}-x_{oD}-u_{D})\tilde{\delta}(y_{D}-y_{oD})\mathrm{d}u_{D} = sf(s)\tilde{p}_{mD} \tag{19}$$

From the perspective of reservoir flow model, the fracture can be discretized into a set of segments, so Eq. (19) is rewritten into a discretized equation in which each segment has a length Δx_{fDi} and flux q_{fDi}. Based on the principle of pressure superposition, the pressure at any point in the fracture can be expressed as:

$$\tilde{p}_{mD}(x_{Dj},y_{Dj}) = \sum_{i=1}^{N} \tilde{q}_{Di}\tilde{p}_{uDj,i}(\beta_{Dj},\beta_{Di},\Delta x_{fDi},x_{eD},y_{eD},s) \tag{20}$$

where $\tilde{p}_{uDj,i}$ represents the pressure disturbance to the j-th segment caused by i-th segment. According to the Green's function[22], the dimensionless pressure solution for uniform-flux segment can be written as

$$\tilde{p}_{uDj,i}(\beta_{wDj},\beta_{wDi}) = \frac{2\pi\Delta x_{fDi}}{x_{eD}}\tilde{H}_{0} + \tilde{F}_{ji} + 4\sum_{n=1}^{\infty} \frac{\tilde{H}_{n}-1}{n\varepsilon_{n}}\cos\frac{n\pi x_{Dj}}{x_{eD}}\sin\frac{n\pi x_{Di}}{x_{eD}}\cos\frac{n\pi\Delta x_{fDi}}{x_{eD}} \tag{21}$$

where

$$\tilde{H}_n = \frac{\cosh\left[\varepsilon_n\left(y_{eD} - |y_{Dj} - y_{wDi}|\right)\right]}{\sinh(\varepsilon_n y_{eD})} + \frac{\cosh\left[\varepsilon_n\left(y_{eD} - |y_{Dj} + y_{wDi}|\right)\right]}{\sinh(\varepsilon_n y_{eD})}$$

$$\tilde{F}_{ji} = \int_{-\Delta x_{fDi}}^{\Delta_{fDi}} \left(\frac{2\pi}{x_{eD}} \sum_{n=1}^{\infty} \frac{1}{\varepsilon_n} \cos\frac{n\pi x_{Dj}}{x_{eD}} \cos\frac{n\pi u_D}{x_{eD}}\right) du_D$$

$$\beta_{wDj} = (x_{Dj}, y_{Dj}) \quad \varepsilon_n = \sqrt{sf(s) + n^2\pi^2/x_{eD}^2}$$

$$f(s) = \omega + \frac{a_g D_D (1-\omega)}{s + D_D}$$

For simplicity, Eq. (20) is rewritten as the corresponding matrix:

$$\begin{pmatrix} \tilde{p}_{mD1} \\ \vdots \\ \tilde{p}_{mDN} \end{pmatrix} = \begin{pmatrix} \tilde{p}_{uD1,1}(\beta_{wD1},\beta_{wD1}) & \cdots & \tilde{p}_{uD1,N}(\beta_{wD1},\beta_{wDN}) \\ \vdots & & \vdots \\ \tilde{p}_{uDN,1}(\beta_{wDN},\beta_{wD1}) & \cdots & \tilde{p}_{uDN,N}(\beta_{wDN},\beta_{wDN}) \end{pmatrix} \times \begin{pmatrix} \tilde{q}_{D1} \\ \vdots \\ \tilde{q}_{DN} \end{pmatrix} \quad (22)$$

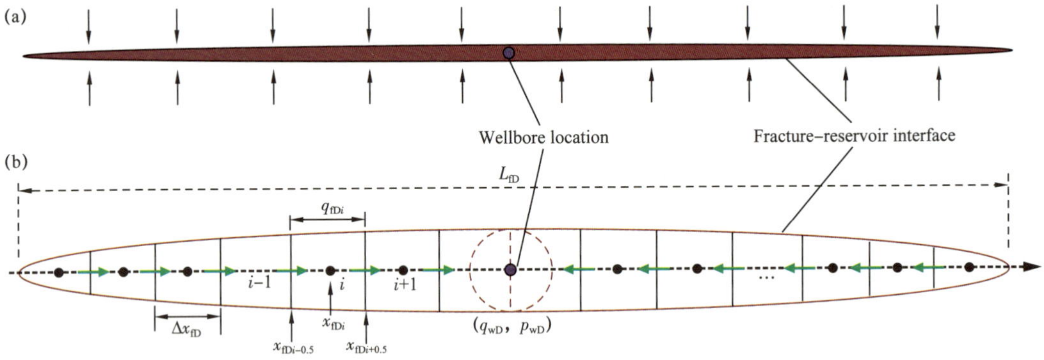

Fig.3 Schematic of the fracture domain including coupled reservoir-fracture model (a); variable width fracture model (b)

1.3 Analytical solution for variable-conductivity fracture

In 1D coordinate of fracture model, the fluid flow from the reservoir into hydraulic fracture, and then flow towards wellbore along fracture panel. As shown in Fig. 3a, the flux density along fracture is given by $q_f(x, t)$, and the rate extracted from wellbore is denoted as $q_w(t)$.

To account for variable conductivity fracture shown in Fig. 3b, the pseudo-pressure equation is described in the following dimensionless form:

$$\frac{\partial}{\partial x_{Dhf}}\left(C_{fD}(x_{Dhf})\frac{\partial \tilde{p}_{fD}}{\partial x_{Dhf}}\right) - 2\pi \tilde{q}_{fD}(x_{Dhf}) + 2\pi\left[\tilde{q}_{wD}(t_D)\delta(x_{Dhf}, x_{Dwhf})\right] = 0 \quad (23)$$

where δ is the Dirac function, and dimensionless variable conductivity is written as

$$C_{fD}(x_{Dhf}) = C_{fDmax}\left(-x_{Dhf}^2 + 2x_{Dhf}\right) \quad (24)$$

Equation (23) is a pseudo-linear differential equation because the conductivity C_{fD} is a function of x_{Dhf}. Here coordinate transformation equation is introduced to make the pseudo-linear problem linear, which is expressed as

$$\xi_D(x_{hfD}) = \hat{C}_{fD} \int_0^{x_{hfD}} C_{fD}^{-1}[p_{fD}(x_D)]dx_D \tag{25}$$

and $\hat{C}_{fD} = \dfrac{L_{fD}}{\int_0^{L_{fD}} C_{fD}^{-1}[p_{fD}(x_D)]dx_D}$. Incorporating Eq. (25) and boundary element method [23] to deal with Eq. (23), the solution for variable conductivity fracture can be derived as

$$\tilde{p}_{wD} - \tilde{p}_{fD}(\xi_D) = \frac{2\pi}{\hat{C}_{fD}}\tilde{q}_{wD}G(\xi_D,\xi_{wD}) - \frac{2\pi}{\hat{C}_{fD}}[I(\xi_D) - I(\xi_{wD})] \tag{26}$$

where G is the integral of Heaviside unit step function, and the Fredholm integral function is defined as

$$I(\xi_D) = \int_0^{\xi_D} d\zeta \int_0^{\zeta} \tilde{q}_{fD}(\varsigma)d\varsigma \tag{27}$$

To take the additional pressure drop due to the convergence of fluid into the horizontal wellbore from transverse fractures, the convergence flow skin S_c is introduced to modify Eq. (26), which is rewritten as

$$\tilde{p}_{wD} - \tilde{p}_{fD}(\xi_D) = \frac{2\pi}{\hat{C}_{fD}}\tilde{q}_{wD}G(\xi_D,\xi_{wD}) - \frac{2\pi}{\hat{C}_{fD}}[I(\xi_D) - I(\xi_{wD})] + \tilde{q}_{wD}S_c$$

$$S_c = \frac{2h_D}{\hat{C}_{fD}L_{fD}}\left[\ln\left(\frac{h_D}{2r_{wD}}\right) - \frac{\pi}{2}\right] \tag{28}$$

Eq. (28) can be also expressed in the matrix form:

$$\begin{pmatrix}\tilde{p}_{wD}\\ \vdots \\ \tilde{p}_{wD}\end{pmatrix} - \begin{pmatrix}\tilde{p}_{fD1}\\ \vdots \\ \tilde{p}_{fDN}\end{pmatrix} = \frac{2\pi}{\hat{C}_{fD}}\begin{pmatrix}\tilde{q}_{wD}\gamma_1\\ \vdots \\ \tilde{q}_{wD}\gamma_N\end{pmatrix} - \frac{2\pi}{\hat{C}_{fD}}\begin{pmatrix}\alpha_{1,1} & \cdots & \alpha_{1,N}\\ \vdots & & \vdots \\ \alpha_{N,1} & \cdots & \alpha_{N,N}\end{pmatrix} \times \begin{pmatrix}\tilde{q}_{fD1}\\ \vdots \\ \tilde{q}_{fDN}\end{pmatrix} + S_c\begin{pmatrix}\tilde{q}_{wD}\\ \vdots \\ \tilde{q}_{wD}\end{pmatrix} \tag{29}$$

where $\gamma_i = G(\xi_{Di},\xi_{wD})$, $\alpha_{j,i} = RS_i + RT_i$,

$$G(x,x_0) = \begin{cases} x - x_0 & (x \geq x_0) \\ 0 & (x < x_0) \end{cases},$$

$$RS_i = \begin{cases} \xi_{Dj}\Delta\xi_{Dj} - 0.5(\xi_{oD,i}^2 - \xi_{oD,i-1}^2) & (i < j) \\ \xi_{Dj}(\xi_{Dj} - \xi_{oD,j-1}) - 0.5(\xi_{Dj}^2 - \xi_{oD,j-1}^2) & (i = j) \end{cases},$$

$$RT_i = \begin{cases} (\xi_{wD} - \xi_{Di})\Delta\xi_{Dj} & (i < 1+\Xi) \\ \left(\xi_{wD} - \sum_{i=1}^{\Xi}\Delta\xi_{Di}\right)\left(0.5\xi_{wD} + \sum_{i=1}^{\Xi}\Delta\xi_{Di}\right) & (i = 1+\Xi) \end{cases}$$

2 Model calculation and verification

According to the continuity condition that the reservoir pressure equals to the fracture pressure on the fracture-reservoir interface, the relationship is expressed as:

$$\begin{cases} \tilde{p}_{fD}(x_{hfD}) = \tilde{p}_{mD}(x_{oD} + x_{hfD}, y_{oD}) \\ \tilde{q}_{fD} = \tilde{q}_{mD} \end{cases} \quad (30)$$

It is worth noting that the relation between flux-density of fracture segment and rate of wellbore in the transformed domain is:

$$\tilde{q}_{wD} = \sum_{j=1}^{N} \tilde{q}_{fDj} f_{cj} \quad (31)$$

where $f_{cj} = \Delta \xi_{Dj}/\Delta x_{Dj}$. The element of $\tilde{p}_{uDj,i}(\beta_{wDj}, \beta_{wDi})$ in the matrix of Eq. (22) is modified as $f_{cj}\tilde{p}_{uDj,i}(\beta_{wDj}, \beta_{wDi})$.

According to coupled condition [Eq. (30) and Eq. (31)], combining the solution for the reservoir [Eq. (22)] and the solution for the fracture [Eq. (29)] provides the closed solution for flux-density distribution along fracture and production rate of the well. The unknown variables can be expressed in vector form:

$$\begin{aligned}(\tilde{q}_{fD1,1}^{(1)}, \tilde{q}_{fD1,2}^{(1)}, \cdots, \tilde{q}_{fD1,N}^{(1)}, \cdots, \tilde{q}_{fDn_f,1}^{(1)}, \tilde{q}_{fDn_f,2}^{(1)}, \cdots, \tilde{q}_{fDn_f,N}^{(1)}, \tilde{q}_{wD}^{(1)}, \tilde{G}_{pD}^{(1)}, \cdots, \\ \tilde{q}_{fD1,1}^{(n_w)}, \tilde{q}_{fD1,2}^{(n_w)}, \cdots, \tilde{q}_{fD1,N}^{(n_w)}, \cdots, \tilde{q}_{fDn_f,1}^{(n_w)}, \tilde{q}_{fDn_f,2}^{(n_w)}, \cdots, \tilde{q}_{fDn_f,N}^{(n_w)}, \tilde{q}_{wD}^{(n_w)}, \tilde{G}_{pD}^{(n_w)})\end{aligned} \quad (32)$$

The cumulative production is obtained by using Laplace transformation, which is given by $\tilde{G}_{pD} = \tilde{q}_{wD}/s$.

It is worth noting that the pseudo-time is a function of spatial and temporal variables. The pseudo-time can be regarded as single integral of average reservoir pressure over time, as presented in Eq. (33), and can be calculated by incorporating material balance equation [Eq. (34)].

$$t_a = t \beta_m [p_{avg}(t)] \quad (33)$$

$$\frac{p_{avg}(t)}{Z_{avg}(t)} = \frac{p_i}{Z_i}\left(1 - \frac{G_p}{\text{OGIP}}\right) \quad (34)$$

The detailed information on iteration algorithm is provided in the work of Ye and Ayala [25].

Afterwards, to verify the accuracy of the semi-analytical model in our work, the transient response solutions were compared with other semi-analytical model and the results of numerical simulation. For the semi-analytical results presented by Poe et al [26], the variation of fracture conductivity is described by four constant-length zones and each zone is assigned a different conductivity. As shown in Fig. 4, our pressure solutions are consistent with Poe's results, validating the calculation accuracy of our model in simulating pressure in one fracture with variable width.

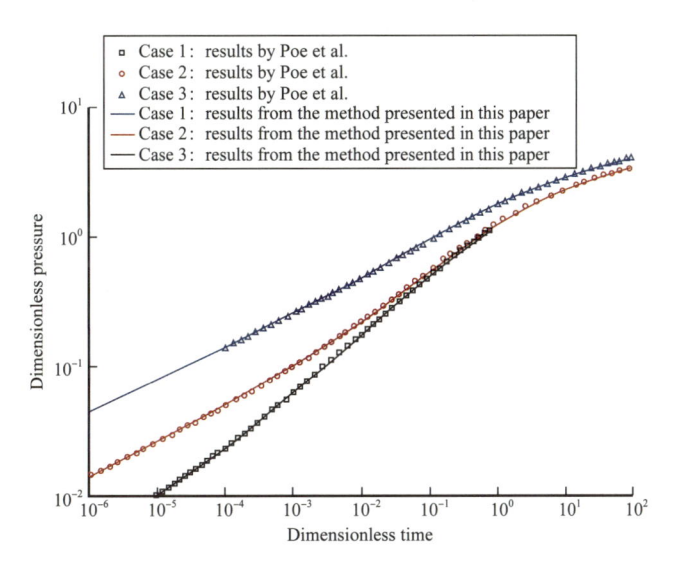

Fig. 4 Comparison of transient pressure solutions from our model with solutions from Poe et al

Commercial numerical simulator was used to further verify the reliability of our model in simulating interwell interference. The parameters used in the model were: p_i=45.0MPa, p_{wf}=6.5MPa, h=20m, s_o=85%, c_t=4.35×10^{-4}MPa^{-1}, k_m= 0.01×10^{-3}μm^2; μ_o=1mPa·s, B_o=1.0m^3/m^3, r_w=0.019m; n_f=30, $x_e × y_e$=1500m×1600m, x_f=114.95m, w_f=0.0127m, ϕ_f=35%. To generate accurate rate response, the local-grid-refinement (LGR) was used to refine the grids around fracture panel as shown in Fig. 5. Since ϕ_f=35 % and w_f=0.0127m, the equi-valent fracture porosity calculated is 4.4%, the equivalent fracture width is 0.1m, and the equivalent fracture permeability is calculated using equation 12×[($K_f w_f$)/($K_m x_f$)]×10^{-3}μm^2. As presented in Fig. 5, the permeability distribution within fracture satisfies Eq. (2).

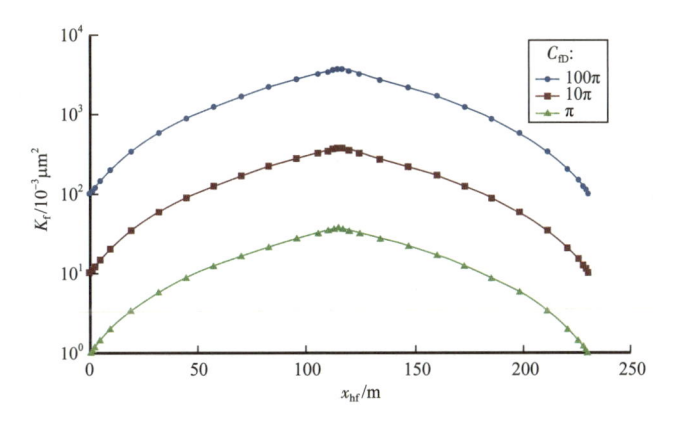

Fig.5 Sketch of numerical model of the well pad with three MFHWs

Due to the ideal assumption of even well configuration, all the wells have the same production performance (Fig. 5). Fig. 6 shows the dimensionless production performance from numerical simulation and the method presented in this paper, it can be seen the single well production and cumulative production from the two methods tally well with each other, verifying the adaptability of the method proposed in this paper in cases with multiple wells and multiple fractures.

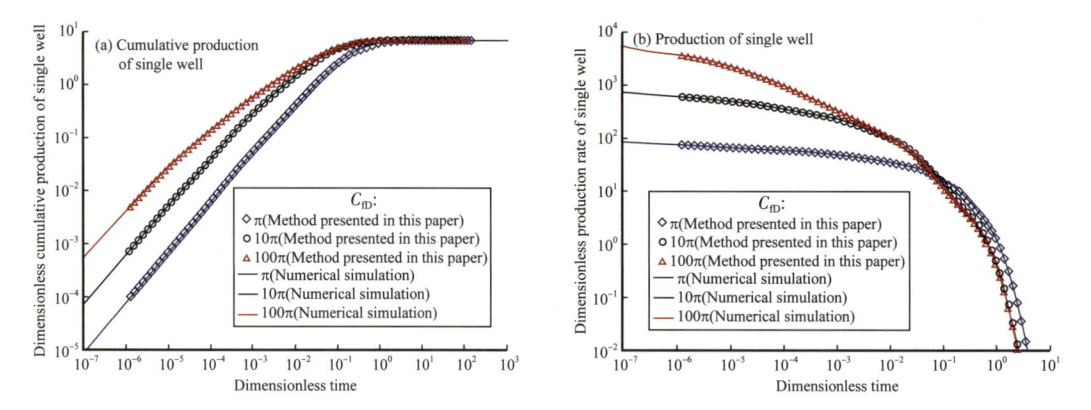

Fig. 6 Comparison of dimensionless production and cumulative production from the two mehtods

3 Optimization of fracture parameters and well spacing

As mentioned before, the production performance of well pad is affected by various factors, including fracture dimensions (width, length and height), number of wells, number of fractures, and fracture/well arrangement. To simplify the issue, we assume an ideal homogeneous configuration, with wells and fractures in even distribution and same in dimensions. The fractures occupy the same sub-drainage area ($1/n_w n_f$ of the total drainage area). Thus the factors affecting production performance of the well pad can be reduced to length and conductivity of the fracture.

At a given mass of proppant, the fracture length and fracture width are competing for the proppant. The maximum productivity index would be achieved on the best compromise between length and conductivity, known as UFD. The classical UFD theory is originally proposed for single fracture with uniform conductivity in the pseudo-steady state (PSS) flow conditions[27-28], and the objective function is the productivity index in the PSS condition, while not (cumulative) rate in transient conditions.

In this paper, an extension of UFD theory to account for the geometry of multiple MFHWs is presented, and the proppant number is modified as

$$N_{prop} = \frac{4x_{ef}}{3y_{ef}} \frac{n_f C_{fDmax} I_x^2}{n_w} \qquad (35)$$

The extended UFD could consider time dependency of transient response covering both transient and PSS conditions. The fixed proppant number in the UFD is regarded as technical constraint. With further consideration of economic constraint, a nested optimization method containing outer-optimization and inner-optimization shells is further developed to optimize an overall economic objective.

3.1 Inner-optimization shell

The inner-optimization shell incorporating the UFD is used to optimize fracture dimensions. In this work, fracture dimensions mainly refer to fracture conductivity and fracture length. By designing appropriate total fracture length and conductivity matching with inflow and outflow relations between fracture and reservoir, the productivity can be maximized. We take a single fracture as an example to discuss the optimization process of fracture dimensions. The fracture dimensions are assumed as 400m × 50m × 20m, where the fracture length is represented by fracture-penetration ratio ($I_x = L_f/x_e$), and the fracture width is represented by dimensionless conductivity (C_{fD}).

The optimization results of fracture dimensions with proppant index in Fig. 7 show when proppant volume is not considered, EUR denoted by dashed line increases monotonously with the increase of I_x and C_{fD} till reaches the maximum value. At this point, the fracture full penetrates the drainage area and reaches infinite conductivity. When the constraint of proppant volume is considered, the cumulative production and fracture dimensions have optimal values, which are denoted by discrete points in the Fig. 7. The maximum dimensionless cumulative production increases with the increase of proppant index. More interestingly, the optimum fracture dimensions are different at different times as seen from Fig. 7 (a), 7 (b).

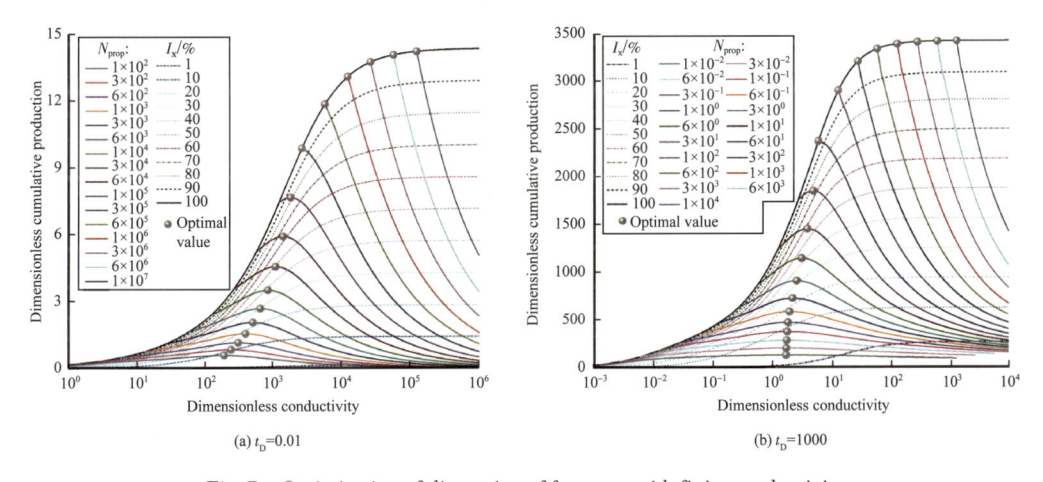

(a) $t_D = 0.01$

(b) $t_D = 1000$

Fig. 7 Optimization of dimension of fractures with finite conductivity

To determine the effect of production time on the optimum results of fracture dimension, the optimized fracture dimensions with time were calculated (Fig. 8). As shown in Fig. 8 (a), as expected, EUR_{max} is an increasing function of t_D for $N_{prop}=0.1$, while C_{fDopt} is a decreasing function and approaches a limit. From Fig. 8 (b) to Fig. 8 (d), we can see that the larger the proppant index, the larger the dimensionless cumulative production at the same time will be, the wider the variation range of optimum dimensionless conductivity in the same time interval, and the larger the value the dimensionless conductivity approaches will be. For example, as presented in Fig. 8 (b) for $N_{prop}=1$, when t_D is in the range of 0.01–1000, the corresponding $C_{fD,\ opt}$ is in the range of 1–50, and the limit value of $C_{fD,\ lim}$ is approaching 1.62. In contrast, in Fig. 8 (d) for $N_{prop}=1000$, the variation range of $C_{fD,\ opt}$ is 15–400 with $C_{fD,\ lim}=15$.

- 229 -

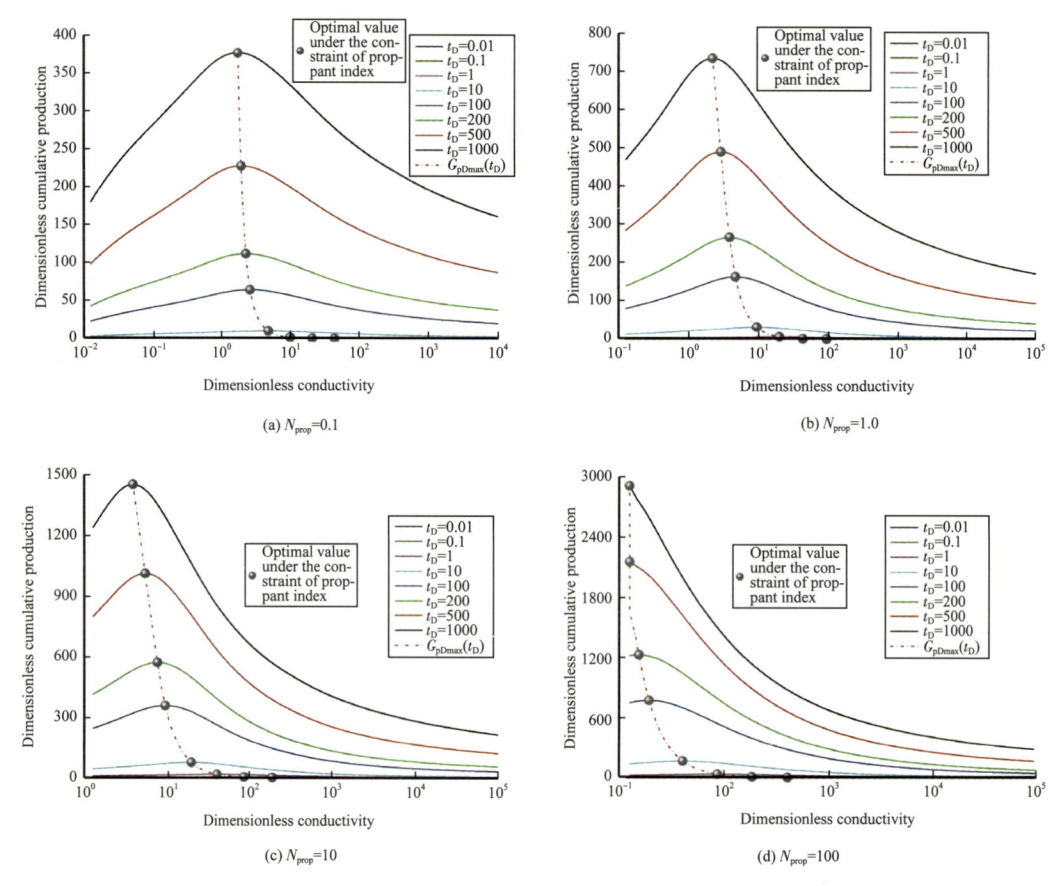

Fig. 8 Optimization of fracture dimensions under finite conductivity at different proppant indexes

(a) $N_{prop}=0.1$

(b) $N_{prop}=1.0$

(c) $N_{prop}=10$

(d) $N_{prop}=100$

Fig. 9 shows the variation patterns of maximum EUR and optimal fracture dimensions at different proppant indexes and different times. Fig. 9 (a) shows a positive correlation between $C_{fD, opt}$ and N_{prop}, which means that the $C_{fD, opt}$ decreases with the decrease of N_{prop} till reaches a limited value and then keeps constant basically. The bigger the value of t_D is, the larger the variation range of N_{prop}. For example, when $t_D=1000$, $C_{fD, opt}$ approximates a constant in the range of $N_{prop}=10^{-8}-10^{-1}$. However, when $t_D=0.01$, N_{prop} is in the range of $10^{-8}-10^{-7}$. Besides, both $C_{fD, opt}$ and t $I_{x, opt}$ increases with N_{prop}. Once $I_{x, opt}$ is fixed at unity, $C_{fD, opt}$ would show a linear relation with N_{prop}. Fig. 9 (b) shows the same results, but highlights the effect of proppant index on the optimum results under different dimensionless fixed period of time. For a given N_{prop}, $C_{fD, opt}$ decreases with time and approaches a constant value. The constant value of $C_{fD, lim}$ is positively correlated with N_{prop} (e.g., $C_{fD, lim}=1.62$ for $N_{prop}=10^{-2}$, while $C_{fD, lim}=175$ for $N_{prop}=10^3$). Put it in another way, when dimensionless fixed period of time is shorter, the optimum design indicates a fracture with shorter length and higher conductivity; when dimensionless fixed period of time is longer, the opposite is the case. It is worth noting that the characteristic value of $C_{fD, lim}$ is 1.62 in the case of small N_{prop}, which is consistent with the results of Volko and Economides[27], proving the method presented in this paper is reliable.

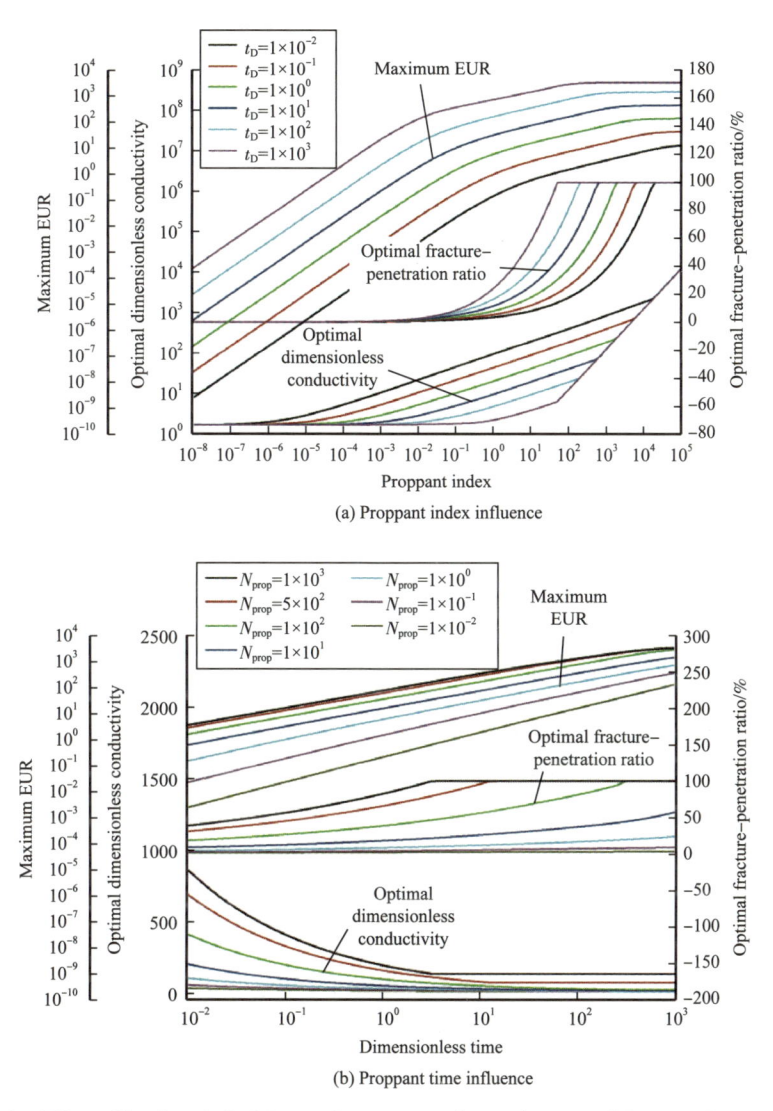

Fig. 9 Effect of fixed period of time and proppant index on the optimal fracture dimensions

3.2 Outer-optimization shell

Since the number of wells (n_w) and the number of fractures per well (n_f) jointly affect the achievable EUR, the outer-optimization shell is proposed to determine the optimal designs of n_w and n_f.

Assuming the drainage area is fixed, the geometrical shape affects the well performance and the optimum fracture dimension. The drainage-area aspect ratio describes the geometrical shape, which is defined as $\lambda = y_e/x_e$. Noted that aspect ratio is also determined by n_w and n_f ($y_e = y_{ef}/n_f$, $x_e = x_{ef}/n_w$). Fig. 10 shows the effect of λ on the production performance. In this case, the fractures are assumed to fully penetrate the pay in the plane, and all fractures have infinite conductivity. From Fig.10, the maximum EUR of a fixed drainage area is a constant ($0.5x_{eD}y_{eD}/\pi$), irrelevant to the ratio of λ. The initial rate is higher and the maximum EUR is reached in a shorter duration for drainage area with more pronounced elongation when λ is smaller. In fact, the more elongated the drainage area, the more efficiently it can be drained, since the corresponding fracture length is larger, thus making it easy to drain the whole drainage area.

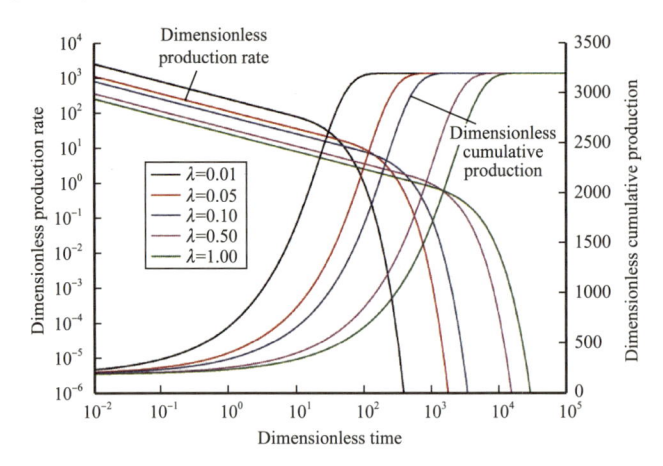

Fig. 10　Effect of drainage-area aspect ratio on performance

When the proppant number is taken into account, Fig. 11 shows the effect of λ on maximum EUR and the optimal fracture dimensions when t_D is small. Fig. 11 (a) shows that EUR_{max} increases with N_{prop} until reaches the peak. The smaller the λ (small fracture space and large well space), the larger the maximum dimensionless cumulative production will be, and at this point, the fractures penetrate the whole formation and reach infinite conductivity [Fig. 11 (b)]. When the proppant index is smaller (less than 10^3), for larger λ values, the corresponding optimum conductivities are the same, but the optimum fracture-penetration ratio is larger, and the flow resistance is smaller, so the corresponding maximum dimensionless cumulative production is larger [Fig. 11 (a)]. It must be highlighted that the production characteristics reflected by the template in Fig. 11 is related to the production time; when the time is long enough, the peak values of EUR_{max} for all λ values are equal; when the time is short enough, λ has little effect on type curves (shown in Fig. 11) in the case of smaller N_{prop}.

3.3　Nested optimization

The development effect of multi-stage fracturing horizontal well is optimized by increasing the contact area between fractures and formation, reducing interwell and inter-fracture interferences, balancing the inflow and outflow between fracture and formation. When these seepage relationships reach balance, the production effect is the best. In this study, half of the well pad is assumed to be 1500m × 1500m × 20m, with the proppant volume as constrain, and the NPV as the objective function, a nested method was used to optimize the parameters. The detailed workflow entails the following steps.

Step 1: Define input variables, including reservoir/well properties, fluid properties, proppant type and the targeted production period.

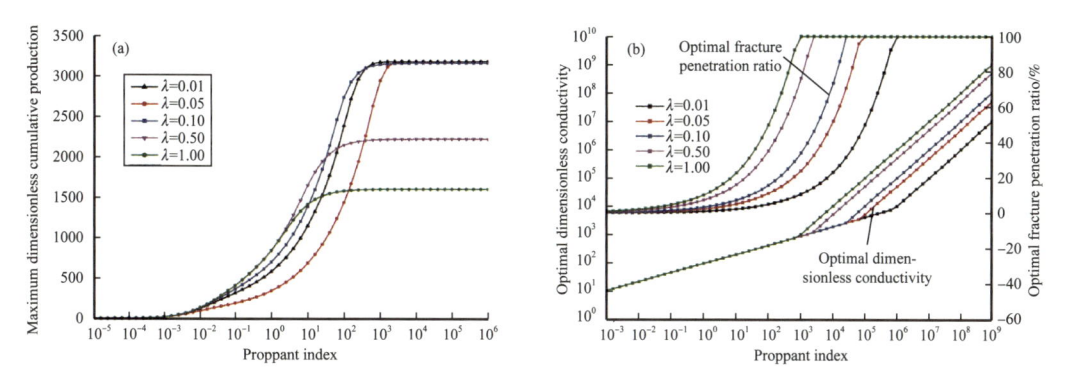

Fig. 11. Effect of fracture space/ well space ratio on maximum EUR (a) and optimal fracture dimensions (b) when t_D=1000

Step 2: Define three key variables to be optimized, including number of wells n_w, number of fracturing stages per well n_f, and proppant volume per fracture, M_p.

Step 3: Calculate the optimum fracture dimensions and maximum cumulative production at different number of wells, fracturing stages and proppant volumes by using the UFD method.

Step 4: Calculate the corresponding NPV.

Step 5: Based on Powell global-optimization algorithm, return to Step 2 till the maximum NPV is achieved, and the horizontal well and fracturing parameters at this point are the best design parameters.

The NPV is calculated as

$$NPV = \sum_{ii=1}^{n_{year}} \frac{(G_{p,ii} - G_{p,ii-1})P_{gas}}{(1+i_r)^{ii}} - \left[FC + \sum_{k=1}^{n_w} \left(C_{well} + \sum_{k=1}^{n_f} C_{fracture} \right) \right] \quad (36)$$

In this case, the selected dimensionless fixed period of time is set to be 1000, and total mass of proppant of multiwell pad is assumed to be fixed, as the actual fracturing scale is limited by engineering conditions. Fig. 12 shows a 3D plot of the total EUR_{max} variation with the combination of n_w and n_f. It can be seen from Fig. 12, EUR_{max} increases with the increase of wells and fracturing stages monotonously. When the proppant index is smaller [e.g. N_{prop}=0.1 in Fig. 12 (a)], EUR_{max} gradually increases in the same magnitude with the increase of wells and fracturing stages [Fig. 12 (a)]. When proppant index is bigger [e.g. N_{prop}=1000 in Fig. 12 (b)], the increase magnitude of EURmax decreases relativley. When the fractures are more than 30 and wells are more than 5, the EURmax hardly increase anymore [Fig. 12 (b)]. Therefore, in the case with the EURmax as the objective function and the fracturing scale as the constraint, there is only optimal fracture dimensions but no optimal well space (number of wells) and fracture space (number of fractures) for a well pad.

Substituting the development index results into the economic evaluation model, the optimization was repeated with NPV as the objective function, the results are shown in Fig. 13. There are obvious extreme points in Fig. 13, indicating there are optimum well space and fracture space in this case. Similarly, the corresponding optimization results of fracture dimensions are shown in Fig. 14. It can be seen with the increase of n_w and n_f, although the EUR$_{max}$ increases, the costs also go up. In other words, when the incremental NPV caused by EUR increase is offset by the cost increase caused by creating more wells and fractures, the economic benefit would get worse gradually. Therefore, under the dual constraints of fracturing scale and economic benefit, there are optimum n_w, n_f and fracture dimensions for a well pad, providing optimization room in development program design.

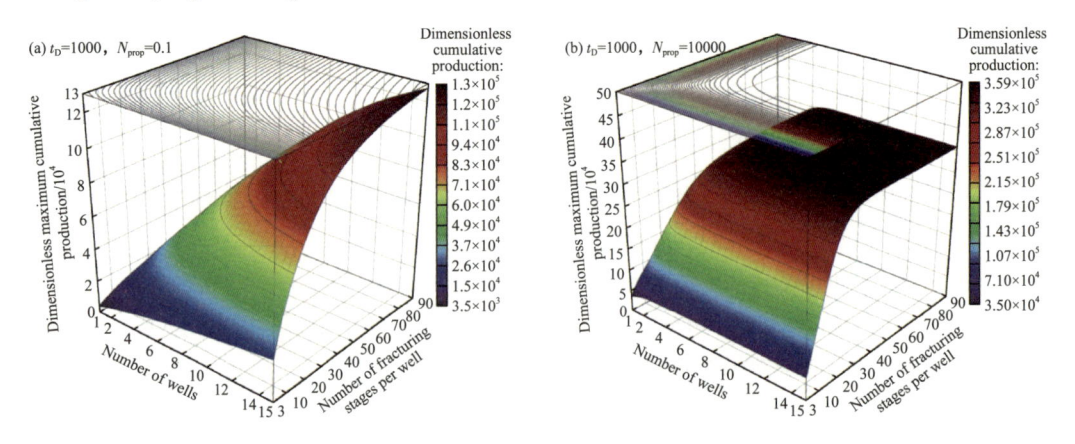

Fig. 12 Maximum EUR vs. number of wells and number of fractures per well

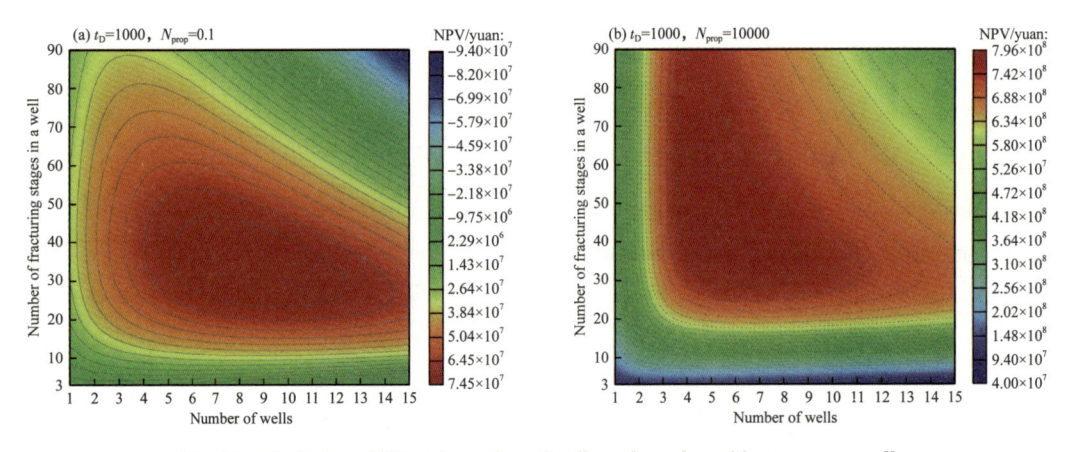

Fig. 13 Variation of NP with number of wells and number of fractures per well

It can be seen from Figs. 13 and 14, when the proppant index is smaller (N_{prop}=0.1), the optimal well space is smaller ($n_{w,opt} \approx 8$), the optimal fracture space is wider ($n_{f,opt} \approx 35$), the corresponding optimal fracture-penetration ratio is smaller ($I_{x,opt} \approx 11\%$), and the fracture conductivity is smaller ($C_{fD,opt} \approx 1.60$). When the fracturing scale is larger ($N_{prop}=1 \times 10^4$), compared with the case with small fracturing scale, the optimal well space increases ($n_{w,opt} \approx 6$), the optimal fracture space decreases ($n_{f,opt} \approx 50$), the fracture penetration rate increases significantly ($I_{x,opt} \approx 72\%$), but the optimal conductivity is still fairly low ($C_{fD,opt} \approx 3.54$).

The specific results of nested optimization are related to input parameters, such as geometry of the well pad, development duration, economic parameters, complexity of fracture networks and constraint conditions (i.e., mass of proppant per fracture, well or well pad etc.). However, these general statements can serve as reference in the more-detailed optimization works.

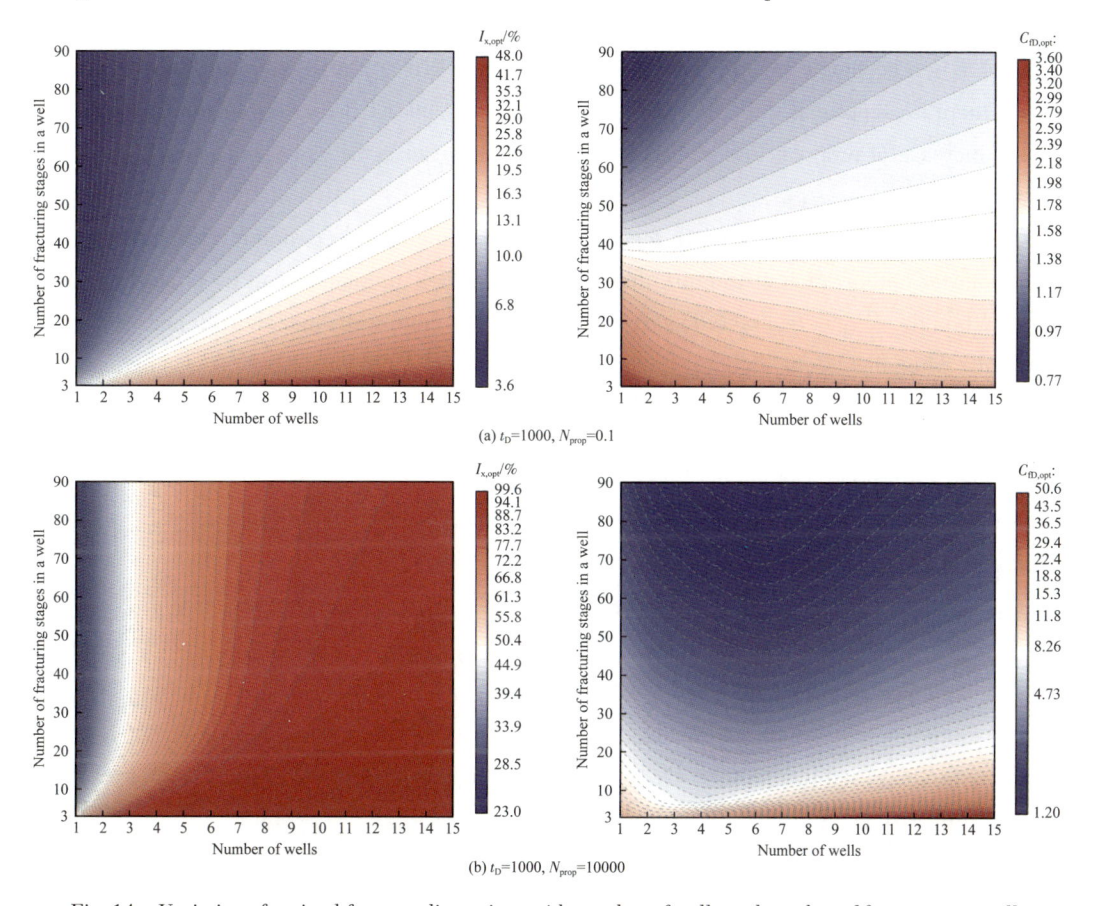

Fig. 14 Variation of optimal fracture dimensions with number of wells and number of fractures per well

4 Conclusions

In this study, we established a general semi-analytical model to study the production performance of a well pad with multiple MFHWs. Dimension transformation is incorporated to change the flow equation of the variable conductivity fracture into the equation of the constant conductivity fracture, and a universal approach is developed to solve the coupled reservoir-fracture flow model. We have reach some general conclusions from the study:

(1) A comprehensive optimization approach has been developed on the basis of semi-analytical model integrating modules for production-performance, extended UFD, and NPV estimation by a mathematical global optimization algorithm.

(2) In the inner-optimization shell, there exists a best com promise between fracture conductivity and fracture length under the constraint of N_{prop}. As N_{prop} and t_D increase, the EUR_{max} increases monotonously, the optimal fracture penetration rate also increases. With the decrease of

proppant index and increase of time, the optimal conductivity of fracture decreases gradually and approaches to a limited value ($C_{fD,\ lim}$=1.62). Besides, the ratio of fracture space to well space also affects the results of the optimal fracture dimensions. In the case with larger N_{prop}, the design with larger well space and smaller fracture space is better.

(3) When economic benefit isn't considered, the EUR_{max} increases with the increase of n_w and n_f, there is no optimal wells and fracturing stages. But the EUR increase magnitude decreases with the increase of proppant index, when the n_w and n_f reach a critical value, the EUR doesn't increase anymore.

(4) In the nested optimization accounting for technical and economic constraints, there exists room for optimization of fracture design when the NPV increment can be offset by increase in cost. When the N_{prop} is relatively lower, the design with more wells and short-length fractures is suggested; when the N_{prop} is higher, the design with less wells and longer fractures is better.

Nomenclature

a_g—gas desorption coefficient, dimensionless;
B_g—gas volume factor, m³/m³;
b_{am}— slippage factor, dimensionless;
C_{fD}—dimensionless fracture conductivity;
$C_{fracture}$— cost of fracturing one stage, Yuan;
C_{well}—the cost of drilling one well, Yuan;
c_t—formation compressibility factor, Pa⁻¹;
c_g—gas compressibility factor, Pa⁻¹;
D_D—dimensionless diffusivity coefficient;
D_g—gas diffusivity coefficient in matrix, m²/s;
E—Young modulus, Pa;
FC—the total fixed cost, Yuan;
G_p—cumulative production, m³;
$G_{p,\ ii}$—the ii-th year cumulative production, m³;
h—formation thickness, m;
I_x—fracture penetration ratio;
i_r—annual interest rate, dimensionless;
k_m—reservoir permeability, m²;
k_f—fracture permeability, m²;
L_f—fracture length, m;
L_{ref}—reference length, m;
M_p—mass of proppant, g;
M_g—molecular weight, g/mol;
m_f—pseudo-pressure in fracture, Pa;

m_m—pseudo-pressure in matrix, Pa;
n_f—number of fractures per well;
n_{year}—production time, a;
n_w—number of wells in a well pad;
OGIP—original gas in place, m^3;
P_{gas}—gas price, Yuan/m^3;
p—pressure, Pa;
p_i—initial pressure, Pa;
p_{net}—net pressure, Pa;
p_{wf}—bottomhole pressure, Pa;
q_w—production rate, m^3/s;
q_{ref}—reference production rate, m^3/s;
q_f—influx density, m^2/s;
$\tilde{q}_{fDn,i}^{\langle w \rangle}$—$i$-th segment of n-th fracture in w-th well;
R_g—gas constant, J·mol^{-1}·K^{-1};
R_m—matrix radius of gas desorption, m;
r_w—wellbore radius, m;
s—Laplace variable, dimensionless;
T—Temperature, K;
t—time, s;
t_a—pseudo-time, s;
V_{prop}—volume of proppant, m^3;
V_E—gas concentration in the center of spherical matrix, m^3/m^3;
V—gas concentration, m^3/m^3;
v_m—flow velocity in matrix, m^3/s;
v—Poission's ratio, dimensionless;
w_f—fracture width, m;
x, y—spatial variable, m;
x_{hf}—coordinate in fracture, m;
x_f—fracture half-length, m;
x_{of}, y_{of}—location of fracture tip, m;
x_{whf}—intersection between fracture and wellbore, m;
x_{ef}—reservoir width, m;
x_e—width of sub-drainage area, m;
y_{ef}—reservoir length, m;
y_e—length of sub-drainage area, m;
Z_g—gas deviation factor, dimensionless;
Z_i—initial deviation factor, dimensionless;

ρ_g—gas density, g/m^3;

μ_g—gas viscosity, $Pa \cdot s$;

ρ_{prop}—proppant density, g/m^3;

φ_{prop}—fracture porosity;

ξ_D—dimensionless transformed spatial variable;

φ_m—reservoir porosity;

∇p_m—pressure gradient, Pa/m;

λ_m—dimensionless viscosity-compressibility ratio;

β_m—dimensionless rescaling factor;

γ_m—nonlinear modified factor;

δ—Dirac function;

ζ, ξ, u—general spatial variable, such as x, y;

ω—storage ratio, dimensionless.

Subscripts:

D—dimensionless;

f, hf—hydraulic fracture;

i—initial pressure condition;

i, j—number of fracture segment;

m, n—number of fracture;

opt—the optimal;

max—the maximum;

lim—the limited;

sc—standard condition.

Superscripts:

<>: number of wells;

−: Laplace symbol

References

[1] Al-Rbeawi S. Productivity-index behavoir for hydraulically fractured reservoirs depleted by constant production rate considering transient-state and semisteady-state conditions [J]. SPE 189989, 2018.

[2] Yao J, Sun H, Huang Z Q, et al. Key mechanical problems in the development of shale gas reservoirs [J]. SCIENTIA SINICA Physica, Mechanica & Astronomica, 2013, 43(12): 1527-1547.

[3] Sahai V, Jackson G, Rai R. Effect of non-uniform fracture spacing and fracture half-length on well spacing for unconventional gas reservoirs [J]. SPE 164927, 2013.

[4] Wei Y S, Wang J L, QI Y D, et al. Optimization of shale gas well pattern and spacing [J]. Natural Gas Industry, 2018, 38(4): 129-137.

[5] Wang X D, Luo W J, Hou X C, et al. Transient pressure analysis of multiple-fractured horizontal wells in boxed reservoirs [J]. Petroleum Exploration and Development, 2014, 37(11): 43-52.

[6] Chen C C, Raghavan R. A multiple-fractured horizontal well in a rectangular drainage region [J]. SPE 37072, 1997.

[7] Chen Z, Liao X, Zhao X. A practical methodology for production-data analysis of single-phase unconventional wells with complex fracture geometry [J]. SPE 191372, 2018.

[8] Fang W C, Jiang H Q, Li J J, et al. A numerical simulation model for multi-scale flow in tight oil reservoirs [J]. Petroleum Exploration and Development, 2017, 44 (3): 415-422.

[9] Yu W, Wu K, Zuo L H, et al. Physical models for inter-well interference in shale reservoirs: Relative impacts of fracture hits and matrix permeability [J]. URTEC 2457663, 2016.

[10] Li L L, Yao J, Li Y, et al. Productivity calculation and distribution of staged multi-cluster fractured horizontal wells [J]. Petroleum Exploration and Development, 2014, 41 (4): 457-461.

[11] Sun H D, Ouyang W P, Zhang M, et al. Advanced production decline analysis of tight gas wells with variable fracture conductivity [J]. Petroleum Exploration and Development, 2018, 45 (3): 455-463.

[12] Li H T, Wang J C, Li Y, et al. Deliverability evaluation method based on volume source for horizontal wells by staged fracturing [J]. Natural Gas Industry, 2015, 35 (9): 1-9.

[13] He Y W, Cheng S Q, Li S, et al. A semianalytical methodology to diagnose the location of underperforming hydraulic fractures through pressure-transient analysis in tight gas res ervoir [J]. SPE Journal, 2017, 22 (3): 924-939.

[14] Liang T, Chang Y W, Guo X F, et al. Influence factors of single well's productivity in the Bakken tight oil reservoir [J]. Petroleum Exploration and Development, 2013, 40 (3): 357-362.

[15] Li B, Jia A L, He D B, et al. Productivity analysis and completion optimization of fractured horizontal wells in low- permeability tight gas reservoir [J]. Journal of Central South University (Science and Technology), 2016, 47 (11): 3775-3783.

[16] Yu W, Sepehrnoori K. An efficient reservoir-simulation approach to design and optimize unconventional gas production [J]. SPE 165343, 2014.

[17] Jiang R Z, Liu M M, Xu J C, et al. Application of genetic algorithm for well placement optimization in Sulige gasfield [J]. Natural Gas Geoscience, 2014, 25 (10): 1603-1609.

[18] Adibifard M, Tabatabaei-Nejad S, KHODAPANAH E. Artificial neural network (ANN) to estimate reservoir parameters in naturally fractured reservoirs using well test data [J]. Journal of Petroleum Science and Engineering, 2014, 122: 585-594.

[19] Ertekin T, King G R, Schwerer F C. Dynamic gas slippage: A unique dual-mechanism approach to the flow of gas in tight formation [J]. SPE Formation Evaluation, 1986, 2: 43-52.

[20] Ozkan E, Raghavan R, Apaydin O G. Modeling of fluid transfer from shale matrix to fracture network [J]. SPE 134830, 2010.

[21] Xiao L, Zhao G. Study of 2-D and 3-D hydraulic fractures with non-uniform conductivity and geometry using source and sink function methods [J]. SPE 162542, 2012.

[22] Gringarten A C, Ramey H J. The use of source and Green's functions in solving unsteady-flow problems in reservoirs [J]. SPE 3818, 1973.

[23] Bao J Q, Liu H, Zhang G M, et al. Fracture propagation laws in staged hydraulic fracturing and their

effects on fracture conductivities [J]. Petroleum Exploration and Development, 2017, 44 (2): 281–288.

[24] Wang X D, Li G H, Wang F. Productivity analysis of horizontal wells intercepted by multiple finite-conductivity fractures [J]. Petroleum Science, 2010, 7: 367–371.

[25] Ye P, Ayala H L F. A density-diffusivity approach for the unsteady state analysis of natural gas reservoirs [J]. Journal of Natural Gas Science and Engineering, 2012, 7 (1): 22–34.

[26] Poe B, Shan P, Elbel J. Pressure transient behavior of a finite-conductivity fractured well with spatially varying fracture properties [J]. SPE 24707, 1992.

[27] Valko P P, Economides M J. Heavy crude production from shallow formations: Long horizontal wells versus horizontal fractures [J]. SPE 50421, 1998.

[28] Bhattacharya S, Nikolaou M, Economides M J. Unified fracture design for very low permeability reservoirs [J]. Journal of Natural Gas Science and Engineering, 2012, 9: 184–195.

四川盆地威远区块典型平台页岩气水平井动态特征及开发建议

位云生，齐亚东，贾成业，金亦秋，袁 贺

（中国石油勘探开发研究院）

摘 要 位于四川盆地长宁—威远国家级页岩气示范区范围内的威远页岩气田（以下简称为威远区块），同一平台上气井的生产动态特征存在着较大差异，目前对于其页岩气井产气量的主控因素和开发工艺措施的有效性认识尚未明确。为此，以威远区块PT2平台的6口水平井为例，针对气井生产动态存在的差异，从钻遇优质页岩段的长度、水平段轨迹倾向、压裂段长度、改造段数、加砂量及井底积液等方面进行分析，明确了影响威远区块页岩气水平井产气量的主要因素，进而提出了有针对性的开发措施建议。研究结果表明：（1）优质页岩段钻遇长度是气井高产的物质地质保障，水平压裂段长度、改造段数/簇数和加砂量是主要的工程因素；（2）页岩气井生产早期均为带液生产且水气比较大，当产气量低于临界携液流量时，井底积液对产气量和井口压力的影响不容忽视；（3）建议低产井应采用小油管生产（油管内径小于等于62mm），对于上半支低产井，应及早采取橇装式排水采气工具和措施以释放气井产能，对于下半支低产井，则应放压生产，防止井底过早积液。

关键词：页岩气；产量主控因素；生产动态；临界携液流量；开发措施；四川盆地；长宁—威远国家级页岩气示范区；威远页岩气田

平台化钻井、工厂化多段大规模压裂改造已成为目前页岩气效益开发的核心技术[1-3]。威远页岩气田（以下简称为威远区块）处于长宁—威远国家级页岩气示范区的范围内，构造位置属于四川盆地川中隆起区的川西南低陡褶带，为一大型的穹隆背斜构造[4]，构造相对单一，地层分布稳定，一定范围内储层特征差异较小，但区块内甚至同一平台井间生产特征差异较大，目前对于该区气井产量的主控因素和开发工艺措施的有效性认识尚不明确。为此，笔者选取该区块内的一典型平台——PT2平台的6口水平井，通过研究各水平井地质与工程参数、生产规律，确定单井产量的主控因素，并结合实际生产动态分析，提出了开发建议。

1 地质特征与工程参数

1.1 地质特征

威远区块志留系龙马溪组页岩地层自下而上划分为龙一段和龙二段，其中龙一段为主要目的层，龙一段自下而上又划分为龙一$_1$亚段和龙一$_2$亚段，龙一$_1$亚段自下而上又细分为4层，即龙一$_1^1$、龙一$_1^2$、龙一$_1^3$、龙一$_1^4$层，含气量均较高，但龙一$_1^4$层岩石脆性矿物含量（硅质、碳酸盐、黄铁矿）较低（表1），可压性相对较差，因此确定龙一$_1^1$、龙一$_1^2$、龙一$_1^3$这3层为优质页岩段。奥陶系五峰组储层在威远区块不发育。

表1 威远区块龙一$_1$亚段各小层岩石脆性矿物含量统计表

层位	统计井	
	威202	威204
龙一$_1^4$	49%	47%
龙一$_1^3$	66%	74%
龙一$_1^2$	67%	58%
龙一$_1^1$	82%	66%

威远区块属海相深水陆棚沉积环境，区域储层地质参数（优质页岩厚度、TOC、孔隙度、脆性指数等）在一定范围内稳定分布，横向变化较小[5]，如威202井与威204井井间距离为22km，但两口井优质页岩段厚度分别为13.9m和16.7m，横向变化率仅0.13m/km。在一个平台范围内，储层地质特征横向上基本一致。

1.2 工程参数

威远区块建产区为背斜构造的一翼，水平井钻井方向几乎与埋深等值线垂直，因此普遍存在上半支井上倾、下半支井下倾的情况；上半支井工程施工难度较大，成功率较低，且测试产量普遍低于下半支井，如图1所示；将目前气井产量和临界携液流量进行对比，认为2017年10月之前投产的89口平台井中有63口井存在不同程度的井底积液，且大部分为上半支井。以威远区块PT2平台6口井为例，该平台PT2-1、2、3井为上半支水平井，PT2-4、5、6井为下半支水平井，各井主要压裂改造参数如图1所示。另外，PT2-1、2、4井采用的压裂液类型为滑溜水，PT2-3、5、6井采用的压裂液类型为滑溜水+线性胶；PT2-1、2、3井采用的支撑剂类型为树脂覆膜砂，PT2-4、5、6井采用的支撑剂类型为陶粒。

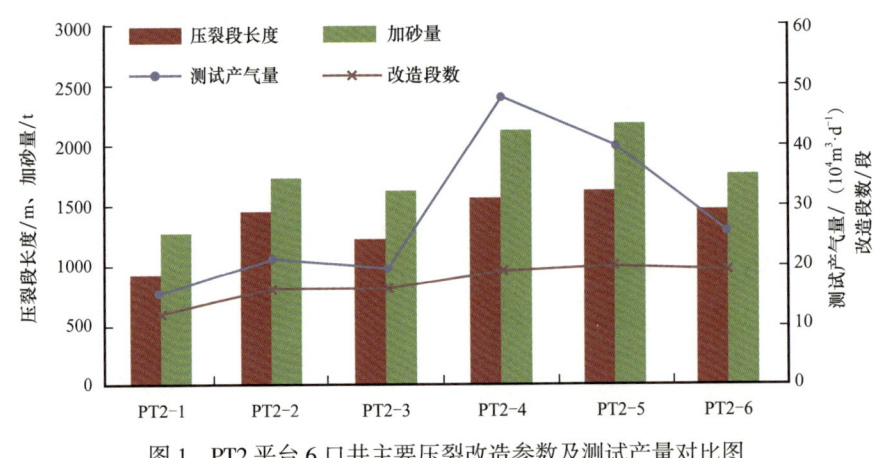

图1 PT2平台6口井主要压裂改造参数及测试产量对比图

2 生产动态特征分析

2.1 气井生产动态

由于钻遇层位与工程参数的差异，威远区块开发平台上、下半支井的产量与压力变化差异较大[6-10]，同时，由于页岩气井压裂入地液量较大，气井在投产早期较长一段时间内为带液生产，返排液对气井产量会产生较大影响[11]。以PT2平台水平井为例，上半支3口水平井（PT2-1、2、3井）初期稳产气量和井口压力均较低，初期稳产气量介于（4.0~6.3）×10^{4}m^3/d，平均为4.8×10^{4}m^3/d，初期套压介于23.3~28.9MPa，平均为25.8MPa；产气量月递减率介于10.2%~19.4%，平均为14.7%；关井后压力恢复速率较慢，介于0.39~0.52MPa/d，平均为0.48MPa/d。下半支水平井中PT2-4、5井初期稳产气量和井口压力均较高，达到20×10^{4}m^3/d以上，初期套压在40MPa上下，且产气量递减较慢，其月递减率在9%左右；PT2-6井虽然初期稳产气量仅6.5×10^{4}m^3/d，但关井压力恢复速率较快，达0.96MPa/d。该平台PT2-1、4井的生产曲线如图2所示。

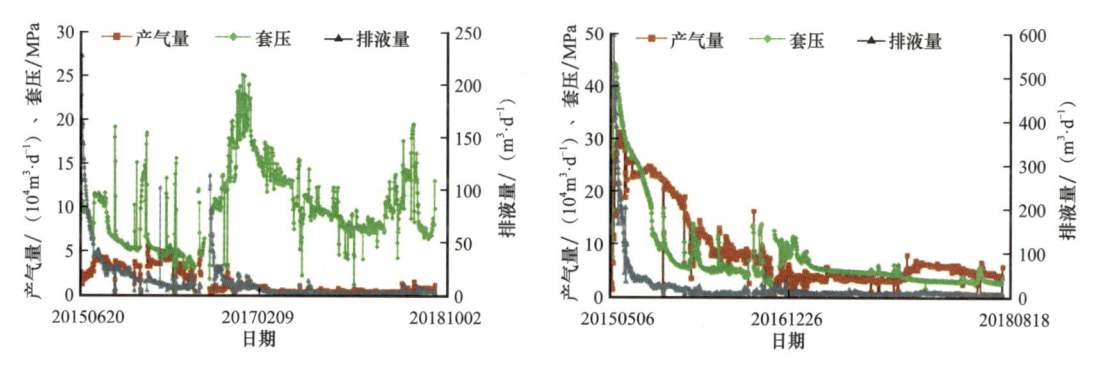

图2 PT2-1（左）、4井（右）生产曲线图

2.2 临界携液流量

由于页岩气井采用大液量进行压裂改造，因此在生产早期均为带液生产，且水气比较大，由于普遍采用油层套管进行生产，由此带来的携液问题不容忽视。本文选用李闽模型[12-14]，以井口为参考点进行临界携液流量的计算。

PT2平台水平井产出气体相对密度为0.5684，水密度为1030kg/m^3，油层套管内径为114.3mm，井口温度平均为310K，气水界面张力取0.06N/m。根据李闽模型，计算不同套压下的临界携液流量，连同PT2平台6口水平井的套压和产气量数据，共同绘制在图3。

2.3 生产动态特征差异原因分析

通过对PT2平台6口水平井优质页岩段的钻遇情况、压裂实施参数、实际生产动态及临界携液流量计算结果进行对比，分析各井产量及压力存在差异的主要原因。

（1）由于钻遇储层条件、压裂规模存在明显差异，导致气井生产动态存在明显差异。上半支PT2-1、2、3井的初期产气量和套压均较低，原因在于：①优质页岩段钻遇长度介于526~937m，平均仅707m，优质页岩段钻遇率介于40.4%~63.3%，平均仅52.2%，可见这3口水平井钻遇的优质页岩段较短；②压裂段长度介于920~1440m，改造段数介于12～16段,加砂量介于1265～1718t,平均加砂量1533t,支撑剂类型为树脂覆膜砂,由于压裂规模小造成储层改造体积小、泄流区渗透性的改善效果有限,造成气井初期产气量和套压低。同时,下半支PT2-6井的初期产气量和套压也较低,该井压裂段长度1450m,改造段数19段,加砂量1766t,支撑剂类型为陶粒,但该井优质页岩段钻遇长度仅77m,优质页岩段钻遇率仅5.1%。综合判断认为PT2-6井压裂规模较大,泄流区渗透性的改善效果较好,但由于靶体钻遇的页岩段大部分为非优质页岩段,储层物质基础较差,因此造成气井初期产气量和套压均低。另外,下半支PT2-4、5井钻遇优质页岩段长、压裂规模较大,使得气井产气量较高,同时井底无积液。PT2-4、5井优质页岩段钻遇长度分别为1486m和1578m,优质页岩段钻遇率均高达90%以上,压裂段长度分别为1565m和1600m,压裂段数分别为19段和20段,平均加砂量为2160t,支撑剂类型为陶粒,由于钻遇优质页岩段较长,且压裂规模较高,初期产气量和套压均较高,产气量都高于临界携液流量。

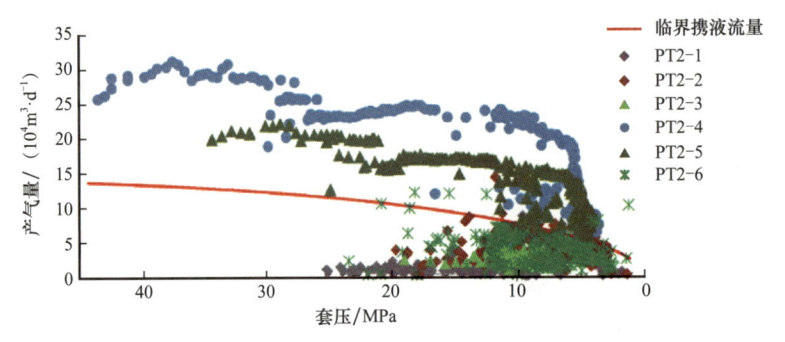

图3 不同套压下产气量与临界携液流量对比图

(2)由关井套压恢复速率推断,PT2-1、2、3井泄流区渗透性较差,又由于产气量一直低于临界携液流量,导致井底存在积液,且上半支PT2-1、2、3井由于轨迹上倾导致井底积液位于水平井跟端附近(图4),进一步增加套压恢复难度;下半支PT2-6井泄流区渗透性较好,尽管该井产气量一直低于临界携液流量,但由于轨迹下倾,积液位于水平井趾端附近(图4),因此套压恢复较容易。

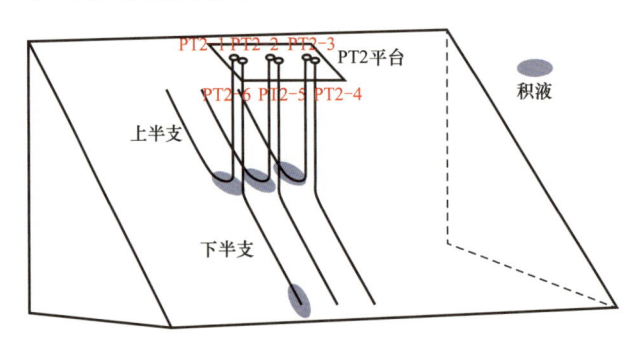

图4 PT2平台6口水平井轨迹和积液示意图

3 单井产气量的主要影响因素

针对页岩气井产气量受到地质与工程因素的共同影响,很多学者已进行了相关探讨[15-20]。其中，储层横向特征变化较小，垂向特征差异较大，是页岩气井开发效果差异的主要因素。威远区块水平井目标靶体位置不断优化、下移，目前的靶体位置为龙马溪组底部的龙一$_1^1$小层碳质页岩段，含气量和压力系数最高，通过水平井多段大规模压裂改造，可获得较高的单井产量。但实际钻井过程中，由于目标层位厚度较薄（5m左右）、地层倾角的小幅变化和钻井导向的精度限制，实际水平井靶体位置有一定变化。另外，从试采动态监测结果看，垂向最大水力压裂缝缝高约40m，有效支撑缝缝高介于10~15m。综合判断，认为压裂改造后的水平井垂向上可以动用优质页岩段。因此，本文采用优质页岩段钻遇长度作为分析影响单井产气量的主要地质因素。

在钻遇优质页岩段的前提下，水平井压裂段长度决定了沿水平井筒方向打开储层的范围。在压裂段内，通过优化段数/簇数，并采用"千方砂+万方液"进行大规模压裂改造，构建复杂的裂缝网络系统，形成有效的气体渗流通道，从而增大泄气面积，提高改造区储层渗透性[9]，获得高产气井。可见，水平井压裂段长度、改造段数/簇数和加砂量是影响水平井产气量的主要工程因素。

从PT2平台6口井的数据来看，优质页岩段钻遇长度、水平井压裂段长度、改造段数、单井加砂量与测试产气量之间有明显的正相关关系（图5）。同时，上半支井的产量低于下半支井，这是威远区块的一个普遍规律。原因在于以下3个方面：（1）威远区块上半支水平井轨迹上倾，井眼轨迹的控制和随钻导向整体效果不如下半支水平井；（2）射孔枪和压裂分段桥塞是通过电缆下入井下，主要靠水动力输送，在上半支水平井易造成完井和压裂工具下入困难，因此，导致上半支水平井的井筒完整性和压裂段数不及下半支水平井；（3）由于滑溜水携砂能力有限，采用滑溜水与陶粒的压裂液与支撑剂组合，上半支水平井支撑剂泵送较为困难，造成加砂量一般比下半支水平井少。另外，支撑剂类型对产量的影响还需要更多井的数据进行论证。

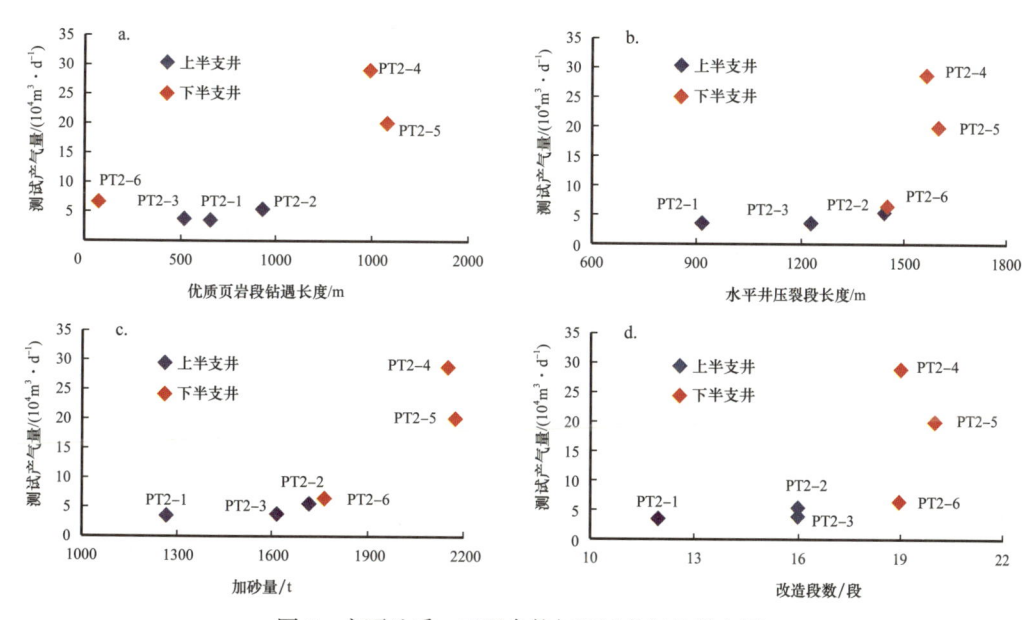

图5 主要地质、工程参数与测试产气量散点图

4 开发措施建议

基于PT2平台6口水平井钻遇优质页岩段的情况、压裂规模及生产动态特征，提出该区块的开发措施建议。

（1）由于页岩气井钻完井及压裂施工费用较高，需采用平台化布井模式，现场施工采用"大兵团、工厂化"的作业模式，同时加强对各技术环节的把控，提升施工效率，尽可能降低单井投资。对平台上的每口井，都应将水平井靶体位置控制在优质页岩段内，同时保证水平井井筒完整性和压裂段长度，且尽可能增加改造段簇数和加砂规模，提高单井产气量。如PT2-4、5井，由于钻遇优质页岩段长，且钻完井及压裂作业均成功实施，单井累产气量预计可达到$1\times10^8 m^3$左右。

（2）针对低产井，建议采用小油管（油管内径小于等于62mm）生产；对于平台上半支低产井，应尽早采取排水采气措施，如PT2-1、2、3井井底积液位于水平段跟端附近，由于压裂液量是有限的，可采用橇装式排水采气工具和措施[21-22]，释放气井产能，恢复气井产量；对于平台下半支低产井，应放压生产，使气井尽量保持较高的产气量，防止井底过早积液，如PT2-6井，井底积液位于水平段趾端附近，实施排水采气工艺难度较大，建议进行放压生产。

以上措施建议已在威远区块的开发中得到推广应用，指导了后续类似平台的有效开发。今后，还需在现场录取并完善更多的动态资料以更全面、深入地分析页岩气井生产动态特征，进一步优化威远区块页岩气的开发技术。

5 结论

（1）实现页岩气井的高产，优质页岩段钻遇长度是物质保障，水平压裂段长度、改造段数/簇数和加砂量是主要的工程因素。

（2）页岩气井的产气量除受钻遇储层品质及压裂改造效果的影响外，还受地层中返排压裂液的影响，返排液体对井筒举升提出了要求。若气井产气量低于临界携液流量，井底存在积液，在分析井口产量和压力时不容忽视。

（3）对于低产井，建议采用小油管生产（油管内径小于等于62mm），对于上半支低产井，应及早采取橇装式排水采气工具和措施以释放气井产能，对于下半支低产井，应放压生产，防止井底过早积液。

参 考 文 献

[1] 邹才能, 董大忠, 王玉满, 等. 中国页岩气特征、挑战及前景（二）[J]. 石油勘探与开发, 2016, 43（2）：1-13.

[2] 王红岩, 刘玉章, 董大忠, 等. 中国南方海相页岩气高效开发的科学问题[J]. 石油勘探与开发, 2013, 40（5）：574-579.

[3] 李文阳, 邹洪岚, 吴纯忠, 等. 从工程技术角度浅析页岩气的开采[J]. 石油学报, 2013, 34（6）：1218-1224.

[4] 王玉满, 董大忠, 李建忠, 等. 川南下志留统龙马溪组页岩气储层特征[J]. 石油学报, 2012, 33（4）：

551-561.

[5] 贾爱林, 位云生, 金亦秋. 中国海相页岩气开发评价关键技术进展[J]. 石油勘探与开发, 2016, 43(6): 949-955.

[6] Wang J L, Jia A L. A general productivity model for optimization of multiple fractures with heterogeneous properties[J]. Journal of Natural Gas Science and Engineering, 2014, 21(2): 608-624.

[7] 王军磊, 贾爱林, 位云生, 等. 有限导流压裂水平气井拟稳态产能计算及优化[J]. 中国石油大学学报(自然科学版), 2016, 40(1): 100-107.

[8] Wang J L, Yan C Z, Jia A L, et al. Rate decline analysis of multiple fractured horizontal well in shale reservoirs with triple continuum[J]. Journal of Central South University, 2014, 21(11): 4320-4329.

[9] 王军磊, 贾爱林, 何东博, 等. 致密气藏分段压裂水平井产量递减规律及影响因素[J]. 天然气地球科学, 2014, 25(2): 278-285.

[10] 王晓泉, 张守良, 吴奇, 等. 水平井分段压裂多段裂缝产能影响因素分析[J]. 石油钻采工艺, 2009, 31(1): 73-76.

[11] 廖开贵, 李颖川, 杨志, 等. 产水气藏气液两相管流动态规律研究[J]. 石油学报, 2009, 3(4): 607-612.

[12] 李闽, 郭平, 谭光天. 气井携液新观点[J]. 石油勘探与开发, 2001, 28(5): 105-106.

[13] 李闽, 孙雷, 李士伦. 一个新的气井连续排液模型[J]. 天然气工业, 2001, 21(5): 61-63.

[14] 李闽, 郭平, 张茂林, 等. 气井连续携液模型比较研究[J]. 西南石油学院学报, 2002, 24(4): 30-32.

[15] 谢军, 赵圣贤, 石学文, 等. 四川盆地页岩气水平井高产的地质主控因素[J], 天然气工业, 2017, 37(7): 1-12.

[16] 张晓明, 石万忠, 徐清海, 等. 四川盆地焦石坝地区页岩气储层特征及控制因素[J]. 石油学报, 2015, 36(8): 926-939.

[17] 郭彤楼, 张汉荣. 四川盆地焦石坝页岩气田形成与富集高产模式[J]. 石油勘探与开发, 2014, 41(1): 28-36.

[18] 李庆辉, 陈勉, Wang FP, 等. 工程因素对页岩气产量的影响——以北美Haynesville页岩气藏为例[J]. 天然气工业, 2012, 32(4): 54-59.

[19] 贾成业, 贾爱林, 何东博, 等. 页岩气水平井产量影响因素分析[J]. 天然气工业, 2017, 37(4): 80-88.

[20] 雷征东, 覃斌, 刘双双, 等. 页岩气藏水力压裂渗吸机理数值模拟研究[J]. 西南石油大学学报(自然科学版), 2017, 39(2): 118-124.

[21] 张书平, 吴革生, 白晓弘, 等. 橇装式小直径管排水采气工艺技术[J]. 天然气工业, 2008, 28(8): 92-94.

[22] 叶长青, 熊杰, 康琳洁, 等. 川渝气区排水采气工具研制新进展[J]. 天然气工业, 2015, 35(2): 54-58.

页岩气水平井生产规律

郭建林[1]，贾爱林[1]，贾成业[1]，刘 成[2]，齐亚东[1]，位云生[1]，赵圣贤[3]，

王军磊[1]，袁 贺[1]

（1.中国石油勘探开发研究院；2.中国石油浙江油田公司；3.中国石油西南油气田公司）

摘 要：随着中国页岩气开发工作的深入和开发规模的加大，适时开展气井生产数据分析、评价历年已投产井生产规律是评价阶段开发效果和制订下步开发部署的依据和保障。为此，本文通过对北美地区 6 个页岩气开发区块和中国长宁—威远和昭通区块的页岩气水平井生产数据进行分析，并采用双曲—指数混合递减模型建立不同区块的归—化产量递减曲线，探讨了页岩气井初始产量和产量递减率的变化趋势，最后建立了页岩气水平井 EUR 快速评价方法。研究结果表明：（1）不同区块在开发初期页岩气水平井平均初始产量均呈现逐年上升的趋势，但初始产量分布存在差异，而后气井平均初始产量变化呈平台式，后期则呈逐年下降的趋势；（2）双曲—指数递减模型应用于中国和北美地区页岩气水平井的产量递减分析具有较好的适应性，同一区块不同批次的气井产量递减规律相似；（3）与北美地区 5 个页岩气区块相比，长宁区块前 3 年产量递减比例依次为 55%、38%和 33%，与 Fayetteville 区块接近，威远区块生产初期产量递减比例依次为 63%、46%和 37%，明显高于其他区块；（4）生产时间为 20 年，气井 EUR 与第 1 年累计产气量（Q_1）正相关，一般为 Q_1 的 2～5 倍，Woodford 区块气井 EUR 与 Q_1 比值最高，长宁区块与 Barnett、Eagle Ford、Fayetteville 和 Haynesville 区块相当，威远区块则相对较低。

关键词：页岩气；水平井；生产规律；初始产量；递减率；北美地区；中国国家级页岩气示范区

页岩气与常规天然气的渗流规律明显不同，对页岩气井进行生产规律的研究难度大，主要表现在以下 3 个方面：（1）页岩气在储层中的赋存状态分为吸附态和游离态，吸附于有机质和黏土矿物颗粒表面为吸附气，游离气则游离于基质孔隙或裂隙中；（2）由于页岩孔隙一般为纳米级，常规渗流理论不能准确描述页岩气的运移规律；（3）页岩储层基质渗透率极低，通常在纳达西级，大规模体积压裂是页岩气井获得工业产量的必要增产措施，压裂后形成的人工裂缝网络、天然裂隙和层理中流体的流动存在着多尺度耦合问题[1-2]。目前，针对页岩气井渗流特征与生产规律的研究主要包括以下 3 个方向：（1）开展页岩气渗流模型研究，通过简化吸附和解吸附方程，建立页岩气在人工裂缝与基质间耦合流动的渗流模型[1-3]，由于受页岩有机质中气体吸附状态难于表征、体积压裂后形成的人工裂缝网络形状不规则等问题的影响，导致以 Langmuir 等温吸附理论为基础、简单裂缝网络为假设条件的渗流模型难以描述在气藏温度、压力条件下复杂天然裂缝—人工裂缝网络中气体的实际渗流状态；（2）以物质平衡方程为理论基础，通过流动状态分析对不稳定渗流理

论模型进行修正和改进,实现压裂水平井的生产数据分析,评价页岩气井产能[4-5],但由于页岩气井在生产过程中需经历构型扩散、吸附/解吸与表面扩散、体相气体传输等多个复杂流动阶段[6-7]且气井生产制度对流动状态的影响显著,导致流态分析和渗流模型仅适用于单井,难以满足区块开发、批量投产情况下的生产数据分析;(3)以 Arps 经验模型为基础,对产量递减率等参数进行修正和改进,通过生产数据的历史拟合建立产量递减经验公式,明确气井产量与累计产气量随生产时间的变化趋势,从而开展相应的生产规律分析和产能评价[8-10],该方法具有相对简单、快速、实用的特点,在矿场的适用性更强。由此可见,页岩储层是典型的多尺度多孔介质,不同尺度空间中气体赋存方式和传输机理复杂[6-7],明确页岩气井渗流特征和生产规律是页岩气高效开发的关键。

为此,本文通过对北美地区 Barnett、Eagle Ford、Fayetteville、Haynesville、Woodford 和 Marcellus 等 6 个页岩气开发区块和中国首批页岩气示范区——长宁—威远和昭通区块的页岩气水平井生产数据进行分析(图 1),并采用双曲—指数混合递减模型建立不同区块的归一化产量递减曲线,探讨了页岩气井初始产量和产量递减率的变化趋势,最后建立了页岩气井最终可采资源量(EUR)快速评价方法。所取得的研究成果对国内页岩气开发,特别是规模化开发后不同区块页岩气井的生产规律分析与预测提供借鉴,为科学制订页岩气开发部署和决策奠定基础。

1 双曲—指数混合递减模型

Arps 递减模型要求气井以定压生产且达到拟稳态流动或边界控制流动状态,而致密气和页岩气等非常规气井实际生产数据呈早期快速递减、长期处于不稳定流动阶段,难以达到拟稳态或边界流控制的流动状态。Ilk 等[8-10]、Nobakht 等[11]基于统计规律建立了幂律指数递减模型,该方法基于对产量递减率随时间变化规律的研究,用衰减指数定律近似表征产量递减特征,引入新的计算产量递减率随时间变化的关系式,从而建立有别于双曲递减的产量递减模型,但该模型存在递减率表达式量纲不齐次、物理意义不明确以及气井生产后期拟合效果不理想等缺陷;Patzek 等[12]、Felguerosoa 等[13]基于北美地区 Barnett 页岩气井实际生产数据,分析表明多级压裂水平井产气量与生产时间的递减关系呈两段式,在生产早期阶段表现为双曲递减特征,在生产后期阶段表现为指数递减特征;基于前人对非常规油气井产量递减规律的认识,笔者采用双曲—指数混合递减模型[14],对北美和中国页岩气水平井的产量递减率和 EUR 等参数开展分析。

双曲—指数混合递减模型中气井产量递减率和产气量的表达式分别为:

$$D = \frac{D_i - D_\infty}{1 + mD_i t} + D_\infty \quad (1)$$

$$q = \frac{q_i}{(1 + mD_i t)^{\frac{1}{m}\left[1 - \frac{D_\infty}{D_i}\right]} e^{D_\infty t}} \quad (2)$$

其中,D 为气井产量递减率,d^{-1};下标 i 为拟合时间段的初始时间;D_∞ 为生产时间趋于无穷大时的递减率,d^{-1};m 为时间指数,无量纲;t 表示生产时间,d;q 为产气量,$10^4 m^3/d$。

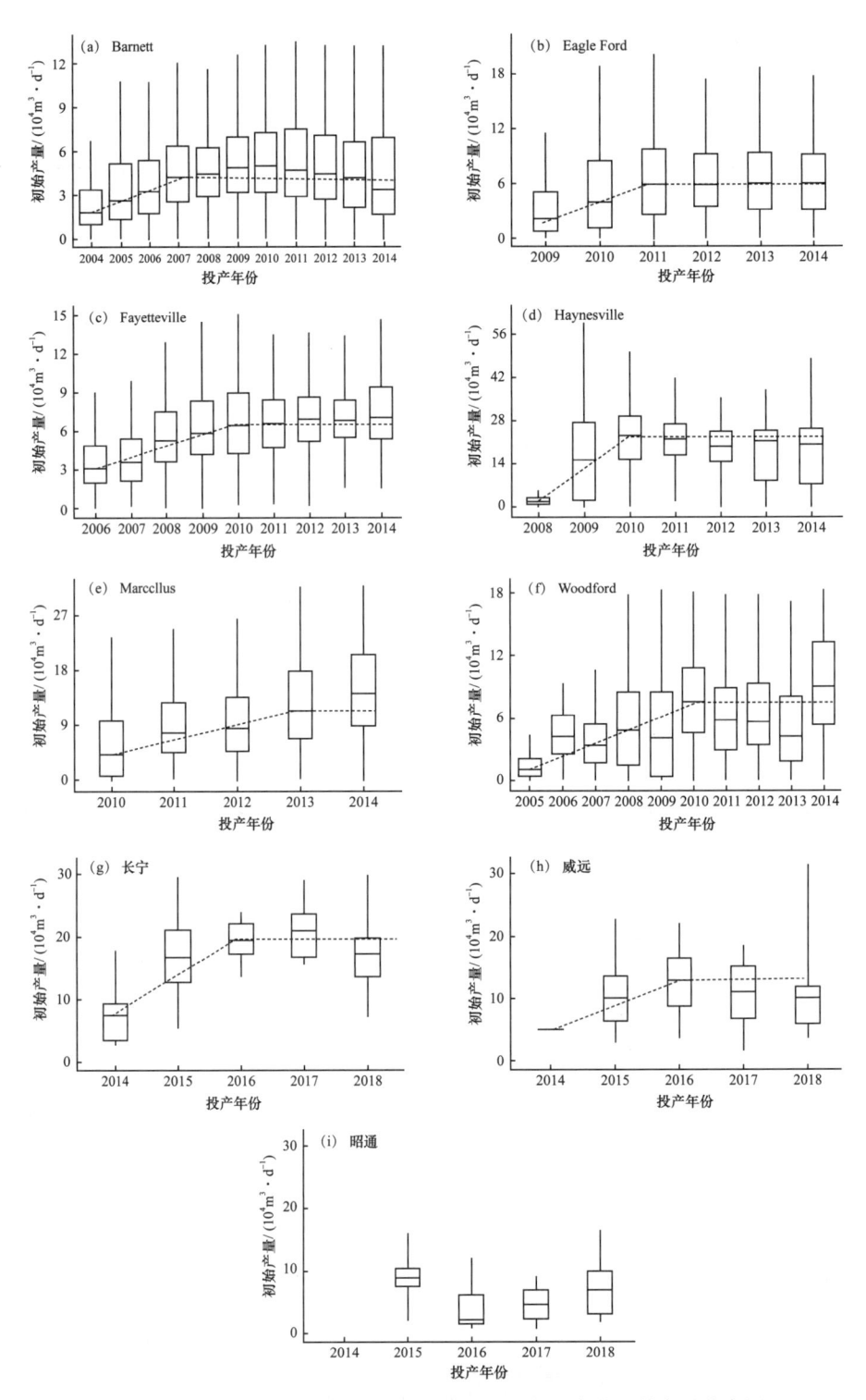

图 1 中国和北美地区主要页岩气开发区块历年投产井初始产量分布图

注：由于公开的生产数据时间周期较长，北美地区页岩气井生产数据的统计截止日期为 2015-01-01；图中 K 线上影线顶点对应最大值，下影线顶点对应最小值，箱体上沿为 75% 累计概率对应值，箱体下沿为 25% 累计概率对应值，箱体中部实线为平均值；(a) 至 (f) 中数据引自本文参考文献 [14]

2 北美地区页岩气水平井生产规律

2.1 气井初始产量

由式（2）可知，气井产量主要由初始产气量和产量递减率共同决定。自 2002 年 Devon 能源公司在 Barnett 页岩气区块试验的 7 口水平井取得成功，水平井开发技术在北美地区页岩气开发中迅速推广应用，推动了美国页岩气产量的快速攀升[3]。北美地区通常将气井压裂返排投产后前两个月的平均产量定义为初始产量，如图 2 所示，北美地区不同页岩气区块水平井初始产量变化均可划分为 3 个特征阶段：第 1 阶段，区块水平井的平均初始产量快速、较大幅度上升，以 Barnett 区块为例，自 2004 年大规模采用水平井开发以后，随着对区块地质认识的逐步深入和水平井开发技术、完井工艺日渐成熟，至 2008 年区块水平井的平均初始产量是开发初期的 2 倍；Barnett、Eagle Ford、Fayetteville、Haynesville、Marcellus 和 Woodford 区块在第 1 阶段的持续时间约 3～6 年。第 2 阶段，区

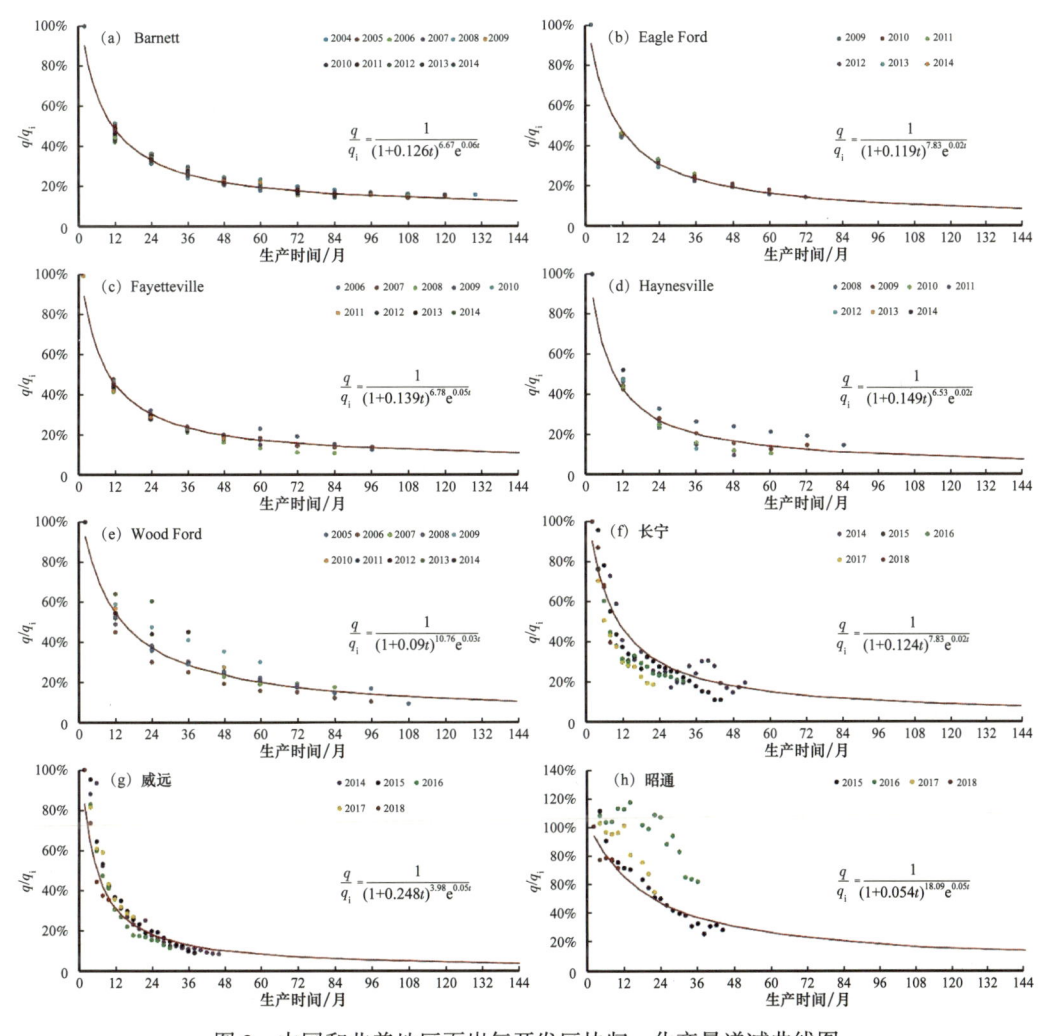

图 2 中国和北美地区页岩气开发区块归一化产量递减曲线图

块水平井的平均初始产量变化呈平台式，表明区块内水平井钻完井工艺趋向成熟、稳定，水平井初始产量提升空间有限，该阶段一般持续时间约7年。第3阶段，区块内水平井平均初始产量呈现下降趋势，表明随着"甜点区"以外布井数量的增加，外围区域储层地质条件相对较差，从而导致外围区域投产井的开发效果相对较差[15-16]。

同时，对不同区块间进行对比，可以看出随着水平井开发技术在Barnett区块试验成功，随后采用水平井规模开发的Haynesville等区块水平井初始产量在区块投入开发的初期阶段均表现出较大幅度的提升。因此，水平井初始产量呈现良好的"学习曲线"形态，即同一区块内随着地质认识的深入和开发技术的完善，初始产量呈现上升趋势；不同区块间随着相关技术的完善和借鉴，后期开发的区块水平井初始产量增幅较前期开发的区块要大。

2.2 产量递减率

自2004年大规模采用水平井开发以来，Barnett地区已公开的生产数据历史达10年，其它区块推广应用水平井开发技术之后，生产历史数据介于5～9年不等，为开展页岩气水平井生产规律分析提供了良好的资料基础[15-16]。为便于表征不同投产年份的气井产量，将月平均产量均一化处理（气井的月平均产量除以其初始产量），同时将历年投产井按年份依次记录，同一区块不同批次的气井产量递减规律相似（图3）；采用本文建立的双曲—指数混合递减模型对月平均产量历史数据进行拟合，得到各区块归一化产量递减曲线（图3）。

为了便于应用区块单井产量递减规律进行相应的分析与评价，本文将上述产量递减曲线按照等时间步长进行离散化，根据初始产量（Q_1）以及每一个时间点较上一个时间点的产量递减比例（g_i），计算任意时间点的产量（Q_n），计算公式如下：

$$Q_n = Q_1 \prod_{i=1}^{n}(1-g_i) \tag{3}$$

其中，Q_n为第n个时间点的气井产量，$10^4 m^3/d$；Q_1为气井初始产量，$10^4 m^3/d$；g_i为气井产量递减比例，无量纲；n为生产时间，d；i为循环因子，无量纲。

如表1所示，中国和北美地区各区块水平井产量逐年递减比例呈现早期高、后期逐渐降低的趋势。

表1 中国和北美地区页岩气开发区块水平井产量逐年递减比例（%）

区块	1年	2年	3年	4年	5年	6年	7年	8年	9年	10年
Barnett	58	28	19	17	12	12	10	10	6	3
Eagle Ford	58	31	28	16	16	16				
Fayetteville	57	33	24	18	13	16	12	3		
Haynesville	59	39	30	18	14	10				
Woodford	47	23	17	14	9	3	10	19	7	
长宁	55	38	33							
威远	63	46	37							

3 长宁—威远和昭通区块页岩气水平井生产规律

四川盆地长宁—威远区块和滇黔北昭通区块是中国首批国家级页岩气示范区，自2014年整体部署水平井进行规模化开发，4年来已累计投产水平井372口，累计产气量超过$100\times10^8m^3$。但与北美地区相比，中国页岩气开发整体尚处于开发初期阶段，水平井生产历史相对较短；开展国内与北美地区页岩气井生产规律的对比分析，是合理评价水平井产能、科学制订区块开发技术政策的基础。

3.1 气井初始产量

长宁—威远和昭通区块历年投产水平井初始产量统计数据表明，不同年份投产气井初始产量的变化呈现与北美地区相似的变化规律，开发初期气井初始产量整体偏低，随着水平井靶体位置优化和井眼巷道控制技术进步，气井平均初始产量呈现上升的趋势，但不同区块初始产量的分布略有差异。如图2所示，长宁区块气井初始产量呈现与Haynesville区块相似的特征，即前3年气井平均初始产量呈线性递增，第5年较前一年略有下降，同期投产气井最大初始产量和最小初始产量变化区间波动较大；威远区块气井初始产量变化曲线显示2015—2018年间平均初始产量分布范围在（10.1~13.2）×10^4m^3/d，平均初始产量变化幅度较小，但同期投产气井初始产量最大和最小值变化范围较大，该区块与Marcellus区块开发初期气井初始产量的大小较接近；昭通区块什么时间段？自投产以来初始产量变化范围介于（1.8~15.7）×10^4m^3/d，历年投产井最大和最小初始产量变化区间范围均较大，且平均初始产量呈现较大波动，2016年和2017年投产井的平均初始产量均低于2015年，该特征与Woodford区块开发初期气井初始产量变化特征类似。

3.2 产量递减率

与北美地区5个页岩气区块相比，长宁区块前3年产量递减比例依次为55%、38%和33%，与Fayetteville区块接近；威远区块生产初期产量递减比例分别为63%、46%和37%，明显高于其他区块；昭通区块不同年份投产井生产方式不同，开发初期（2015年）采用衰竭式生产与控压生产相结合的方式，生产初期产量递减比例分别为33%和28%，从2016年开始投产井均采用控压生产方式，两年来气井月平均产量波动幅度介于初始产量的100%~120%，表现出较好的稳产特征，气井稳产可持续22~34个月；由于投产时间较短，产量递减期规律尚不明确。

4 气井EUR预测

基于气井产量递减模型，全生命周期内气井EUR表达式为：

$$\mathrm{EUR}=\sum_{m=1}^{T_\mathrm{w}}\left[Q_1\prod_{j=1}^{m}\left(1-g_j\right)\right]=Q_1\sum_{m=1}^{T_\mathrm{w}}\left[\prod_{j=1}^{m}\left(1-g_j\right)\right] \quad(4)$$

其中，Q_1为投产第1年气井年产气量，10^4m^3；T_w为气井生产时间，年；m为投产年份，年；j为循环因子，无量纲；g_j为气井产量年递减比例，无量纲。

由式（4）可知，气井EUR与Q_1正相关，将第1年年产量作为基准值，气井生命周期内的累计产气量是产量年递减比例的 Π 函数；不同区块气井 EUR 与 Q_1 的比值随时间变化的幅度不同。由图 3 展示的不同区块气井 EUR 与 Q_1 的比值随时间变化曲线，可实现对不同区块气井的产能 EUR 快速评价。

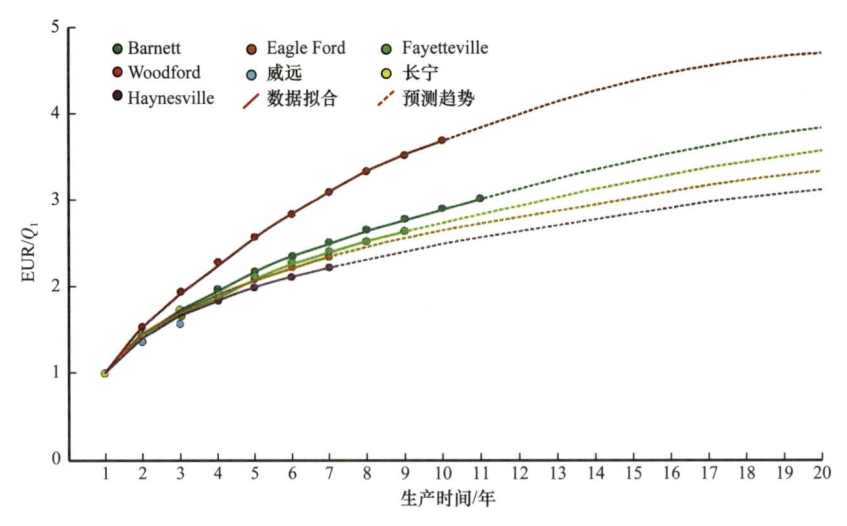

图 3 中国和北美地区页岩气水平井 EUR 与生产时间关系曲线图

对气井全生命周期产气量与第 1 年累计产气量比值与生产时间关系曲线进行分析，可以看出，在 20 年内，EUR 与第 1 年累计产气量正相关，一般为第 1 年累计产气量的 2～5 倍；Woodford 区块气井 EUR 与 Q_1 比值最高，长宁区块与 Barnett、Eagle Ford、Fayetteville 和 Haynesville 区块相当，威远区块则相对较低。

5 结论

（1）不同区块在开发初期页岩气水平井平均初始产量均呈现逐年上升的趋势，但初始产量分布存在差异，而后气井平均初始产量变化呈平台式，后期则呈逐年下降的趋势；

（2）双曲—指数递减模型应用于中国和北美地区页岩气水平井的产量递减分析具有较好的适应性，同一区块不同批次的气井产量递减规律相似；

（3）与北美地区 5 个页岩气区块相比，长宁区块前 3 年产量递减比例依次为 55%、38% 和 33%，与 Fayetteville 区块接近；威远区块生产初期产量递减比例分别为 63%、46% 和 37%，明显高于其他区块；

（4）生产时间为 20 年，气井 EUR 与第 1 年累计产气量（Q_1）正相关，气井 EUR 一般为 Q_1 的 2～5 倍，Woodford 区块气井 EUR 与 Q_1 比值最高，长宁区块与 Barnett、Eagle Ford、Fayetteville 和 Haynesville 区块相当，威远区块则相对较低。

参 考 文 献

[1] 夏阳, 金衍, 陈勉, 等. 页岩气渗流数学模型 [J]. 科学通报, 2015, 60（24）: 2259-2271.

[2] 段永刚, 魏明强, 李建秋, 等. 页岩气藏渗流机理及压裂井产能评价 [J]. 重庆大学学报, 2011, 34（4）: 62-66.

[3] 张东晓, 杨婷云. 页岩气开发综述 [J]. 石油学报. 2013, 34 (4): 792-801.

[4] 王军磊, 位云生, 程敏华, 等. 页岩气压裂水平井生产数据分析方法 [J]. 重庆大学学报, 2014, 37 (1): 102-109.

[5] 徐兵祥. 变产—变压情况下的页岩油气生产数据分析方法 [J]. 天然气工业, 2017, 37 (11): 70-76.

[6] 吴克柳, 陈掌星. 页岩气纳米孔气体传输综述 [J]. 石油科学通报, 2016, 1 (1): 91-127.

[7] 吴克柳, 李相方, 陈掌星. 页岩气有机质纳米孔气体传输微尺度效应 [J]. 天然气工业, 2016, 36 (11): 51-64.

[8] Ilk D. Well performance analysis for low to ultra-low permeability reservoir systems [D]. Texas, U.S.A: Texas A&M University, 2010.

[9] Ilk D, Rushing J A, Perego A D, et al. Exponential vs. Hyperbolic Decline in Tight Gas Sands — Understanding the Origin and Implications for Reserve Estimates Using Arps' Decline Curves [C] // paper 116731-MS presented at the SPE Annual Technical Conference and Exhibition, 21-24 Sept. 2008, Denver, Colorado, USA. https://doi.org/10.2118/116731-MS.

[10] Ilk D, Currie S M, Symmons D, et al. Hybrid rate decline models for the analysis of production performance in unconventional reservoirs [C] // paper 135616-MS presented at the SPE Annual Technical Conference and Exhibition, 19-22 Sept. 2010, Florence, Italy. https://doi.org/10.2118/135616-MS

[11] Nobakht M, Clarkson C R, Kaviani D. New type curves for analyzing horizontal well with multiple fractures in shale gas reservoirs [J]. Journal of Natural Gas Science and Engineering, 2013, 10: 99-112.

[12] Patzek T W, Male F, Marder M. Gas production in the Barnett Shale obeys a simple scaling theory [J]. Proceedings of the National Academy of Sciences of the United States of America, 2013, 110 (49): 19731-19736.

[13] Luis C-F, Ruben J. Forecasting long-term gas production from shale [J]. Proceedings of the National Academy of Sciences of the United States of America, 2013, 110 (49): 19660-19661.

[14] 齐亚东, 王军磊, 庞正炼, 等. 非常规油气井产量递减规律分析新模型 [J]. 中国矿业大学学报, 2016, 45 (4): 772-778.

[15] Saussay A. Can the US shale revolution be duplicated in continental Europe? An economic analysis of European shale gas resources [J]. Energy Economics, 2018, 69: 295-306.

[16] Railroad Commission of Texas Oil and Gas Division. 2014. Annual summary of Texas natural gas 2014. [EB/OL]. (2016-05-01) [2018-06-25]. http://www.rrc.texas.gov/media/33644/annual-gas-summaries-2014.pdf